The Environment of Man:

the Iron Age to the Anglo-Saxon Period

edited by

Martin Jones and Geoffrey Dimbleby

BAR British Series 87

1981

B.A.R.

B.A.R., 122 Banbury Road, Oxford OX2 7BP, England

B.A.R. 87, 1981, "The Environment of Man"

ISBN 9780860541288 paperback
ISBN 9781407323602 e-book
DOI https://doi.org/10.30861/9780860541288
A catalogue record for this book is available from the British Library
This book is available at www.barpublishing.com

CONTENTS

LIST OF FIGURES

Figure		Page

LIST OF PLATES

LIST OF CONTRIBUTORS

Mr Martin Bell
Dept of Geography, Bristol University

Dr Don Brothwell
Institute of Archaeology, London University

Prof. Geoffrey Dimbleby
34, House Lane, Sandridge, St. Albans

Ms Annie Grant
26, Beckwith Road, London SE24

Mr F. J. Green
Test Valley Archaeological Committee, Romsey, Hants

Mr Martin Jones
Botany School, Oxford University

Prof. Dr Udelgard Körber-Grohne
Istitut für Botanik, Universität Hohenheim (Stuttgart)

Prof. Hubert H. Lamb
Climatic Research Unit, University of East Anglia

Dr Richard Macphail
Institute of Archaeology, London University

Mr Mark Maltby
Department of Archaeology, Southampton University

Mr David Miles
Oxford Archaeological Unit

Dr Peter J. Reynolds
Butser Ancient Farm Project, Rookham Lodge, East Meon, Hants

Dr Mark Robinson
Oxford Archaeological Unit

Dr M. L. Ryder
ARC Field Laboratory, Roslin, Midlothian

Dr Judith Turner
Dept of Botany, Durham University

INTRODUCTION

Martin Jones

The environment of man has come to occupy a prominent position in archaeological studies. The close relationship between human activity and changes in climate, vegetation and soils has been recognised, and such changes have become part of the structure within which developments in prehistory are viewed. The better known applications of ecological studies, such as those relating to the emergence of agriculture, belong to periods of prehistory preceeding the Iron Age, and have been formulated within considerable constraints in the form of scarcity of settlement evidence and the relative imprecision of the associated chronological framework.

Students of these earlier periods would delight in the abundant settlement of Iron Age and later date that the recent growth of rescue archaeology has produced. They would be overwhelmed by the rich and well-dated assemblages of biological material recovered from these settlements. The rescue 'boom' has indeed led to the establishment of a number of environmental specialists in archaeological units and universities, the fruits of whose labours are beginning to see the light of day in publication. However, for all the diligent work that has been conducted on relatively favourable material of Iron Age to Anglo-Saxon date, this period seems to lack the general ecological perspective that exists for earlier periods of study. Much of the information produced is still site specific rather than relating to general ecological questions. An understanding of the period as a whole, and indeed the environmental questions themselves are still structured entirely around political events, in particular the magical dates of 43 and 410 A.D. This conservatism of approach is perhaps related to the rate at which material is being produced. The possibility of fresh thought on the structure of environmental studies has disappeared behind boxes of bone stacked to the ceiling, and beneath sacks of earth by the hundredweight, all waiting to be 'processed' by overworked specialists. The perils of this means of accumulating data are well expressed by Maltby in this volume; his assessment of the small fraction of the extant bone data that can be usefully assimilated into reliable ideas on animal husbandry is indeed a cautionary tale.

The clear need for some appraisal of environmental studies in the period in question led to the idea of bringing together a group of researchers in various fields. Each would attempt to synthesise the information they felt relevant to their own field, and suggest models and directions for further research. This was undertaken at a conference in Oxford entitled "The Environment of Man: The Iron Age to the Anglo-Saxon Period". The papers in this volume stem from that conference.

Whilst the option of contributing to the volume was made available to all researchers in the field, through the recently formed Association for Environmental Archaeology, the initial selection of contributors inevitably reflects my own ideas about potentially fruitful directions in environmental research. Professor Dimbleby rightly observes in his contribution that a volume with the same name could include an entirely different series of papers. It is therefore appropriate to outline my sentiments here.

Man is not a particularly amenable organism for scientific study, and general principles of ecology have in the past been developed on the basis of other, more amenable species. The value of studying the ecology of man lies not in revealing these general principles, but in revealing the features that are distinctive, the ecologically unusual aspects of the relationship between man and his environments. This unusual nature of human ecology must to a great extent lie in the considerable influence played by social factors in human development and behaviour. The period under study is unique in offering both a well defined long-term record of substantial ecological change, together with a well defined long term record of substantial social change. Other periods may present one or other of these records, but periods that present both with similar chronological precision are hard to find. I would suggest that it is in the opportunity of comparing these records, of collating variations in man-plant and man-animal relationships with variations in social status and structure of settlement hierarchy within a developing society, that the importance of this period to the study of human ecology lies.

If these sentiments are reflected in the general structure of the volume, they are by no means the uniform dogma of the contributions. The various authors have taken individual and often contrasting approaches to the subject, and the volume has benefited as a result. By way of introduction to the papers, I have summarised what strike me as the salient themes that emerge from the contributions.

In the section entitled "The Context of Environmental Studies", Reynolds and Miles view the direction of environmental studies from the standpoint of two related fields that have also come to prominence in the study of this period, experimental archaeology and landscape archaeology. Both authors emphasise the importance of research questions in the design of an archaeological strategy, and lament the lack of this element in past work. Reynolds considers in detail the transformations that occur between the initial elements of an ecosystem, and the resultant debris that is recovered by archaeologists. This question is also raised by Bell and Maltby, Miles draws attention to the problems inherent in the concepts of 'the site' and 'complete excavation', and there was a general consensus at the conference that a proper study of the environment of man should involve the examination of artefactual and environmental data from beyond the confines of conventional sites. Reynolds, for example, laments the lack of a comprehensive excavation of a 'celtic field'. Bell demonstrates admirably that such exercises can be successfully mounted.

Bell's contribution falls in the second section entitled "The Environmental Setting". The three papers on climate, vegetation and soils are not to be thought of as strictly equivalent syntheses, or to exclude consideration of human intervention as might be inferred from the section heading. Lamb draws together available evidence on climatic change, and his paper illustrates how broad a range of data, geological, biological, archaeological and historical, may be drawn on in environmental studies of this period. Climatic changes in Britain are reviewed in the context of changes further afield, and in places related to contemporary historical and cultural developments. Turner deals with one specific form of data, pollen, and stresses the factors involved in the selection of suitable sites for analysis, and in the interpretation of associated radiocarbon dates. As well as outlining the national trends during the period, both Lamb and Turner draw attention to regional diversity. The potential for studying regional diversity and the importance of adequate regional coverage was alluded to by several authors. Bell, dealing with the specific question of valley sedimentation, considers the implications of his data, both in the interpretation of primary evidence from field survey, and in the subsequent understanding of concurrent changes in settlement pattern indicated by such surveys. The latter aspect is also explored by Jones in the following section.

In this section, entitled "The exploitation of Environmental Resources", Jones and Green deal with evidence for crop husbandry in the period. Jones collates the evidence from plant macrofossils with evidence of agricultural implements, buildings and farm layout, and attempts to isolate periods of change. A model is then proposed to relate these changes to concurrent developments in settlement hierarchy. Green looks in more detail at the interpretation of crop plant remains from a specific region, giving greater consideration to methodological problems than Jones, and stresses the internal diversity and complexity of the crop record. Maltby's paper on animal husbandry deals specifically with the evidence from preserved bone. The assemblages for which useful quantitative data exists are identified, and the form, abundance and age structures of the key domesticates are considered, both in relation to different periods of time and to different types of settlement context. Certain clear trends emerge, particularly in relation to age structure, and Maltby concludes by indicating the kinds of questions that may usefully be directed to the data, and the methological improvements that are necessary for the answers to be forthcoming. Maltby draws attention to the considerable diversity of information that can be derived from vertebrate remains, a point which is amplified in the remaining papers in this section. Grant considers the social implications of the occurrence of wild animal bones in periods that are predominantly agricultural. Ryder draws attention to a form of animal sub-fossil less familiar than bone, fragments of wool and associated fibres, and relates their occurrence to the micro-evolution of sheep in the period under study. Brothwell reviews the forms of disease that manifest themselves in the bone material, and discusses how this may be related to such topics as the micro-evolution of domestic stock, health and nutrition, and the treatment of animals in captivity.

Three authors describe specific case-studies in the section entitled "The Ecology of Settlement". Each study is conducted within a long-term archaeological programme, and the three papers display various ways in which an environmental study may be incorporated into such programmes. Körber-Grohne describes an intensive study of a single example of a type of site that occurs all along the North German coast. Changes in the environment are related to the initiation, development and decline of that individual settlement in a manner that may be extended to other similar sites. Robinson deals with a number of sites in the upper Thames Valley, scattered in space and time, and formulates a picture of the environment of each site. The changing environment is then analysed in terms of the ecological differences between each site. Macphail also deals with a number of sites, but confines his attention to a particular point in the stratigraphic sequence of each site, corresponding to the so-called "dark earth" in late Roman settlements. The deposition of the layer itself is investigated from an ecological point of view, and implications about the use of the settlements in late Roman Britain discussed. All three authors argue on the basis of a very broad range of ecological data, pollen, phytoliths, molluscs, insects, soils, etc., in each case leading to a far more powerful argument than would have been derived from one form of data alone. Körber-Grohne's discussion of plant associations rather than individual species also brings strength to her case.

Each of the contributors was asked to consider the relevance of the chronological framework based on the Roman Occupation to their own subject matter. Whilst overall differences between the Iron Age, Romano-British and Anglo-Saxon periods are recognised by various authors, the timing of the various changes is occasionally brought into question. Green isolates the 3rd and 2nd century B.C. as a transitional period in terms of the Wessex crop record, whereas Jones would locate a transitional period earlier in the millennium for Southern Britain as a whole. Miles, Jones and Robinson place emphasis on the late Iron Age as a key period of change, and there was a general consensus at the conference that more ecological investigation of the period might lead to the backdating of many supposed Roman innovations. A great deal of change in late Roman Britain is evidence from the papers of Jones, Green, Maltby, Grant and Macphail. The volume in general includes a number of points about this period and the ensuing Dark Ages. It is interesting to compare Turner's cessation of woodland clearance and Grant's resurgence in the exploitation of wild animals with Jones' sustained intensification of lowland agriculture, Maltby's sustained improvement in Stock rearing, and Macphail's horticultural activity in the towns. This is clearly a period to which environmental archaeology has exciting contributions to make.

These are the themes that present themselves most clearly to me. I am grateful to Professor Dimbleby for casting a more experienced eye over the proceedings in his 'concluding remarks'. The papers are densely packed with new information and ideas, and many other themes will undoubtedly emerge in due course. The execution of the conference and this volume have already been a source of great stimulus to me, and I should like to acknowledge the many people who allowed them to happen. The conference was organised under the auspices of Oxford University Department of External Studies and the Oxford Archaeological Unit, and my thanks go especially to Trevor Rowley, Shirley Herman and Richard Chambers for their help. Throughout the development of the conference nd volume, and especially during the editing, the extensive and careful guidance given by Professor Dimbleby has been invaluable. My thanks also go to all those who by attending the conference made both it and this volume possible, and in particular to the contributors themselves, whose tasks of synthesis were never easy. To those who spent on this task so much of their precious time-off from the boxes of bones and sacks of earth, only to be rewarded by my bullying phone-calls and heartless editing, may I extend my gratitude.

Martin Jones,
Oxford, November 1980

NOTE ON DATING

Throughout this volume, the following convention has been used:

i) 'bc' and 'ad' are used to denote dates measured in radiocarbon years, calculated using the conventional Libby value of 5568 years for the half-life of radiocarbon.

ii) 'BC' and 'AD' are used to denote dates measured in calendar years.

THE CONTEXT OF ENVIRONMENTAL STUDIES

1. SOCIAL LANDSCAPES: PATTERN AND PURPOSE?

David Miles

'The great majority of the English settlers faced a virgin country of damp oak-ash forest, or beech forest on or near the chalk; and what was not thickly forested was likely to be cold, high, mist-wrapped moorland, or water-logged wet heath, drowned marshes or estuary saltings, or sterile thin-soiled, dry heath.' (Hoskins 1955: 44)

This evocative picture was conjured up over twenty five years ago in that classic of historical geography The Making of the English Landscape. Since then the scene has changed dramatically thanks to the work of archaeologists, historical geographers, and paleoecologists. It would be unsafe to say however that the picture has yet been brought into sharp focus. Still less understood are the human societies which lay behind a man-made landscape which was already ancient, complex and widely exploited when the Saxons arrived in Britain. No longer can we go along with the view current in 1955 that 'the English landscape as we know it today is almost entirely the product of the last fifteen hundred years, beginning with the earliest Anglo-Saxon villages in the middle decades of the 5th century'.

British archaeology has a long, venerable though only intermittently vigorous tradition of fieldwork, running from Camden to Crawford, and in the past few decades carried chiefly on the shoulders of the various Royal Commissions. Since at least 1844 when Lord Albert Conyngham, 'dressed in an exploring costume', supervised the 'excavation' of eight barrows watched by two hundred spectators in their carriages, most archaeologists have been carried away by the glamour of digging, encouraged by museums in search of the 'morbid or the aesthetic'. Fieldwork in the sense defined by Taylor (1972) has usually taken a back seat.

The 1930s saw something of a renaissance in field archaeology with the surveys of Crawford, Fox, Daniel, Grinsell and others. The decade also saw the impressive results of aerial photography by fliers such as Major Allen. The aerial photographs were a mixed blessing; rather than stimulating field work on the ground they possibly stifled it. Only a few people such as Crawford and C. W. Phillips in the Fens saw the implications for expanding archaeology beyond the supposedly unique site to the landscape as a whole. The Fenland project in particular, culminating in The Fenland in Roman Times (ed. C. W. Phillips 1970) was a model of inter-disciplinary co-operation, pioneered at Cambridge by Grahame Clark and Harry Godwin, Star Carr being a good example. Unfortunately it was not a model that proved catching.

The emphasis on excavation lay very much behind the British rescue boom of the late '60s and early '70s. At the same time the observation of motorways and gravel pits and the increasing use of aerial photography generated a growing awareness that archaeological sites were not rare and isolated phenomena. On the M5 in Gloucestershire and Somerset, for example, about 130 sites were found along the road's fifty yard wide corridor through some of southern England's more intractable claylands; where only ten sites in total had been known before there was now an average of two sites per mile. Similar increases were found in the main river valleys as a result of aerial photography. In the Nene Valley, in Northamptonshire and Huntingdonshire where 36 sites were known in 1931, there were 434 by 1972.

In the 1940s the anthropologist Clyde Kluckhohn (1940) voiced his famous criticism of American archaeologists 'I should like to record an overwhelming impression that many students (of Middle American prehistory) are but slightly reformed antiquaries ... there seems a great deal of obsessive wallowing in detail of and for itself ... proliferation of minutiae is not its own justification'. The criticism could equally be applied to much of the excavation that was carried out in the 1970s in Britain. Sites were dug because they were threatened; there were individual discoveries but the overall results were disappointing.

In the light of the growing evidence that archaeological sites were not rare, lip service was paid to total excavation and landscape archaeology. Both were well-meaning and in a sense progressive, a reaction against the key-hole excavation of highly selective 'important' sites. Of course the total excavators were predisposed to 'relevate' some data at the expense of others (this useful neologism, first coined by the physicists Bohm and Schumacher, is used by them to point out the way scientific theory (often unexpressed) will push certain things into the foreground and 'irrelevate' other form of evidence, leaving it ignored. (See Thompson 1979: 21). One excavation of the early '70s, for example, was noted at the time for the meticulous three-dimensional recording of all its finds—except animal bones. Total excavation tended to concentrate on the production of large-scale site plans, though even in this respect with the use of heavy machinery for site clearance the adjective was of dubious application.

Landscape archaeology was another attempt to approach the mirage of totality; to see sites in context rather than in isolation. It grew partly out of the much admired historical geography school of Hoskins and Finberg, combined with the ecological strain which ran through British archaeology via the fieldwork tradition and fuelled by the tremendous increase in data in the late '60s and '70s. Landscape archaeology had a romantic and aesthetic quality which appealed both to the environment generation and the traditional lovers of the British countryside, the National Trust joiners and the Country Life readers. It was always perilously close to being a new form of antiquarianism; collection and classification without the study of process. Books on the subject (e.g. Taylor 1974, Aston & Rowley 1974) tended to be basic manuals of 'how to do it' rather than 'why do it?' or 'what to do with it when you've got it'.

The growth of interest in the landscape and conservation was behind the sudden burgeoning of Sites and Monuments Records in Museums and Planning Departments. While their creation was and is wholly admirable they often seemed to encourage a train-spotting mentality: at one time becoming the first county to have 15,000 sites seemed to be a leading priority. As in landscape archaeology site collection could become an end in itself; and even that was not always done very well. Today the best Sites and Monuments Records are in the Cultural Resource Management business, adivising local authorities on protection, but few instigate systematic and intensive site location or analysis in their own right.

The flagships of British field archaeology are the Royal Commissions for Historic Monuments, RCHM, for England, Wales and Scotland. It may be more accurate to call them royal barges as their progress has been described by one of their deck-hands as 'leisurely' (Taylor 1978). There is more than a hint in the title of Taylor's article that the institution he works for is something of a sacred cow itself. Certainly its aims, methods and forms of publication have until recently avoided overt criticism since it started on its regal way in 1908. In that time surveys of only seventeen English towns or counties have been published and not all of them completely. Two pioneering RCHM surveys highlighted the threat to archaeological sites A Matter of Time (1960) and Monuments Threatened or Destroyed (1963). They were evidently ahead of their time for all the impact they had. Surveys of Newark (1961) and Peterborough New Town (1969) were also threat-orientated.

On the whole though RCHM surveys seems to have been produced in a vacuum, free from theoretical, methodological or practical concerns. Their obvious aim is to provide basic data with relatively little comment and thus avoid the whims of current fashion. This in the Commission's work has been normative rather than progressive; trapped in an undefined and invisible cocoon of ageing attitudes. The contents of the volumes are not as unchanging as their regal bindings suggest. Hertfordshire (1910) bears the stamp of its time, as does Dorset: East (1975) or Northamptonshire: Central (1979). The Dorset: East volume is curiously old-fashioned for the mid '70s: only 32 Iron Age and Romano-British settlements are recorded, including hillforts in an area of over 110 square miles and the vast majority are on the chalk-lands. The clay and Bagshot Sands are almost empty of settlement. The same pattern emerges in Dorset : North; 22 prehistoric and Romano-British settlements, all on chalk between the 200 and 400 foot contour. Opulently produced surveys which fossilise this kind of biased information are of dubious value.

The latest RCHM volume Northamptonshire: Central reflects the recent explosion in site numbers, 200 late prehistoric/Romano-British sites in total. 15 are concentrated around Brafield-on-the-Green, but, in contrast, only one site is recorded in the whole of the north west part of the survey area. The density of late prehistoric occupation in some parts at least, is a factor which has only recently come to be appreciated. Still we are quite rightly told in the earlier Northamptonshire volume (RCHM 1975, XXIV) 'The value of distributional analysis is open to question since the distribution pattern is controlled by the methods by which the evidence has been collected. Thorough examination of every feature of the landscape and of current alterations to the terrain, and also of air-photographic cover under varying conditions over a long period, is necessary before the extent of early occupation can be fully understood. Such is a task for the future'.

The main reason for the large increase in sites in Northamptonshire (aside from aerial photography) is the activity of a small number of dedicated and skilful amateur fieldworkers. For the most part professional archaeologists spend their time cataloguing known sites or digging them. As a result field archaeology's techniques have not improved much on John Aubrey's, and its practice in Britain is not usually integrated into the growing body of archaeological theory.

While appealing for more and better fieldwork Taylor (1978) questioned the growing interest in survey work in the mid to late '70s. 'It is, it seems to me, largely the result of the bursting of the rescue bubble following the restraints on excavation'. There is an element of truth in this, but it should not be allowed to cloud a genuine need, nor to encourage the unfortunate divisions that already exist in archaeology, between fieldworkers, 'rugged' diggers and theoreticians. Archaeologists attempting to create a regional research framework in which to operate have not had the decades necessary to await the docking of the RCHM in their particular pond; sites with planning permission for redevelopment can have disappeared by the time RCHM volumes appear (for example in Gloucestershire's Cotswold water park) and many archaeologists feel that unsystematically collected data are of little use to them anyway for tackling questions of current interest (Cherry & Shennon 1978).

As Renfrew (1978: 164) has rightly said 'Rural archaeology in Britain is becoming a total disaster'. In contrast urban archaeological problems have been tackled more coherently, with implications studies, both local and national, better fund raising and long term urban excavation units. The formation of priorities has proved relatively straightforward in spatially and temporarily limited areas with a strong sense of identity like Winchester, Oxford, Norwich and York. The archaeology of the rural areas has presented a much more open-ended problem: a data base constantly growing at one end as a result of unsystematic discoveries but eroded at the other by modern agriculture, quarrying and suburban expansion, combined with a lack of specified priorities and means of tackling them. Many ill-conceived rural surveys produced in the '70s reflected the uncertainty of their aims (Cleere 1978). As a result surveys became unpopular with the DoE and the baby was thrown out along with the murky bath water.

A rapid glance over the Current Research Projects of any edition of American Antiquity will show that in the United States survey is assumed to be the logical first stage in the implementation of a regional research design. This is partly the result of the nature of federally funded projects (Schiffer & Gumerman 1977) but also the widespread notice eventually taken of Binford's warning 16 years ago 'Under current programmes of salvage archaeology and increased foundation support for archaeological research we are being given the opportunity to study major regions intensively. In spite of the opportunities currently available, it is my impression that very little thought has been given to research design.' (Binford 1964).

In large relatively unexamined parts of the United States, particularly in the south west with its dry conditions and low modern population, field survey is the obvious first element in regional analysis. Conditions are often better for the surface location of sites than in Britain; even so the possibility of obtaining a complete inventory has been questioned (House & Schiffer 1975: 41). The same doubts have been raised in Britain (Taylor 1972) but American archaeologists have taken a more positive and systematic approach to the problems of site definition and location, survey intensity, sample size and the variables which can effect the discovery pattern (see the delightfully onomatopoeic Plog, Plog & Wait 1978).

In recent years British archaeologists have attempted to modify the framework in which they operate to develop coherent regional programmes of the kind advocated by Binford (1964). The definition of regions 'which can be expected to have supported cultural systems' is no easy task for the archaeologist. Systems will have fluctuated over time: the hill-fort territory of the early Iron Age varying from that of the later Iron Age oppidum for example. A pragmatic strategy resorted to is the one advocated by Cherry & Shennan (1978: 103) 'The alternative to specifying systems of interest and then studying them is the one usually adopted: we take a block of land and study bits of changing systems within it ... however we specify our sytem of interest, its boundaries are bound to by fuzzy and arbitrary, so there is no harm in looking at an arbitrary segment; furthermore, if our interest is, as we believe it should be, in processes of change and evolution through time, then taking a particular area as the object of study cuts the Gordian knot of changing system boundaries.'

After the initial gathering of known data the logical first stage in this kind of project is to select a random sample of the area in question for field survey (e.g. Schadla-Hall & Shennan 1978). Unfortunately this is exactly the phase which is missing from most excavation based work. In spite of the growing interest in the landscape the idea of the unique site has persisted and with it a reluctance to generalise. On the other hand there has been a welcome if limited growth of regional studies: in Dartmoor (Wainwright, Fleming & Smith 1979), various river valleys and most recently the collaboration of the RCHM (England) with the Wessex Archaeological Unit on Salisbury Plain. There has also been a willingness not only to blurr the boundaries of ancient systems but also those of modern archaeologists. The Oxford Archaeological Unit operating within the Upper Thames region has transcended county boundaries to implement a research strategy specifically aimed at the study of Iron Age and Romano-British settlement and the processes affecting change in the first millennia BC/AD (Jones & Miles 1979).

There is general agreement that 'rescue is research or it is nothing' (Renfrew 1978: 178) (except perhaps on the part of some of archaeology's administrators). The present failure is to integrate the resources as coherently as possible. In particular theorists, field archaeologists, excavators, environmentalists and cultural resource managers need to co-ordinate regional research designs so that our raw material can be both preserved and collected in a way which will provide a basis for present and hopefully future problem solving. The need for teamwork is so obvious as to be practically a cliché but the fragmentation into separate institutions and isolated specialists is currently our biggest handicap.

The study of late prehistory, the Roman and early medieval periods in Britain has been bedevilled by historical specialization. A series of invasions by early Iron Age Celts, Belgae, Romans and Saxons have provided neat divisions and ready-made explanations. The limitations of the invasion hypothesis were highlighted by Clark (1966), but nevertheless the divisions effectively remained, institutionalised in British archaeology. The movement away from invasion based explanations has been most pronounced among prehistorians and a modified systemic approach has percolated into some of the textbooks (Cunliffe 1974), though even with the more progressive of these (Bradley 1978) the arrival of the Roman legions in AD 43 has generally terminated their speculations.

A subject of traditional interest has been the so-called continuity question. This has most often exercised the minds of those concerned with the 5th century AD in Britain. It has often been approached at a basic and not especially informative level: 'Continuity is currently thought of as continuous exploitation of resources in a given area, ranging from the use of a single field or house through to that of major sectors of Britain' (Rahtz 1978: 35). However it has occasionally seduced researchers around the Maginot lines of period based archaeology. Even so there has sometimes been a strange lack of dialogue between, for example, historical geographers who have postulated Iron Age hillforts as the centre of multiple estates extant in the Saxon landscape and the prehistorians of the Iron Age with their Thiessen polygons.

If medievalists have pushed backwards to explain the origins of their period, then prehistorians have begun tentatively to invade Roman Britain, in the belief perhaps 'that Roman Britain is too important to be left to the Romanists' (Cunliffe 1979: 359), several of whom have 'made clear their dislike of the intrusion of sociological and economic concepts in their subject' (Jones & Miles 1979: 315). If to the Classical Historian Roman Britain is only a minor province, to the anthropologically-minded archaeologists it is a rich testing ground for theories of tribal development, the growth of urbanisation and change, the impact of an imperial world system and many other questions. Kemp (1976) has described the prehistorians response 'to some variant of a Parkinson law about techniques developing to fill the space created by particularly mute evidence'. It is no bad thing that the techniques are now beginning to be applied in the more testable proto-historic and historic periods where there is a growing body of settlement data.

Iron Age, Roman and Saxon studies have been bedevilled by over-simplistic settlement type-fossils: Little Woodbury, Roman villas, Sutton Courtenay have all illustrated the past willingness to argue on the basis of the particular and the partial. The CBA's <u>Rural Settlement in Roman Britian</u> (Thomas 1966) emphasised the need for detailed regional studies to shed light on the complexity of past societies; Hogg (1966) for example showed that the Highland Zone cannot be regarded as a uniform entity and the same is certainly true of the lowlands. Fourteen years ago this type of detailed work was in its infancy, and since it has grown more slowly than we might have wished. To tackle the problems raised a longterm effort is needed on the part of a team of researchers combining specialised skills. At the same time the aims of regional research need to be agreed on in outline and spelled out (see the SARG project Gumerman 1971; Cherry & Shennan 1978b) and set against a wider background: Friedman & Rowlands (1977: 204) have emphasised 'As production of exchange seems to be a constant factor in evolution we must deal with a system larger than the local political unit, whether it be tribe or state, if we are to understand its conditions of existence and transformation.'

Cunliffe (1978, fig. 13) has proposed a simple model for the development of Iron Age society in which three significant phases are identified during which population expansion can be correlated with socio-economic change. The model has been extended in outline into the Roman period (Cunliffe 1979: 363-84) with a suggestion of a 2nd-3rd century population decline and subsequent late 3rd-early 4th century recovery. This may have been followed by a further late 4th century decline, a phenomenon which is currently being examined in Romano-British towns (see Macphail this volume). The population model of two steps forward, one step back and two forward again is an appealing one (Braudel 1967; Wynne Edwards 1966). Population should not be seen as a prime mover: a feedback process involving population, commercial trends, political control, agricultural productivity, redistribution of production, belief systems and uncontrollable factors such as climate and the slowness of transport or the chronic power shortage must all play a part. But population is a vital factor about which there has been much speculation, some of it meriting Braudel's criticism (1967: 16) 'Historians who persist in the old way of explaining Western demographic movements by the fall in urban mortality, the advance in hygiene and medicine, the decline in smallpox, the numerous drinking water supply systems, the devisive fall in infant mortality, plus a general fall in the mortality rate and a younger average age of marriage, should abandon it as inadequate and outmoded'.

Systematic sampling of regional settlement patterns is likely to be the best means of tackling the population problem: at present our statistics are woefully inadequate. These settlement patterns should also give rise to testable hypotheses relating to models such as those proposed by Cunliffe (op. cit.). In the Thames Valley attempts are being made to construct a hierarchy of regional settlement in which initial trends are apparent. The middle Iron Age, for example, is characterised by hillforts, settlements or greater than extended family size on the valley slopes and higher ground, (Parrington 1978) while on the flood plain there are seasonally occupied herdman's compounds (Lambrick & Robinson 1979). This range of sites does not characterise the late Iron Age by which time there is an increase in single enclosed farmsteads on the gravel terraces (Miles ed. 1981 forthcoming), banjo enclosures on the Cotswolds slopes, extensive trackway systems associated with enclosures and new central places at the junction of communications routes. The radical changes in the late Iron Age settlement layout may reflect the social and economic changes proposed by Cunliffe and observed also on the Continent (for example in Gotland, Lindquist 1974).

It is probable that these transformations in Celtic society in Southern Britain made possible the subsequent Romanisation of that part of the province (Groenman van Waateringe 1980) just as the internal problems of Roman Britain opened the way for Saxon settlement. If the bow-wave of pre-Conquest Romanisation had its greatest impact in the south-east of Britain, the ripples may have extended much further. Expansion and intensification in the late Iron Age has been detected as far north as Cumbria, followed by contraction in the middle period of Roman occupation (Higham 1978). Late Iron Age expansion has also been detected in north eastern England from pollen diagrams

(Turner 1979). The Roman Conquest as such does not seem to have had a major impact, nor were there dramatic changes in the 5th century AD.

In spite of the problems proto-historic Britain is well supplied with settlement data, even in the so-called destruction zone. A more systematic approach to this evidence would lay the basis for better quality excavation and environmental work and more feedback between specialists. If prehistorians have had to rely on the somewhat mechanistic and deterministic techniques such as site catchment analysis, proto-historians have an opportunity to test hypotheses of more complex hierarchies raised by different specialists (Flannery 1976). Even within a close knit team specialist interests will and should develop, but some basic common themes might discourage the field archaeologist who simply puts dots on a distribution map, the excavator who digs 'because it is there', the environmentalist who just provides species lists and the kleptomaniac theorist who indulges in untestable fantasies. Our common aim should be to understand the development of social landscapes.

Bibliography

Binford, L. R. (1964). A consideration of archaeological research design American Antiquity 29 425-441.

Bradley, R. (1978). The Prehistoric Settlement of Britain. London: Routledge & Kegan Paul.

Braudel, F. (1967). Capitalism and Material Life 1400 - 1800. London: Fontana (English edition 1974).

Cherry, J. & Shennan, S. (1978a). Sampling Techniques and Regional survey strategies. In: New Approaches to our past: an archaeological forum (eds. Darvill, T. C. et al.) Southampton pp. 101-126.

Cherry, J. & Shennan, S. (1978b). Sampling Cultural Systems: some perspectives on the application of probablistic regional survey in Britain. In: Sampling in Contemporary British Archaeology (eds. Cherry, J. F., Gamble, C. & Shennan, S.) BAR 50 17-48.

Clark, J. G. D. (1966). The invasion hypothesis in British archaeology Antiquity 40 172-189.

Cleere, H. (1978). A Survey of Surveys. In: New Approaches to our Past: an archaeological forum (eds. Darvill, T. C., et al.) Southampton pp. 89-100.

Cunliffe, B. W. (1974). Iron Age Communities in Britain. (revised 1978) London: Routledge & Kegan Paul.

Cunliffe, B. W. (1978). Settlement and Population in the British Iron Age: Some Facts, Figures and Fantasies. In: Lowland Iron Age Communities in Europe. (eds. Cunliffe, B. & Rowley, T.) BAR 48 3-24.

Cunliffe, B. W. (1979). Some Concluding Thoughts. In: Invasion and Response: the Case of Roman Britain (eds. Burnham, B. C. & Johnson, H. B.) BAR 73 359-365.

Flannery, K. V. (1976). Empirical Determination of Site Catchments in Oaxaca and Tehnacan. In: The Early Mesoamerican Village (ed. Flannery, K. V.) New York: Academic Press pp. 103-116.

Friedman, J. and Rowlands, M. J. (1977). Notes towards an Epigenetic Model of the Evolution of Civilization. In: The Evolution of Social Systems (eds. Friedman, J. and Rowlands, M. J.) pp. 201-276.

Groenman van Waateringe, W. (1980). Urbanization in the North West Frontier of the Roman Empire. In: Roman Frontier Studies 1979 (eds. Hanson, H. W. & Keppie, L. J. F.) BAR I.S. 71 1037-44.

Gumerman, G. J. Ed. (1971). The Distribution of Prehistoric Population Aggregates. Prescott College Anthropological Reports No. 1.

Higham, N. J. (1978). Continuity Studies in the first millennium AD in North Cumbria. Northern History 14 1-18.

Hogg, A. H. A. (1966). Native Settlement in Wales. In: Rural Settlement in Roman Britain (ed. Thomas, C.) CBA Research Report 7 28-38.

Hoskins, W. G. (1955). The Making of the English Landscape London: Hodder and Stoughton.

House, J. H. & Schiffer, M. B. (1975). Archaeological survey in the Cache River Basin. In: The Cache River Archaeological Project: an experiment in contract archaeology. Arkansas Archaeological Survey, Research Series 8 37-53.

Jones, M. & Miles, D. (1979). Celt and Roman in the Thames Valley: Approaches to Culture Change. In: Invasion and Response: the Case of Roman Britain (eds. Burnham, B. C. & Johnson, H. B.) BAR 73 315-325.

Kemp, B. J. (1976). Review of Mathematics and Computers in Archaeology by J. E. Doran & F. R. Hodson. Journal of Archaeological Science 3 188-9.

Kluckhohn, C. (1940). The Conceptual Structure in Middle American Studies. In: The Maya and their Neighbours (ed. Hay, C. L. et al.) pp. 41-51.

Lambrick, G. & Robinson, M. (1979). Iron Age and Roman riverside settlements at Farmoor, Oxfordshire. CBA Research Report 32.

Lindquist, S. O. (1974). Comments on Iron Age Farms in Southwest Norway: the development of the Agrarian Landscape on Jaeren. Norwegian Archaeological Review 7: 1 47-52.

Miles, D. Ed. (1981 forthcoming). Archaeology at Barton Court Farm, Abingdon. CBA Research Report.

Parrington, M. Ed. (1978). The excavations at Ashville Trading Estate, Abingdon (Oxfordshire) 1974-76. CBA Research Report 28.

Plog, S., Plog, F. & Wait, W. (1978). Decision making in modern survey. In: Advances in Archaeological Method and Theory (ed. Schiffer, M. B.) 383-421.

Rahtz, P., (1978). The Concept of Continuity in Settlement Studies Medieval Village Research Group: Annual Report 26 35-6.

Renfrew, A. C. (1978). Archaeology and Society in 1978. In: New Approaches to our Past: an archaeological forum (eds. Darvill, T. C. et al.) 157-178.

Schadla-Hall, R. T. & Shennan, S. J. (1978). Some suggestions for a sampling approach to archaeological survey in Wessex. In: Sampling in Contemporary British Archaeology (eds. Cherry, J. F., Gamble, C. & Shennan, S. J.) BAR 50 87-104.

Schiffer, M. B. & Gumerman, G. J. (1977). Conservation Archaeology: a guide for cultural resource management studies. London: Academic Press.

Taylor, C. C. (1972). The study of settlement patterns in pre-Saxon Britain. In: Man, Settlement and Urbanism (eds. Ucko, P. J. et al.) 109-118.

Taylor, C. C. (1978). Looking at Cows-Field archaeology in the 1980s CBA Report 28 63-75.

Thomas, C. Ed.) (1966). Rural Settlement in Roman Britain. CBA Research Report 7.

Thompson, M. (1979). Rubbish Theory: the creation and destruction of value Oxford: OUP.

Turner, J. (1979). The Environment of North East England during Roman times as shown by pollen analysis. Journal of Archaeological Science 6 285-290.

Wainwright, G. J., Fleming, A. & Smith, K. (1979). The Shaugh Moor Project: First Report. Proceedings of the Prehistoric Society 45 1-33.

Wynne-Edwards, V. C. (1962). Animal Dispersion in relation to Social Behaviour. Edinburgh: Oliver & Boyd.

2. NEW APPROACHES TO FAMILIAR PROBLEMS

Peter J. Reynolds

This conference could properly be called a 'Convocation of Appendices' and as such it makes a most important step forward in the practice and philosophy of archaeology today. It was, therefore, a great privilege to be invited to give the opening lecture of this conference and to attempt to introduce the theme as expressed in the title 'New Approaches to Familiar Problems'. Traditionally archaeology and in a sense more specifically archaeological reports, have concentrated upon the hard evidence, the real tangible features of pits and post-holes, gulleys, ditches and banks, artefactual material and their seemingly inevitable typological analysis. Indeed so common are these last that the standard impression one gains from flicking through almost any archaeological report are the interminable pages of drawings of sherds, their blackened sections leaning in serried ranks in not dissimilar fashion to the military process of a right marker. Not, it must be added that such pages are without value. Rather one questions proportional values. Within an environment pottery is undoubtedly significant but hardly diagnostic. Similarly even the so-called hard evidence is elusive enough. While there is a concentration upon what is available to analyse almost to the point of cloaking a lack of understanding with massed data banks, there is also an assumption that 'hard evidence' can be cursorily dealt with upon virtue of its clarity and unambiguity. The communication concentrates upon the interpretation, the selection and more often than not entirely subjective presentation of that which is regarded by the excavator as significant. Against this background, with all its inevitable shortcomings, is the recognition of the 'other material products' of archaeological excavation. Today we subsume this material under the umbrella term of 'environmental evidence'. Under this umbrella term we can find the bone assemblages, carbonised seed lists, soil analyses, pollen analyses and allied materials. Under normal circumstances this material is lodged at the end of the report, all in due order as a series of appendices. Commonly it is not referred to in the main text of the report other than to indicate its presence as a specialist contribution. There are, of course, a multitude of reasons for this practice not least of which is the difficulty of communication between excavator and specialist. It is not unusual, for example, for an assemblage of bones to be sent to the specialist, all duly marked with an incomprehensible series of letters and numbers, with a covering note 'Herewith the bones from Nother............ please analyse and send report by..........' (date coincidental with publishers deadline). In reality the bone material from a site is an embarrassment. Liaison between excavator and specialist is commensurate with the former's sympathy for the material and the latter's commitment. The end product regularly comprises a list comprehensible to another bone specialist and generally of remarkably little value beyond a simple statement of presence and absence. Environmentally we are no further forward at all. The same can be said of seed lists, soil samples and even radiocarbon dating results. All seem to be regarded as addenda rather than fundamental.

Over the last ten years this pattern of activity has been subject to a gradual and occasionally almost imperceptible change. The advent of improved data recovery, wet and dry sieving machines, the unit 'environmentalist', the institution of sub-departments in Universities to deal specifically with faunal and floral materials and the establishment of comparative collections have all served to build upon the early work of the pioneers. There is undoubtedly a trend in process. Site oriented archaeology gave way to catchment areas, locational archaeology to the archaeology of a landscape—effectively nothing new in that most of these aspects have been examined in the past.

The particular trend which concerns us here, however, is a new one and with it there lies a great deal of danger. Environmental archaeology is already accepted as a comprehensible phrase with connotations and denotations which have not yet been fully thought out or appreciated. The danger lies in the overswift adoption of a new technique. Regularly the result of such a rapid acceptance, once it is realised that enormous and even prohibitive limitations are inherently present in the material evidence, is disillusionment and subsequent rejection. Environmental archaeology, a phrase which is likely to do more harm than good is even now in its infancy. Its importance is founded in the belief that it will provide the answers to function, to operation, to land management and site organisation. In effect these answers are the prime objectives of excavation and consequently environmental archaeology is but an integral part of excavation and needs to be viewed as such and no more. It is, as a pursuit, no more important than the results it will yield in giving greater substance to the basic archaeological data. There is a case to be made that it will inspire new questions and the ensuing techniques to satisfy those questions. This is an objective to be actively pursued in that the scale of our ignorance of the past is but underlined by the results already obtained.

In outlining this caveat, my purpose is to encourage the initial establishment of a balance of disciplines avoiding the normal and incidentally time wasting and expensive excesses of extremes which usually presage a compromise. In philosophical terms compromise is the worst of all worlds. In archaeological terms compromise would lead to less and less being achieved. By introducing as a compatible level the disciplines of the specalist to complement the work of the excavator the prime objective must be the highest standards of excellence. One seed impression does not equal a harvest and two different varieties does not equal crop rotation!

There is a temptation, since analysis of environmental material lends itself to statistical analysis, to utilise numbers as a definitive rather than an informative function. It is too simple to disregard the macrostate of a prehistoric environment by detailed analysis of the microstate of an environmental sample. It is vitally important to retain the context and sample size in balance and accord appropriately a tentative conclusion. From the results of environmental analyses an abundance of 'acceptable data' will accrue which unless checked at the outset will allow yet another new branch of archaeology, the predictive statistication, to destroy by selective randomisation the whole purpose of the analyses.

The positive objectives of analysis of environmental material are to complement the physical excavation results. In so doing, and this includes the refining and testing of techniques of both data recovery and subsequent analysis, the aim is to increase our understanding of prehistoric and historic settlements. Even today our knowledge is woefully slim but the opportunity to increase that knowledge, provided it is handled with care, is present. It would be a tragedy should 'environmental archaeology' prove to be just another fashionable phase and phrase.

The introduction to this paper is based in large measure upon the research findings over the last eight years at the Butser Ancient Farm Research Project. The objectives of the project are published elsewhere (Reynolds 1979) but in brief it is an open air scientific research laboratory devoted to prehistoric agriculture and archaeology. In effect it is an environmental research station. Its purpose is to replicate the processes and practices as evidenced by the archaeological data in order to test fully the hypotheses advanced based upon those data. The testing process itself reveals not only inadequacies in the data and its acquisition but also in the hypotheses in such a way as not only to proffer systems for improvement but also by actually carrying out replicated tests over a period of time to learn by observation of previously unexpected and unpredicted occurrences and thus to frame specific questions to test further these new data. The rangeand scope of the research programme at the farm are considerable and there has been great pressure brought to bear to publish the results of these programmes almost as soon as they are achieved. This pressure has generally been resisted because of the reasons outlined in the introduction. Because the research work is primarily involved with agriculture, even now it is too soon to publish any firm results and the most positive observations can only be regarded as trends. A period of a mere eight years in an agricultural research programme is pitifully short. It is a truism to say that no one year is like another and over the last eight years the climate, the major variable affecting agricultural practice and conditions, has been subjected to the wildest extremes. It is quite remarkable how common have been the superlatives amongst those most conservative of men, the meteorologists. The hottest, driest, coldest, wettest since records began typified the weather reports. While the extremes experienced have been of the greatest value in that they have occurred and their effects have been monitored, not enough time has elapsed to allow for an overall statistical validity. A minimum period would seem to be some twenty to twenty-five years. This reluctance to publish data, however, is creating its own problem in that in due course the sheer quantity of data will be almost prohibitive. Even in the following examples drawn from the research programmes current within the Ancient Farm it is important to stress that for the most part any results quoted are interim and subject to overall change as time goes by, and secondly to underline that unless stated elsewhere the programmes are restricted to a specific geological and climatological zone.

To illustrate the work of the Ancient Farm and to show how it is inextricably involved in environmental research I have selected four essentially disparate topics. Three of these are drawn from the core research programmes, the fourth is a specific experiment carried out for a third party utilising the laboratory facilities and expertise available at the Ancient Farm. They involve respectively the storage of grain in underground pits, the performance of prehistoric type cereals against arable weed competition and the improbability that the majority of carbonised grain recovered from archaeological excavations is of significance in gross harvesting terms, the erosion pattern of an experimental domestic earthwork and finally the testing of a so-called Romano-British Grain Drier. The methodology of experimental work as carried out at the Ancient Farm is published elsewhere (Reynolds 1979). The only necessary element of that methodology which needs to be underlined here is the conviction that so far as possible the human element must be disregarded within the experimental milieu. Concentration is entirely focused upon function and the interaction of one element with another and given the equipment available the collection of as much data as possible even though their relevance may be in some doubt at the time. It has been the early experience of experimental work at the Farm that selectivity of data collection denied subsequent and unforeseen questions. This argument could with advantage be applied to excavation procedure.

Pits

There is little doubt but that the major archaeological vestige of the Iron Age period throughout the chalk lands and the sand and gravel terraces of southern England is the pit (Bowen and Wood 1968). Its sheer frequency as a feature implies a specific technology which utilises the pit for a number of different and varied purposes. Because of its predominance it has been the subject of considerable interpretative effort with various blanket explanations being offered ranging from pit dwellings (Wheeler 1943) though to grain storage (Pitt Rivers 1887, Bowen and Wood 1968). This last has gained widespread acceptance though not until relatively recently has that acceptance been supported by empirical evidence. Not until experiments were carried out in the nineteen sixties (Bowen and Wood 1968, Reynolds 1967, Reynolds 1969) was the process of grain storage in an underground pit accepted by the agricultural world. It is, perhaps, worth making the observation that an interpretation which involves a process or function which is beyond the strict limitation of archaeology should satisfy the experts or practitioners of that discipline or process. The do-it-yourself approach has not exactly enhanced experiment in archaeology.

The arguments leading towards the premise that certain types of pits were used particularly for the storage of grain and the results of twelve consecutive seasons of experimental storage of grain in underground silos are available (Reynolds 1978 - unpublished Ph.D. thesis, Univ. of Leicester). In brief the process is relatively simple to understand. Grain during its normal respiration process utilises oxygen and gives off as a waste product carbon dioxide. When placed in a sealed container the respiration cycle continues until the available oxygen is exhausted and the atmosphere has been changed to one charged with carbon dioxide. At this point the grain enters a state of unstable dormancy, the instability being caused by the presence of mirco-organisms fungi and bacteria, capable of surviving and in certain cases thriving in anaerobic conditions. Provided a low temperature is maintained and no further oxygen admitted even the micro-organisms can be kept largely in abeyance. The pit effectively acts as such a sealed container. For brevity reference can be made to Tacitus who described the German practice of sealing underground caches with fimus (Tacitus) and Pliny who described a similar practice in southern Spain. There are no direct classical references to British practice nor indeed, in contrast to continental and African countries, are there any ethnographic historical parallels. In effect the references quoted refer to the impermeable seal necessary for the container to function properly. Controls where pits were left open to the elements have been carried out with the inevitable results (Bowen and Wood 1968).

That Iron Age type pits can be used successfully for the underground storage of grain has now been proved conclusively. Research is nonetheless continuing in order to gain answers to supplementary questions, the most important of which is the assessment of the long term viability of a pit. However, what is of particular interest within the context of this paper is not the efficiency of the process, its economic implications or even the function traces which the process leaves upon the walls of the pit, but rather the waste product of the process itself. Within the experimental programme major attention was focused upon unlined pits where grain was literally stored in an untreated pit. The production of carbon dioxide gas which acts as the main preservative agency is the waste product of the germination of the seed at the interface between stored bulk and the seal over the mouth of the pit and the pit walls. Because the seal is damp in order for it to be impermeable the ideal conditions for germination exist at the interface layer. It is dark, moist and warm and there is abundant oxygen available to accelerate germination. To a lesser extent the same is true of the pit walls where the moisture content of the rock encourages germination. The build-up of carbon dioxide gas to a level which inhibits further germination is relatively swift taking on average no more than three weeks. At this point directly beneath the clay seal the grain has germinated to a depth of some 0.05 m. Similarly the grain has germinated next to the pit wall creating a skin averaging some 0.03 m thick. The skin of germinated grain is a thick matt of intertwined sprouts and rootlets which effectively acts as a further barrier to moisture penetration into the stored grain bulk now in a dormant state. However, the germination process has been arrested before the starch reserves of the majority of seed are fully utilised. The matt of sprouted grain adheres quite firmly to the pit wall during the recovery of of the grain bulk after storage. In itself it is useless, dead, damp and riddled with fungi and bacteria and quite unpalatable even as pig fodder. Clearly it needs to be removed from the pit as further deterioration will take place once it is open to the atmosphere. Naturally it could find its way to a midden but a more practical and simple process is to peel it away from the walls of the pit and burn it in situ. The nature of such a fire, the material itself being damp within the confines of a pit perhaps as deep as 2.00 m, is slow burning and to a large extant a reducing atmosphere. Several tests have been carried out showing that carbonisation regularly takes place in this type of fire. As mentioned above the seeds within this waste skin have germinated pushing out rootlets and sprouts but generally germination has not sufficiently advanced for the starch to have been totally consumed and the seed to have lost its shape. Indeed on average half of the seeds in this matt have only lost the germ element and thus retain their bulk and morphology. Their carbonization is a matter of chance and quite impossible to quantify given the vagaries of such a fire. Naturally enough rootlets and sprouts are utterly destroyed.

Recently analysis of carbonised seed recovered from a so-called Iron Age storage pit revealed exactly this condition (M. Jones, pers. comm.). It is attractive to see this discovery as a clear confirmation of the practice of storing grain not only in underground pits but in unlined pits. It is similarly attractive to use the discovery as a straight vindication of the empirical approach of hypothesis testing. It does illustrate, however, the immense value of close co-operation between specialists. As a final postscript to this section on pits it is worth recording that the matt of germinated grains referred to above does also affect the nature of the pit walls cut into chalk. Close examination of the matt as it is peeled away from the walls reveals tiny granules of chalk rock enmeshed in the rootlet structure. The wall itself after two to three winter seasons of consecutive storage becomes quite smooth even including the interstices in the rock where the rootlets have penetrated in their quest for moisture and stability. This effect has been noted on Iron Age pits by the author on numerous occasions but as yet has not been officially recorded archaeologically. Suffice to comment that the walls of a subterranean structure like a pit should be afforded the same close scrutiny as that lavished upon an upstanding construction.

A logical projection from the storage of grain in an underground silo is the physical growing and harvesting of the cereal crops themselves. For the past eight years and before that for a period of three years extensive cropping trials have been carried out respectively at the Ancient Farm and Avoncroft Museum of Buildings, Bromsgrove, Worcestershire. Broadly the prime objective has been to establish the parameters of crop yield of the prehistoric type cereals against monitored climatic and soil conditions. Ancillary objectives range through stand height variation, arable weed competition, potential indication of sowing seasons, disease resistance and aphid resistance. In direct association with these trials a series of ploughing experiments utilising replica prehistoric ards has been carried out in order to assess the reality of so-called 'ard-marks' recovered by excavation. In practical terms the study involves both the effect of ploughing upon the ard and upon the soil. A further ancillary programme is concerned with sherd/artefact movement in the ploughsoil under prehistoric ploughing conditions. To exemplify all the above, one field designated VII on the demonstration area of the Ancient Farm is used for all of the above trials.

Carbonised Seed

Our purpose here, however, is to examine two factors which have emerged from the trials over the past eleven years and to offer them as a direct contribution toward an analysis of carbonised seed as recovered from Iron Age sites in particular. Ordinarily our prime date for hypothesis formation of cereals grown in the Iron Age stems from two basic sources, seed impressions fired into pottery, presumably by accident, and, more importantly, carbonised seed. As techniques of recovery have improved so the quantities of seed recovered have increased. Yet it is necessary to introduce at the beginning a caveat which is not ordinarily appreciated. From the considerable data bank accruing from the experiments, analysis of emmer wheat (Triticum dicoccum) spikes indicate an average of thirty seeds per spike. Further, accepting the low estimate of a yield of ten hundredweights per acre the potential yield of seeds would be some thirty-two million. When we compare this figure with the average few pounds of carbonised seed the pitifully minute nature of the usual sample is underlined and any statistical analysis of it is brought into serious doubt. Nonetheless from the multitude of palaeobotanical reports now available the vast majority of samples comprise an admixture of cereal and arable weed seed with the latter not infrequently outnumbering the former. It further goes almost without saying that virtually without exception the samples are drawn from excavations—partial, sample oriented or total of settlement sites. It is ironic that at this time despite unquestioned agreement that agriculture formed the basic economy of the prehistoric period, particularly the Iron Age, there have been no excavations other than by fortuitous accident of 'Celtic fields' or outlying land or close investigations of farm organisation (Cf. Bell, this volume). Since it must be logical to accept that settlement is a function of farming rather than the reverse then a prime potential source of information has been largely ignored. There is also little doubt given the increasing numbers of Iron Age settlement sites being found especially in valleys and on the previously unfashionable clay soils that the countryside was effectively shaped at this time. The landscape was in fact a veritable agroscape where agriculture and silviculture were the norm and wasteland at a premium. In all probability the landscape was better tended even than today and presented a 'manicured appearance'. This particular hypothesis will be explored further in due course. It is sufficient here to emphasise that the business of agriculture is to be found in fields and fences. The material recovered from a settlement, already subjected to chance survivial, can at best serve to indicate material taken into the settlement which has potentially been processed prior to its arrival or which has been subjected to processing within the settlement. It is in this context that the samples of carbonised seed need to be viewed. Are they representative of a gross product introduced into the settlement as, for example, the harvest, or are they the waste product of a supplementary process involving the harvest or yet again, despite their being cereal and arable weed seeds, have they anything to do directly with a cereal harvest at all. There are clearly an increasing number of variables in proportion to the comprehension of the agricultural and domestic requirements of an operational economy.

Precise information of harvesting techniques as employed by the Iron Age people as opposed to the Roman systems (White 1970) is difficult to isolate. We do have the comments by Strabo and Diodorus Siculus that the Celtic practice was to reap the ears or spikes of the crops. Support of these statements can be found in the ubiquitous artists' impressions of prehistoric harvesting where the so-called sickle is to be seen neatly cutting off cereal heads. Practice, however, has little in common with imaginative representation. From the observation of some eleven seasons of growing the prehistoric type cereals of emmer and spelt wheats, belief in the determination of the typical small 'sickle' has radically waned to the point of offering an alternative function—this being the splitting of hazel gads to make into thatching spars for which the tool is admirably suited. This is not to say that they cannot be used for cutting wheat ears but rather that they are grossly inefficient and far too slow. The problem lies in the nature of the crop itself. Accepting the classical sources as being a reasonable description of the actual harvesting practice, both at Avoncroft Museum and at the Ancient Farm all the crops have been harvested in this way. Each year replica sickles are provided for the purpose, each year they are unanimously rejected by the reapers. The answer is simply that it is much easier to reap the ears by hand picking.

It was realised from the very first season that the prehistoric cereals have two characteristics which are not ordinarily to be found in modern hybrid wheats. The first major difference lies in the disparity of stand heights achieved by the tillers of the same plant. In the case of both emmer and spelt wheats this disparity can be as much as a metre from the shortest to the tallest spike, while in a modern hybrid it rarely exceeds 0.40 m. The importance of this characteristic will be developed below. The second characteristic is the 'necking' of the prehistoric types. As the ears ripen so they droop gracefully downwards from the main stalk in the fashion of modern barley. The effect of this not only inhibits water retention within the spike and consequent lodging potential, this being truer of emmer than spelt where lodging can be a problem, but also exposes the stalk top to the elements of wind and sun. Once ripe this section of the stalk becomes extremely brittle and is very easy indeed to break off. If allowed to reach maturity the natural result is for the spike to break away from the stalk and seed itself. The skill of farming is, of course, to pre-empt this point by as short a time as possible.

Observations of this disparity of stand height have led to the direct recording of each successive crop. Annually a thousand measurements of each crop variety and treatment are made across random transects with sample points 0.30 m apart. The results of just one year are provided to substantiate the present argument. These can be seen represented in histogram form in figures 2:1 and 2:2. To provide a direct comparison figure 2:3 shows the results from a modern cereal crop grown under exactly the same conditions. The selection of examples is further deliberately pointed in that they are drawn from a field which has been continuously cropped for the past eight years without any added nutrient whatsoever. The farm is, of course, free from all modern herbicides, pesticides and fertilisers.

The prime data comprise an admixture of cereal and arable weed seeds. Throughout the cropping programme the competitive weed flora has been most carefully monitored. In eleven seasons all the harvests have been effectively pure with but two contaminants. By way of explanation the harvest is taken to be the spikes that are collected from the crop. The contaminants, in turn, are arable weeds which can be found within the gathered spikes. The harvesting process consistently follows the classical description. However, the crop itself is regularly infested with an abundant weed flora and during the months of June and July particularly is a riot of colour. The primary enemy of farmers of all periods prior to the introduction of herbicides is charlock (Sinapis arvensis). It is vitally necessary during the early part of the season in April and May to hoe the crop carefully in order to keep at bay this most pernicious competitor. In this century it was not uncommon to hear of crops being ploughed back into the soil because of charlock infestation. Farming by definition is the provision of a preferred habitat for a specific plant. However, once the plant is established and can outgrow its competitors, there is no advantage in further hoeing and indeed there can be direct disadvantages in that too much soil disturbance in the secondary stages of crop development can damage the spreading roots of the crop itself. Consequently at harvest time there is an abundance of other plants amongst the cereal crops. Careful note has been made of those competitors which reach the reaping height. That is, those plants which head out amongst the majority of spikes and which are likely

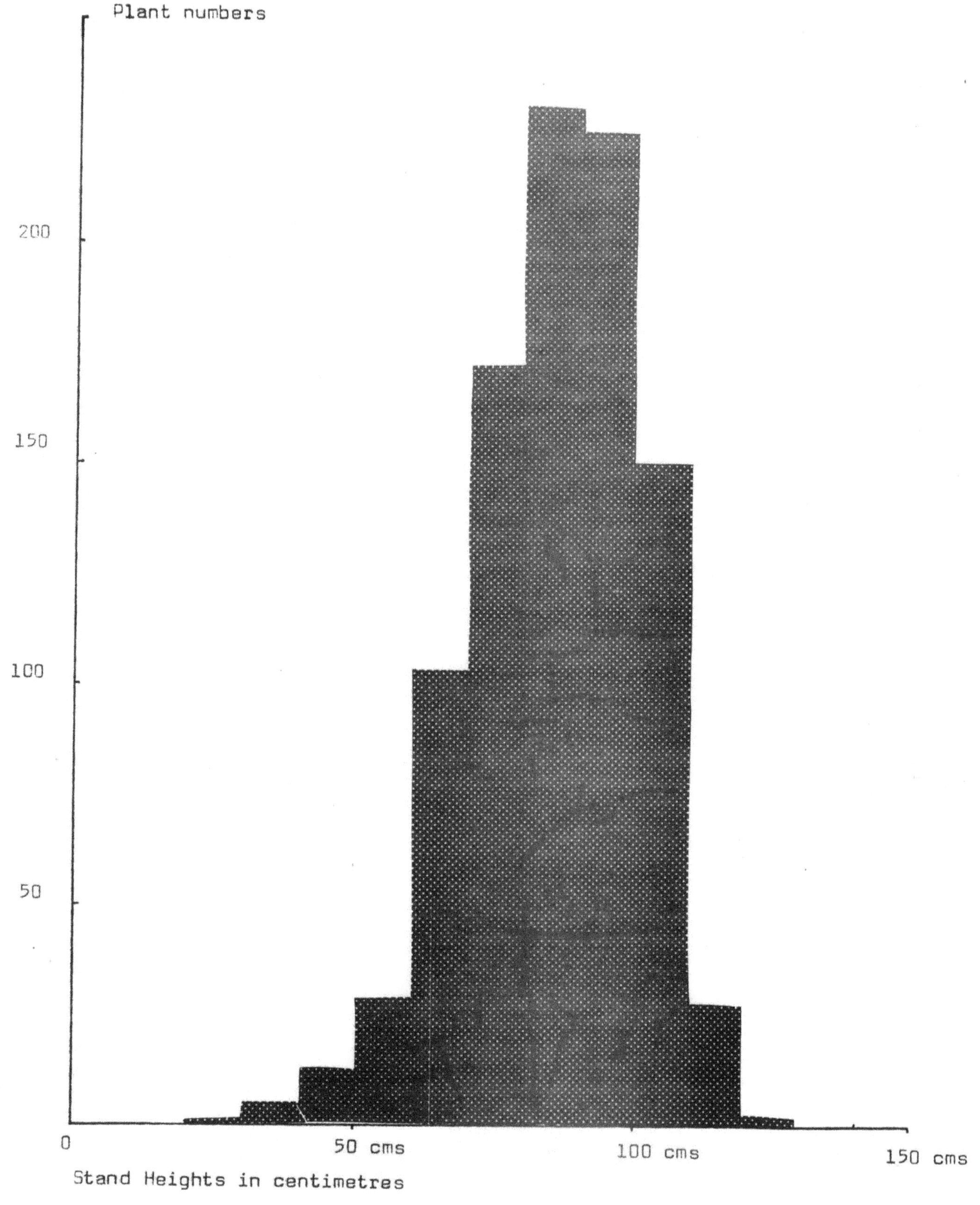

Fig. 2.1 **Stand heights of Triticum dicoccum.** Field II, 1978 **Butser Ancient Farm.**

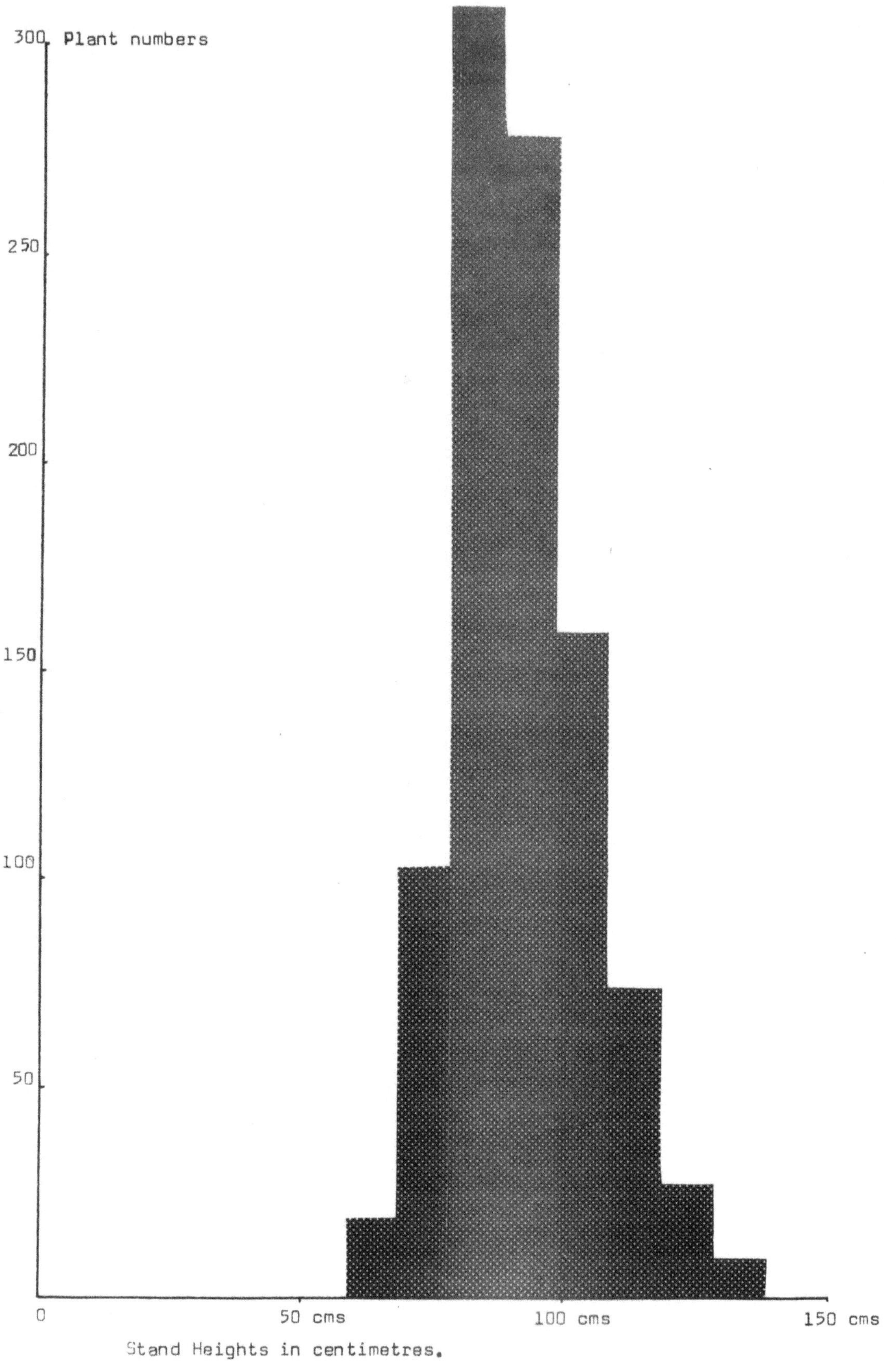

Fig. 2.2 Stand heights of Triticum spelta. Field II, 1978 Butser Ancient Farm.

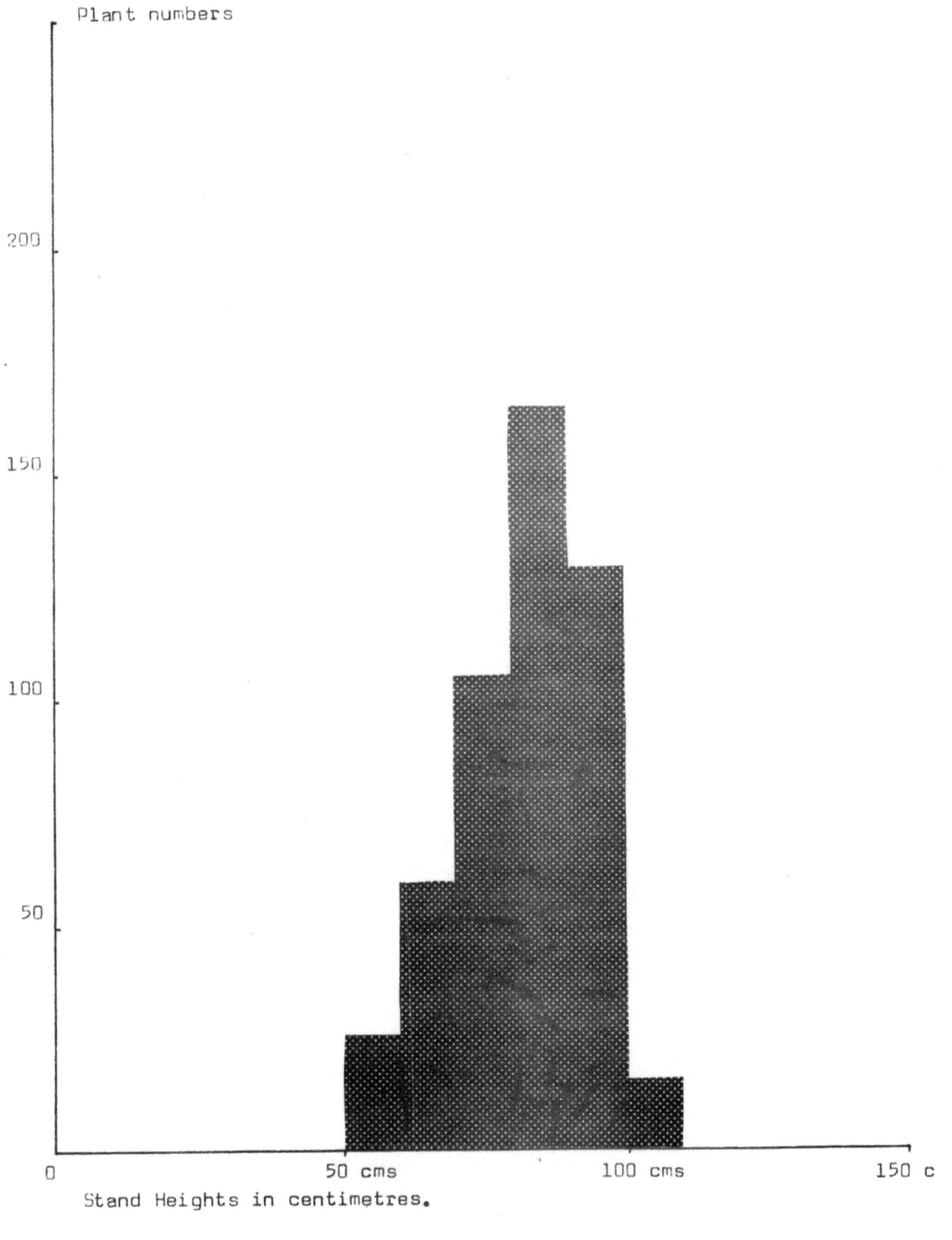

Fig. 2.3 Stand Heights of Triticum aestirum var. Maris Huntsman. Field IV, 1980 Butser Ancient Farm.

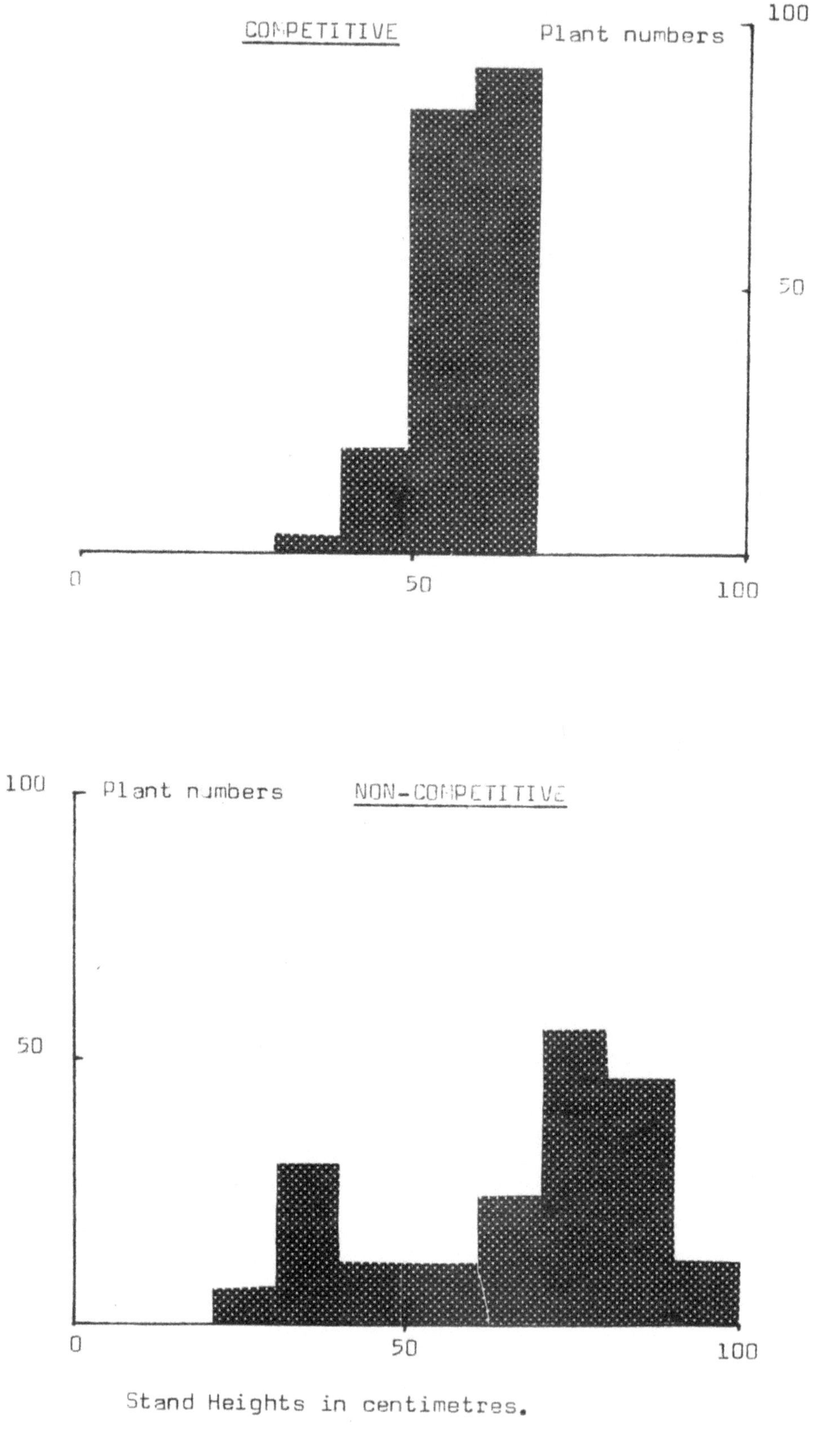

Fig. 2.4 Comparative stand heights of Agrostemma githago in competitive and non-competitive conditions.

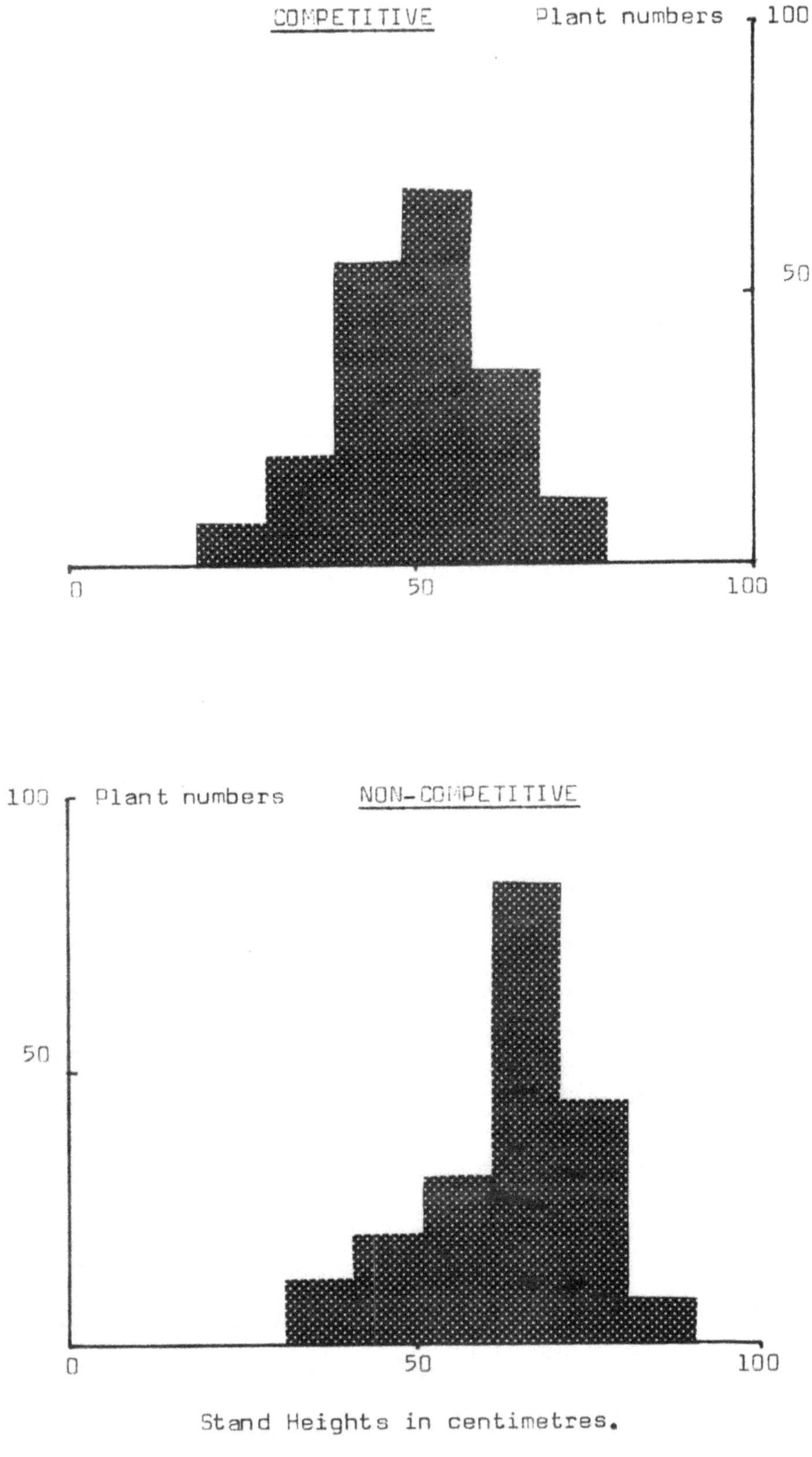

Fig. 2.5 Comparative stand heights of Centaurea cyanus in competitive and non-competitive conditions.

to be collected by accident during the harvesting process. Such plants are effectively few in number and virtually all of them are sufficiently different in nature and form for avoidance to be a matter of course. The most typical on calcareous soils are the thistle family (Cirsium), fat hen (Chenopodium album), poppies (Papaver sp.), charlock (Sinapis arvensis) and hedge mustard (Sisymbrium officinale). On acid soils as well as most of the above red shank (Polygonum persicaria) and pale persicaria (Polygonum lapathifolium) are the most common. The two contaminants referred to above are black bindweed (Polygonum convolvulus) and sowthistle (Sonchus arvensis). The former simply entwines itself up the straw stalks and wraps itself inextricably about the spike. The reaper recognises its presence but can do little about it. It hardly detracts from the crop in so far as it is not a poisonous species. The latter is relatively uncommon as a contaminant and only occurs as such since its seed is wind dispersed and occasionally a seed is trapped amongst the awns of the prehistoric type cereals. It too is unlikely to cause any problems in subsequent utilisation of the harvest. It is rare amongst the carbonised seed record as is only to be expected from its nature but it has recently been identified from an Iron Age site along the course of the M3 motorway (pers. comm. P. J. Fasham). Even though its presence within the harvested spikes is occasional this is not to suggest it is rare amongst the crop.

The purity of the harvest is the direct result of the reaping system. Since the abundant weed flora occupies the bases of the crop and while it does not necessarily detract from its overall performance, it does mask from the view of the reapers those spikes which are in the lowest stand height range. It would be counter-productive to hunt about amongst the base of the crop to recover what amounts to less than one per cent of the total yield. Ironically there is regularly no difference in the overall seed size of these omitted spikes from those of the reaped harvest. Consequently they cannot be identified as being distinct in any specific analysis programme. Similarly the observation needs to be made that carbonisation itself does not affect seeds uniformly and thus predictive assessments of original seed size are impossible.

Throughout the development of the Ancient Farm every effort has been made to supply as accurate a competitive weed flora as possible. In many cases this had led to supplementary research programmes involving individual species which are either threatened or actually extinct from the United Kingdom. In the latter category, for example, thorow-wax (Bupleurum rotundifolium) can only be observed in the fields and propagation plots at the Ancient Farm. The objectives of these supplementary programmes are to study the percentage germination fruiting characteristics and stand heights in competitive milieux of these threatened species and to keep a reserve seed bank in the normal sense as well as the maintenance of living populations. To underline the nature of the competitive arable flora which occupies the base of the crop two typical species, regularly recorded in carbonised seed samples, are selected as particular examples. The first of these, corn cockle (Agrostemma githago) is sufficiently frequent to have been considered as a potential food supply and its protein levels have been analysed (Renfrew 1973) although, in fact, it is poisonous (Forsyth 1954, Renfrew 1973) and spoils bread made from flour which contains it. Similarly it retains its poisonous characteristics even after drying and storage. In Figure 2:4 its stand heights are recorded in competitive and non-competitive conditions. Similarly the results of the performance of cornflower (Centaurea cyanus), another threatened species although non-poisonous, are shown in Figure 2:5. In both cases it can be readily seen that given the harvesting system they are well below the reaping height and are to be found locked within the mass of other arable weeds.

The argument, therefore, is now confused in that to all intents and purposes the norm would have been a pure harvest and yet the carbonised seed samples are anything but pure. Returning to the field situation, once the spikes have been harvested there remains still a second and extremely valuable product to be collected, namely the straw. Because of its relatively low gloss factor the straw from the prehistoric cereals is quite palatable to livestock. Similarly it had great value as thatching material, a fact substantiated by the stand heights alone and it would have been required as bedding material for livestock and not improbably for humans. There is always the alternative that then as now the straw could have been burned in situ but this does not deny the need for straw within the agricultural economy and it is not unreasonable to assume that it was cut, put into sheaves and carted back into the settlement. In any event the field area would have had to be cleared prior to its recultivation. The straw inevitably contained the mass of weed flora within its bulk. In this way, entirely apart from the harvested cereal the arable weeds could have found their way into the settlement. In order to quantify this exercise, each year several of the straw sheaves from the farm are analysed for the seeds they contain after a winters storage within a stack. The results from just one example are given in Figure 2:6. A typical sheaf can be seen in Plate 2:I. Even this analysis does not give a true reflection of the field state in that the seed release mechanisms of the different species of arable weeds are widely diverse. For example, the thistle population since its seeds are wind dispersed is rarely represented. Similarly the seeds of charlock (Sinapis arvensis) are exploded from a pod at the slightest touch of the trigger element. In fact, just prior to harvest one can actually hear the pods exploding as the wind blown stalks of wheat brush against the seed pods. Some simple measurements carried out at the farm have shown that charlock can deliver its seed up to three metres from the parent plant. Both thistles and particularly charlock are dominant weeds of the research crops, a fact which analysis would fail to substantiate. As has been suggested above in reference to sowthistle (Sonchus arvensis) attention needs to be focused upon the characteristics of individual plant species and any analysis needs to take account of these characteristics. A virulent weed in the field need not necessarily appear in quantity in the archaeological record.

However, there is yet the problem of carbonisation for which some account should be made. For carbonisation to occur a fire with a primarily anaerobic atmosphere is required. Undoubtedly some are the product of accidental fires but the following hypothesis seeks to offer a positive alternative to accident. That houses were thatched in the Iron Age is attested by the classical writers and one presumes that straw among other materials was used for this purpose. While it is readily recognised that river weed (Phragmites), heather (Calluna vulgaris) and even bracken (Pteridium aquilinum) could well have been used, the last being a potential explanation of bracken spores identified on calcareous sites, straw undoubtedly was a readily available material product of the

Plate 2.1 A typical sheath of emmer wheat straw.

SHEAF 15. Tr. dicoccum (Emmer Wheat)

Number of straws	1833	Sheaf gross weight	3.6 kg.
Complete spikes	14		
Rachis ends	602		
Seed yield	97		

WEED SPECIES	SEED NUMBERS
Achillea millefolium	900
Anagallis arvensis	73
Centaurea cyanus	1
Chenopodium album	30,000
Cirsium vulgare	60
Convolvulus arvensis	40
Crepis spp.	1330
Fraxinus excelsior	2
Leontodon autumnalis	55
Odontites verna	260
Papaver rhoeas	22
Sinapis arvensis	12
Sisymbrium officinale	911
Sonchus arvensis	403
Stellaria media	105
Veronica persica	30
V. chamaedrys	75

Fig. 2.6 Quantitive analysis for remanent seeds in randomly selected sheaf of emmer wheat straw.

agricultural process. The preparation of straw for thatching requires one most critical operation, it must be cleaned of all other materials including its own leaves, any pithy plant stems, all seeds and especially cereal seeds. The reasons are straightforward. Pithy material rots and by so doing would decrease the bulk and cause the ties to loosen. Any seed material attracts the attention of rodents who in turn may well eat the bundle ties in order to reach the food supplies. Traditionally straw cleaning and the preparation of bundles or yealms which form the thatch require the wetting down of the straw to facilitate the drawing of the straw stalks. The possible implications of the process for the disposition of postholes is discussed elsewhere (Reynolds 1979). The end product of this cleaning process is a pike of damp waste material comprising arable weeds, stems and seeds and waste cereal heads. It is but a small step to hypothesise the disposal of this damp rubbish with a slow burning bonfire. Such a bonfire inevitably has within it an anaerobic atmosphere which will allow carbonisation to take place.

The two factors which form the objective of this aspect of cereal investigation are, in effect, the result of two distinct approaches, each one relying upon the documentary and archaeological evidence available to us. The first is the pure harvest, the direct result of actually growing the prehistoric type cereals in Celtic sized fields and processing them accordingly. The second, the commonly found admixture of carbonised seeds is questioned as being non-representative of a harvest and an alternative hypothesis is raised. If the first, the pure harvest, is a valid hypothesis then the second cannot be representative of that harvest. Throughout it must be emphasised that one is dealing with possibilities and at best probabilities and it would be foolish to deny other hypotheses provided that they can be similarly validated. In effect both the above hypotheses are in a sense offshoots of the main research purpose, that being to establish probable parameters of yield for the prehistoric cereal types given different soil types, treatments and bio-climatic zones. These results will be published in due course as indeed will the full statistical records from which examples have been drawn above.

The third element of this paper refers to a series of experiments carried out at the Ancient Farm on behalf of Mr. C. Partridge of Hertford and funded by the Robert Kiln Trust (Reynolds and Langley 1980). The purpose was to reconstruct and subsequently test the function of a structure initially identified as a corn drier. The project was inspired by its discovery at Foxholes Farm, Hertford when the foundations were found to be virtually intact and standing to a height of approximately a metre. Indeed this substructure was in such a good state of preservation it has been removed in its entirety to Hertford Museum for conservation and display (Partridge 1976). The area plan and elevation of the structure Figure 2:7 shows quite clearly why it has been placed within the overall category of buildings interpreted as corn driers. The evolution of this interpretation is not entirely without interest. Despite a wide divergence in flue design their determination as corn driers was inititated in the early part of this century (Cocks 1921). Further confirmation was provided by Goodchild (1943) when he surveyed T-shaped corn drying ovens in Roman Britain. The interpretation was inspired by the regular discovery of carbonised seeds from flue and stoke hole areas of such structures and a few oblique classical references (Pliny, Virgil). The interpretation gathered so much support that the 'corn drying oven' is probably the most easily identifiable agricultural structure found in Roman Britain (Morris 1980). The motive behind the reconstruction was, in a sense, not so much to see whether the structure worked but rather how efficiently and economically it dried the grain. While this particular structure is unusual in its basic design, in order to balance the research programme a second structure was built based directly upon an example of the traditional T-shaped flue (Figure 2: 8). Although the main interpretation has been based upon the discovery of carbonised grain in stoke holes and flue systems, it is not necessarily wise to assign function from this type of material alone. It has been suggested, for example, that fires were and regularly are started by using straw and in view of the previous discussion it is not impossible to see a ready source of carbonised seeds of both cereals and weeds.

The detailed report of the construction of the Foxholes drier and its subsequent testing has already been published (Reynolds and Langley 1980). The completed structure (Plate 2: II) was used to test the hypothesis that it is necessary to contain totally within a structure the heated air emanating from the flue system to dry grain. The conclusions drawn from the extended series of tests, one example of which is given in Table 1 effectively invalidates the hypothesis. The major block to the structure functioning in the suggested manner is the impermeability of the floor. There is abundant evidence of solid floors being used ranging from the stone slab floor of the Atworth Roman Villa corn drier (Shaw-Mellor and Goodchild 1942) to the plank impressions observed on the chalk slurry from the Foxholes Farm excavation (C. Partridge, Pers. comm.). Such floors are, in fact, very poor conductors of heat and rather than allowing hot air to rise through the grain, they simply warm the lowest levels and cause moisture exchange from one level to another. The nature of the floor prevents a significant increase in temperature and even were it possible the grain would in turn be damaged and spoiled by overheating.

A second series of tests has been carried out in 1980 by Mr. P. McCrone of Glasgow University utilising the second T-shape corn drier built at the farm. In contrast to the Foxholes Farm structure there was no evidence of flooring or building material from the excavation which yielded the flue system at Barton Court Farm (D. Miles, pers. comm.). Consequently a rectangular oak plank floor was constructed 4.00 x 3.00 m which covered the bar and shank of the 'T'. Air flow trials showed how the hot air travelled along the flue and bifurcated to the two exit points at the extremities of the 'T'. The edges of the floors were sealed to the ground surface except for two exit points on either side of the stoke hole. Therefore, the hot air would be drawn back under the floor heating its whole area. Thereafter, trials were carried out to explore the variables of superstructure design in case free passage of air over the grain would enhance the drying process. Initially a simple open-framed building was constructed based upon sill beams. There was no convincing post-hole evidence from the excavation. The roof was thatched with heather and ridged with turves. Extensive trials were carried out with this open-sided roofed structure in a variety of weather conditions. All produced similar results, Table 2. In hot dry weather grain was spread onto a plank in the e open air as a direct control. The control predictably dried quite beautifully. The trial resulted as before. A second variable was the use of hurdles to provide a permeable wall structure. This produced similar results.

Plate 2. II The reconstructed Foxholes Farm Corn Drier. 2 m scales.

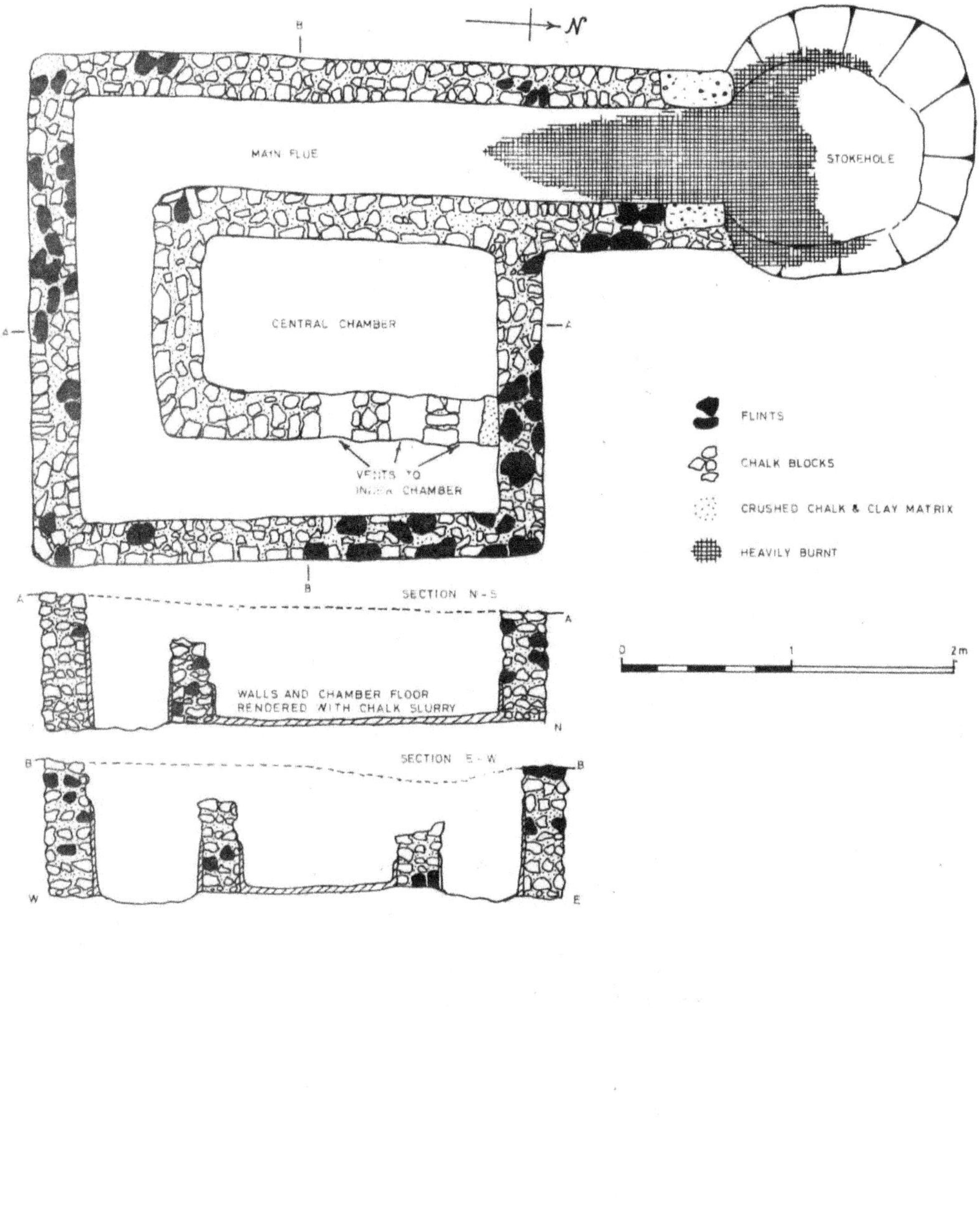

Fig. 2.7 Area plan and elevation of the archaeological evidence for Foxhole Farm Corn Drier (after C. Partridge).

TABLE 1

Thermocouple No.	Temperature in degrees Centigrade at 30 minute intervals (from left to right)													
1	127.0	152.0	172.0	152.0	159.0	184.0	147.0	182.0	167.0	208.0	192.0	127.0	207.0	122.0
2	59.0	62.0	66.5	79.0	78.5	76.0	83.0	80.3	88.0	89.0	83.0	86.0	86.0	88.0
3	49.5	53.0	59.0	61.0	64.5	64.0	72.5	74.0	78.5	88.0	83.0	83.0	88.0	81.0
4	35.0	38.5	42.0	44.5	46.5	48.0	52.0	55.0	55.5	57.0	55.0	59.0	57.0	58.0
5	35.0	38.5	39.0	43.5	43.5	45.0	45.0	48.0	53.0	54.5	54.5	59.0	58.0	60.5
6	29.0	30.0	32.5	35.0	36.0	36.0	42.0	42.0	43.5	45.0	47.5	50.0	49.5	52.0
7	71.0	78.5	86.0	76.0	88.0	93.0	94.0	98.0	98.0	101.0	105.0	83.0	110.0	78.5
		*			*		*		*			*		

Moisture Content Sample	Percentage, at hourly intervals (from left to right)					
1	20.8	22.4	20.3	21.5	23.0	24.75
2	20.4	22.0	21.75	21.0	23.5	26.0
3	18.4	20. 95	19.5	19.5	19.0	20.4
4	19.0	17.7	17.75	18.8	19.0	18.4
5	18.4	19.25	17.75	19.4	18.5	20.0

Temperature Table: Thermocouple positions: 1 in flue above stoke hole, 2, 3 4 & 5 set anticlockwise 0.30 m from corners on floor surface, 6 centre floor surface, 7 flue exit.

Moisture Content Table: Sample numbers correlate in order with floor positions 2, 3, 4, 5 & 6.

General Weather description: Fine dry day with light northerly wind veering south in late afternoon.

* denotes stoking times.

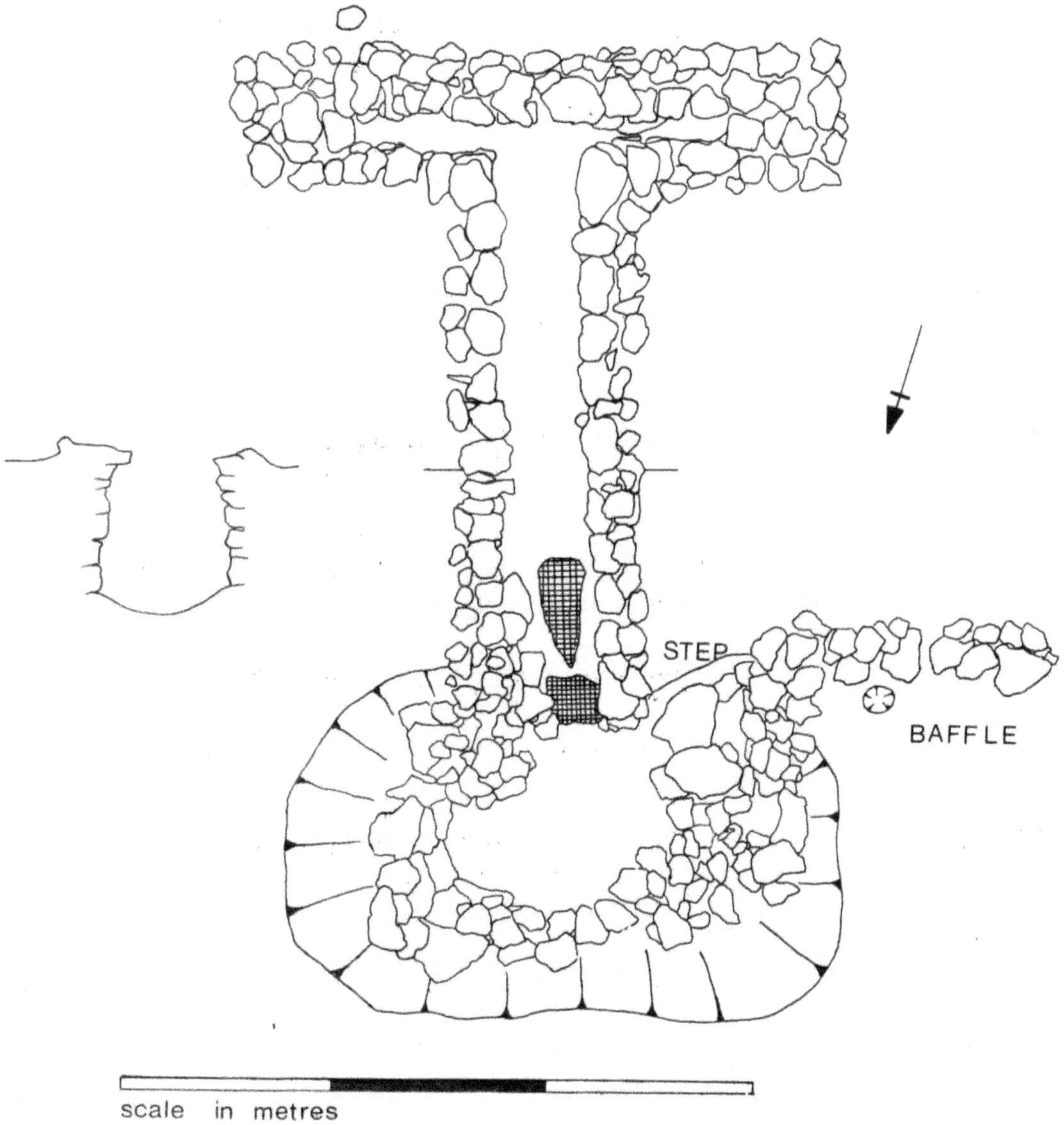

Fig. 2.8 Area plan and elevation of the archaeological evidence for Barton Court Farm Corn Drier (after D. Miles).

TABLE 2

Thermocouple No.	Temperature in degrees Centigrade at 30 minute intervals (from left to right)						
1	190.0	172.0	172.5	152.5	147.0	169.0	174.0
2	90.5	77.0	79.0	78,0	78.0	79.0	86.0
3	88.0	71.0	78.0	77.0	77.0	78.0	82.0
4	76.0	76.0	78.0	77.0	78.0	77.0	76.0
5	61.0	73.5	74.0	74.5	74.5	72.0	74.0
6	71.0	98.0	97.0	97.5	97.5	96.0	94.5
7	96.0	97.5	98.0	98.0	74.5	82.5	89.0
	*				*		

Moisture Content Sample	Percentage, at 30 minute intervals (from left to right)						
1	20.9	19.0	18.0	15.5	15.2	12.5	10.0
2	20.9	19.0	18.0	15.5	12.5	12.0	9.5
3	20.9	19.0	18.0	15.5	13.5	12.5	10.5

Temperature Table: Thermocouple positions: 1 in flue above stoke hole, 2, 3, 4 and 5 set anticlockwise 0.30 m from corners on floor surface, 6 centre floor surface, 7 flue exit.

Moisture Content Table: Sample numbers correlate in order with floor positions 2, 3 and 4.

* denotes stoking times.

Unless the grain is apread so thinly on the floor as to be uneconomic of time and effort, the mechanics of these structures with their solid floors are such that they cannot be considered to be effective for the purpose assigned to them.

Rather than propose a negative, an alternative hypothesis for these structures was raised and explored. The analysis of carbonised barley seeds recovered from the T-shaped 'corn drier' at Barton Court Farm, Abingdon, Oxford showed that they had actually sprouted (M. Jones, pers. comm.). Stimulated by this evidence and further encouraged by the heating patterns experienced on the floors of both reconstructed structures, the hypothesis sought to explore whether barley could be successfully malted therein and that the term 'sprouted' used above could be substituted by the brewers term 'chitted'. Successful trials were conducted and the resultant malt was analysed by Briant & Harman, Consulting and Analytical Chemists of London. The malt was favourably compared to the Amber malts of 1935-45 traditionally used by Guinness. In brief the malt was regarded as quite suitable if a little smoky, for the manufacture of an ale and most importantly, its production was regarded as quite successful.

In conclusion the above trials indicate that the adoption of untested interpretations can prejudice the evidence in such a way that an anomaly such as corn drying in the history of agricultural development can not only be accepted almost without question but also be embraced as a clearly understood phenomenon.

An Experimental Earthwork

The final section of this paper is a brief reference prior to initial publication to perhaps one of the most fascinating elements of the research programme for the archaeologist excavating in the field. Inspired by the experimental earthworks at Overton and Wareham Down (Jewell and Dimbleby 1968) it was decided to incorporate an experimental earthwork within the demonstration area constructed in 1976. Essentially this earthwork differs radically from both Overton and Wareham Down earthworks in that it seeks to explore the silting pattern and plant recolonisation sequence of a 'domestic' earthwork. The distinction is drawn quite clearly in that the earlier earthworks are linear and monumental. In the case of the Butser Ancient Farm Demonstration Area's (BAFDA) earthwork the questions are reduced in scale and complexity and concentrate upon the above two elements. In effect, as with the majority of the research programmes at Butser, this experiment is a pilot scheme.

While the data achieved from the Overton and Wareham earthworks are fundamentally important their relevance is somewhat restricted because of their initial shape and scale and it is extremely difficult to parallel the results to date with the vast majority of ditches and banks which are the subject of excavation. These tend to be of considerably lesser scale and are invariably associated with settlements both large and small. To underline the distinction I refer to these as domestic and the BAFDA earthwork is designed to simulate the typical ditch and bank which were commonly constructed around such settlements. Indeed, its general plan is based directly upon an extant although unexcavated earthwork at East Castle in Dorset (RCHM). The selection was determined by the topographical nature of the site of the demonstration area

Plate 2. III Ditch section east of the BAFDA earthwork. 50 cm scale divisions.

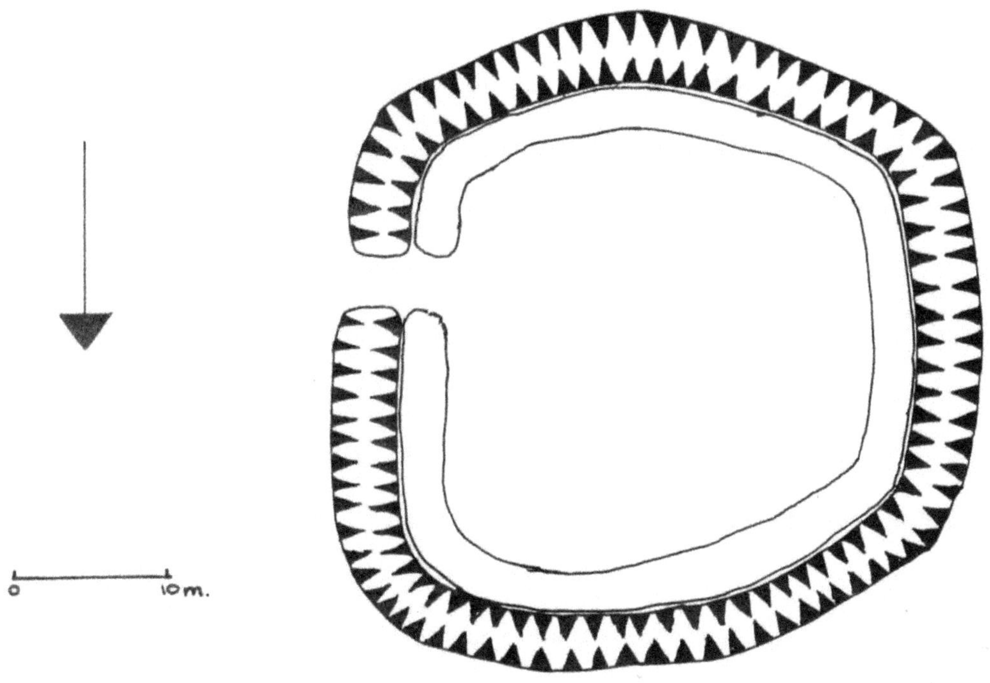

Fig. 2.9 Area plan of experimental domestic earthwork, Butser Ancient Farm Demonstration Area.

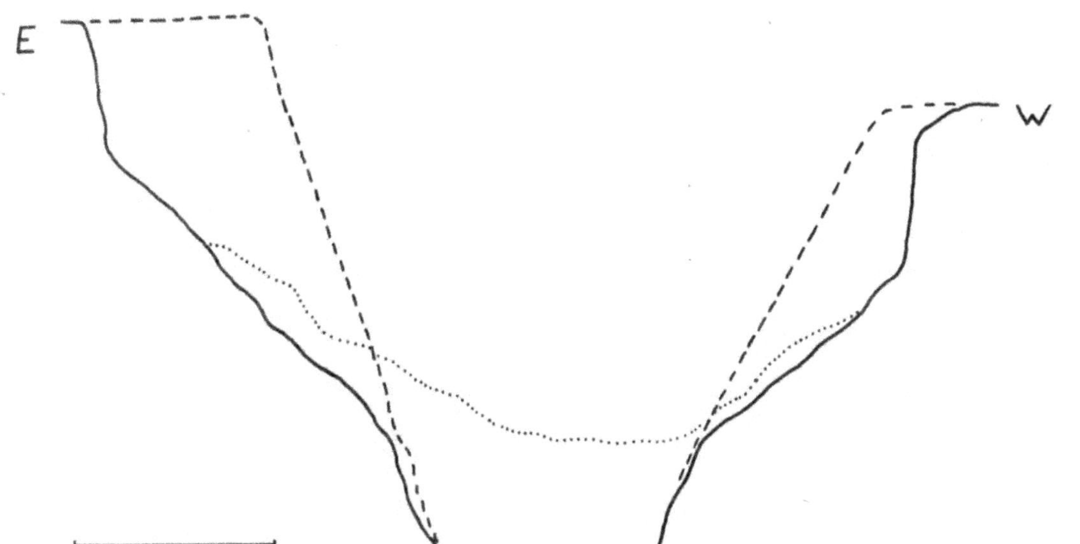

Fig. 2.10 Profiles of sections across eastern arm of the experimental earthwork, Butser Ancient Farm Demonstration Area.

(Figure 2:9). It is rectangular in plan with rounded corners. Its location in the valley bottom allows the north-south line to be its long axis. The ditch was cut with a JCB fitted with the smallest standard triangular shaped bucket available. In practical terms the bucket had a working capacity of between two and three cubic feet. The end product comprised a ditch averaging 1.50 m deep with a flat bottom some 0.30 m wide. The ditch averaged 1.50 m across. It was impossible to standardise this shape around the earthwork as a whole and consequently profiles were taken at 5 m intervals. A berm of 0.30 m was left between the inner edge of the ditch and the simple dump bank. The turf cover from the ditch surface was primarly used as a core to the bank. The bank was not compacted by the machine at all. All shaping was done by hand in order to obtain a consistant profile. No revetting was done on the ditch side of the bank. On the enclosure side a small revetment of turves some 0.30 m high was made. The bank itself was not measured in detail, the primary interest being focussed upon the ditch. The berm width was carefully standardised throughout.

Annually the profiles of the ditch are recorded across the original positions. The ditch is to be sectioned every four years. The plant recolonisation of the bank and ditch sides was carefully monitored season by season. In the late summer of 1980 the first section across the eastern arm of the ditch was excavated with surprising results. The profiles can be seen in Figure 2:10 while the detail of the section appears in Plate 2:III. No less than 0.55 m of build-up had taken place in the base of the ditch with a series of clearly defined layers. All the layers can be tied into specific time/climate periods and none of the source material is derived from the bank at all. Indeed the deposition pattern of the individual layers not only suggests fairly stable periods when silting did not take place allowing a dark 'humic' layer to develop but also the angle of rest suggests the major source of infilling material to come from one particular side. Ordinarily one would assume this to be material derived by slippage from the bank but, in fact, it is derived from the opposite side.

The process involved, once observed, is quite straightforward. Periods of intense cold, for example, the months of January and February in 1978 when for many days the temperature stayed below 0° C, cause the exposed surfaces of the chalk rock in the ditch to fracture. The subsequent and inevitable rainfall loosens the chalk fragments and washes them into the base of the ditch and in so doing undermines the turf and root-bonded topsoil around both the outer and inner perimeter of the ditch. The more intense the cold, the greater the frost penetration and fragmentation, the more deeply undermined the topsoil becomes. Inevitably collapse will take place and again primarily from the outside lip of the ditch by virtue of the disturbance caused by the normal traffic of animals and people. The inner side of the ditch is more favoured in terms of stability not least because it is less subject to traffic but primarily because the berm, from the inception of the earthwork, is a protected habitat for plant growth. Even in the first year verdant growth took over this narrow shelf and effectively acted as a solid barrier and prevented any slippage from the bank into the ditch at all. Thereafter plant colonisation of the bank followed swiftly so that within three years it became stable, bonded by the root action of the herbage. Details of the recolonisation pattern will be published shortly.

Sufficient here to remark that the patterns of occupation are subject to steady change with the final phase of grass colonisation not yet achieved after a period of four years.

Initial analysis of this one section is so remarkable that it offers considerable cause for concern over the interpretation of domestic earthworks. It would further seem to underline the need for more complex experimental earthworks based upon the full range of soil types and located within different bio-climatic zones. A design for such an earthwork has been drawn up and it is hoped to implement at least one in 1981. In general terms it would seem that the single most significant element affecting a ditch and bank is the weather. Therefore, it is necessary to consider the effects of weather upon the orientation of ditches, both archaeological and experimental, in that the silting deposition patterns will inevitably vary. The design selected seeks to oppose at right angles a length of ditch and bank to each of the eight major points of the compass. The length of each arm is to be 20.0 m and, in contrast to the pilot earthwork briefly described above, will be divided into two equal lengths, one with a berm, one without. The resultant octangonal earthwork will not, therefore, simulate any prehistoric earthwork in plan but only in profile. The objectives for such an experiment are essentially two-fold although considerably more data can be garnered than here described. First the plant colonisation of the raw earthwork needs to be monitored carefully and the annual changes of population recorded until stabilisation is achieved. Second the changing profile of the ditch requires annual recording with excavation of sample sections scheduled on a four year basis.

The information gained from such an enterprise will be of the greatest value to archaeologists working in the field in that it will provide a series of standards against which archaeological data can be compared and quantified. Finally both the pilot earthwork at BAFDA and the proposed octagonal earthworks underline the contribution which empirical experiment can make to archaeology in terms of objective interpretation at one extreme, at the other, more logical and cost effective sampling procedures.

I have attempted within the confines of this paper to present but four elements from the research programmes of the Butser Ancient Farm Project. Each seeks to bridge the gap between prime data acquisition by the excavator in the field and the work of the environmental specialist whose potential contribution to the understanding of the material evidence has not yet been fully realised or even appreciated. Indeed they also serve to illustrate the void which exists between them at the present time. The Ancient Farm is, in fact a vast open air research laboratory which is designed quite specifically to integrate the environmental evidence with the archaeological evidence. It is only to be regretted that it is unique in world archaeology and that it has only been in existence for a mere eight years. The four elements are chosen carefully in that each points a different approach yet within itself is an integrated whole. The first, the nature of the waste product of an experimental sequence is allied to the latest analysis of carbonised seed from an archaeological feature, the pit, which initially inspired the hypothesis the experiment sought to test, namely that certain pits were used for the storage of grain. The second element explores the correlation of paleobotanicbl material with documentary references and underlines the dangers of simplistic analysis.

The third explores the recreation of a specific structure, its subsequent testing within the parameters of normal interpretation and brings into question the whole problem of non-validated hypotheses being uncritically adopted into the literature. The fourth element is a direct simulation of a prehistoric feature which embraces all the aspects fundamentally important to archaeologist and environmental specalist alike. The preliminary results from this simulation, which involves human activity the least and relies upon natural phenomena the most, echoes the conclusion drawn from the previous three elements. All four elements are, in fact, almost peripheral to the main work of the Ancient Farm and are offered both as an indication of the environmental resource being built there and as a salutary reminder of the need for far greater integration between excavator and specialist. There are many areas where empirical testing will be quite impossible to achieve but nonetheless in those areas where such testing is feasible then every effort should be made including the provision of financial resources to validate hypotheses. Until integration of purpose and personnel is achieved and until validation of those hypotheses which can be tested are the norm, then archaeology as a discipline will continue to be subject to the trends imposed by fashionable theoreticians.

Acknowledgement

I am indebted to Dr. E. Slater of Glasgow University for reading and commenting upon the manuscript of this paper.

Bibliography

Bowen, H. C. and Wood, P. D. (1968). Experimental Storage of Corn Underground and its Implications for Iron Age Settlements. Institute of Archaeology Bulletin 7 1-14.

Cocks, A. H. (1921). A Romano-British Homestead in the Hambledon Valley, Bucks. Archaeological Journal 71 141-98.

Forsyth, A. A. (1954). British Poisonous Plants. Bulletin 161 Ministry of Agriculture, Fisheries and Food. London: HMSO

Goodchild, R. (1943). 'T'-shaped Corn-Drying Ovens in Roman Britain. Antiquaries Journal 23 148.

Jewell, P. A. and Dimbleby, G. W. (1968). The Experimental Earthwork on Overton Down, Wiltshire, England: the first four years. Proceedings of the Prehistoric Society 32 313-42.

Morris, P. (1979). Agricultural Buildings in Roman Britain. BAR 70 Oxford.

Partridge, C. (1976). Lifting a Roman Corn-Drying Oven. Museums Journal 76.2 69-70.

Pitt-Rivers, A. L. F. (1887). Excavations on Cranbourne Chase. London: Private publication.

Pliny. Natural History 18.18.97.

Renfrew, J. M. (1973). Palaeoethnobotany. London: Methuen & Co. Ltd.

Reynolds, P. J. (1967). Experiment in Iron Age Agriculture. Transactions of the British & Gloucestershire Archaeology Society 86 60-73.

Reynolds, P. J. (1969). Experiment in Iron Age Agriculture. Ibid. 88 29-33.

Reynolds, P. J. (1979). Iron Age Farm: The Butser Experiment. London: British Museum Publications.

Reynolds, P. J. and Langley, J. K. (1980). Romano-British Corn-Drying Oven: An Experiment. Archaeological Journal 136 27-42.

Shaw-Mellor, A. and Goodchild, R. G. (1942). Atworth Roman Villa. Wiltshire Archaeological Magazine 49 46-95.

Tacitus. Germania 16.4.

Virgil. Georgics 1.267.

Wheeler, R. E. M. (1943). Maiden Castle, Dorset. Research Committee Report XII. London: Society of Antiquaries.

White, R. S. (1970). Agricultural Implements of the Roman World. Oxford: Routledge & Kegan Paul.

THE ENVIRONMENTAL SETTING

3. CLIMATE FROM 1000 BC TO 1000 AD

Hubert H. Lamb

Introduction

Climate is for ever changing. The differences which we know from one year to another, and which we sense from one decade to the next, are part of a whole "spectrum" of variations which include significant differences from century to century and from millennium and millennium. Just ten millennia take us back to the end of the last ice age, and from around 8000 to 4000 years ago the climates of the world were mostly warmer than now, in what we now see as the warmest postglacial times.

It is already more than a hundred years since the Norwegian palaeobotanist Axel Blytt (1876) first detected the main outline of the postglacial sequence of climatic regimes in Europe in terms of their temperature characteristics. Since that time, and especially in the last thirty years, many new lines of evidence from both physical and biological science and new techniques of analysis have provided confirmation of the sequence and added many kinds of detail, including much more certainty and precision in the dating of the evidence both by radiometric methods (e.g. radiocarbon) and such techniques as tree-ring chronologies and the counting of year-layers (varves) in lake-beds and ice-sheets. A wide-ranging account of the multifarious types of evidence and analysis techniques has been given by Lamb (1977). Most recently important additions to our knowledge have come from mapping the successive stages of postglacial vegetation history and of the climate itself (Birks et al. 1975; Bernabo and Webb 1977; Wigley 1977). Major efforts have also been put into the collection and meteorological analysis of documentary records of more than a thousand years of past climate and the earliest and longest series of meteorological instrument observations of the last 200 to 300 years (e.g. Manley 1974; Lamb 1977; Ingram et al. 1978; Lamb et al., 1979). Nevertheless, of course, many details must escape our ability to reconstruct the past.

Climate in the Bronze Age

As can be seen from Fig. 3:1, both England and China on opposite sides of the northern hemisphere were enjoying a warmer climate than now until about 1000 BC, although there are some suggestions of unsteadiness after 2000 BC and greater fluctuations from century to century than in most of the preceding four millennia. It is thought that these fluctuations were more pronounced in terms of rainfall than of temperature, some centuries being very dry and others quite wet. It was probably largely due to these fluctuations that, judging from the radiocarbon dating of tree-stumps in the peat, the forest had receded from the exposed places near the Atlantic coasts of Scotland and from the former upper tree line at nearly 800 m on the heights in Scotland,

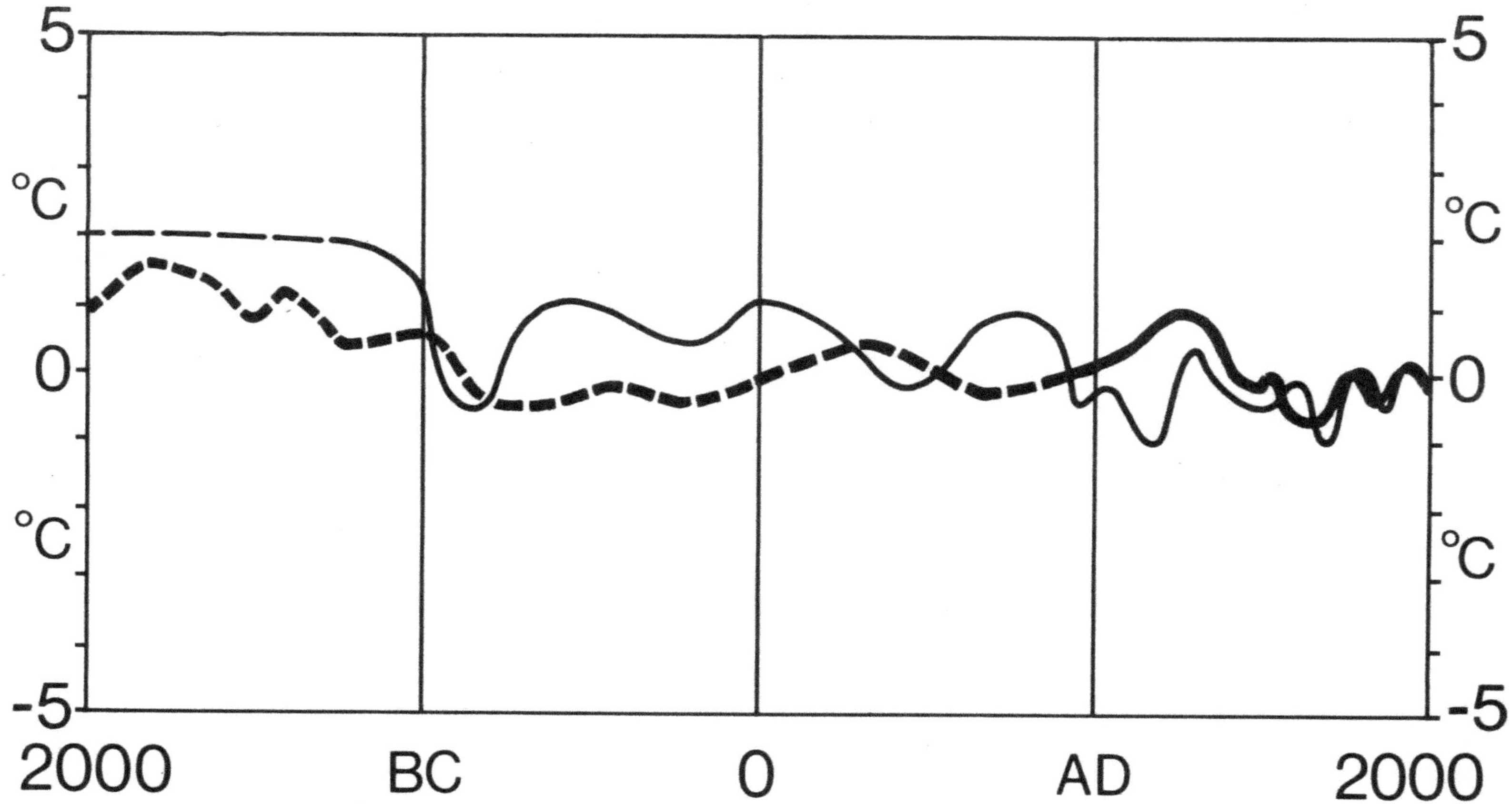

Fig. 3.1 Estimated mean temperatures for the whole year in Central England (bold line) and in China over the last four thousand years, both expressed as departures from the modern all year average.

Wales and northern England around 2000 BC or even a little earlier (see Pears 1972). But soil deterioration and peat growth had begun in many places in the wetter climates of still earlier times, often before 3000 BC: the spread of this tendency in the wetter centuries and the grazing of neolithic man's animals (Moore and Chater, 1969) were also making conditions less favourable for trees on the heights.

Farther south, in Cornwall, Wales and Ireland in the Bronze Age, according to Godwin (1975), forest still extended right to the Atlantic shores and to greater heights on the hills than any present woodlands.

The extension of the forests to the Atlantic edge and the megalithic monuments that were apparently used as observatories from Stonehenge, through Wales, to Lewis and the Orkney Islands, must be held to indicate two other characteristics of the warmer climate phases occurring well after 2000 BC, namely generally lighter winds and freedom from storms, and clearer skies than nowadays. Both these aspects are meteorologically probable (see Lamb 1974) and indicate a greater frequency of anticyclones in these latitudes than now. These conditions are also doubtless related to the evidence of a good deal of seafaring and cultural contacts via the Irish Sea, Hebridean waters and the northern isles as far as Norway, Denmark and the Baltic.

Climate in the early Iron Age

Around 1000 BC one more warm and dry climatic period, lasting a century or two, seems to have occurred; but soon afterwards a sharp decline of the prevailing temperature level and increased storminess seems to have set in. Apart from brief periods—in England the dates were around 250-400 AD and 1100-1300 AD—the temperature levels of 1000 BC and earlier have never been approached since. Fig. 3:1 shows that the timing of the fluctuations in England and China over the last three millennia has been quite different. The timing in England is known to be representative for Europe north and west of the Alps and as far as Iceland and the nearer sector of the Arctic. The amplitude of the temperature fluctuations has been greater in Iceland and the Arctic as far as Spitsbergen than in England. Even in northern Scotland, owing to the advance and retreat of the polar water on the ocean surface (Lamb 1979), the changes of prevailing temperature have probably been twice as great as in England. The sharpened thermal gradient between latitudes 50 and 65° N in the colder times probably meant that in those times some of the windstorms developed to a greater intensity than any experienced in modern times.

The estimated fall of nearly 2° C in over-all mean temperature in England, Ireland, Wales and southern Scotland between about 1000 and 750 BC, presumably shortened the growing season by more than five weeks.

Studies of Tregaron Bog in west Wales (Turner 1965) indicate at the first climax of cold climate between about 750 and 400 BC, when the peat was growing at the rate of 1 cm/4 years, as compared with the average rate for other times of 1 cm/20 years, an extraordinary and unmatched wetness of the climate in that part of the country. By contrast, in the east—in the East Riding of Yorkshire (Holderness and Lincolnshire)—the evidence is rather for a dry climate at that time. It seems therefore that there must also have been an extraordinarily great predominance of westerly winds at that time, unequalled in any other period we know about. It seems reasonable also to suppose that these winds often produced great storms: it was one of the two main periods (the other one was between about AD 1300 and 1800) in which coastal changes around these islands were wrought by blowing sand.

During the wettest period, or periods, in the west during the last millennium before Christ wooden trackways were laid to maintain ways across the Somerset levels and across fens and marshlands in some other parts of the country. Presumably the wetness of the landscape continued to increase for several centuries, as by 350 BC or thereabouts some of these trackways had been abandoned in favour of the use of boats. Around 250 BC conditions were still wet enough in the west for the lake villages to be built at Glastonbury and Meare. (There were shallow lakes or meres there again in another cold climate period in the 16th century AD, which have now disappeared).

In northern England, in the Pennines, the high-level places even up to the summit plateau, up to nearly 800 m, which had been occupied with forts and settlements around 2000 BC—like the Bronze Age tillage up to over 400 m above sea level on Dartmoor—were abandoned by, or during, the last millennium BC. In this last millennium the glaciers in the Alps were advancing from their minimum stands at the end of the warmest postglacial times and in some cases reached advanced positions quite near those of around 1800 AD, while in Scandinavia the Fimbulvinter legend (written down by Snorri Sturlason nearly 2000 years later) probably enshrines a folk-memory of the shock of the increased severity and snowiness of the winter climate in northern Europe, where great glacier advances are now known to have taken place.

Also in the Alps during Hallstatt times the upper limit of the forests sank by several hundred metres (up to 600 m in the case of the spruce), and the formerly thriving gold mining in the Hohe Tauern ceased as the mine entrances became increasingly inaccessible. Most traffic in and across the mountains came to an end (Gams 1937). Around the middle of the millennium the levels of the lakes around the foot of the mountains rose and the lake villages there were finally abandoned (Brooks 1949).

Climate during Roman times

After about 150 BC the climate of Europe was becoming milder again. Ancient Roman writers record the effects (Lamb 1977: 4 and 420) in terms of changes in the southern limit of the beech tree and the northern limits of culture of the vine and olive as well as in the increasing rarity of reporting of severe winters and frozen rivers. Nevertheless, one last tremendous bout of storminess in the North Sea about 120 to 114 BC, with flooding of the coastlands about the German Bight and a permanent eastward shift of the coast of Denmark (Jutland), which set the Celtic and Teutonic peoples migrating (the Kymbrian Flood), was reported by Pytheas.

The climate continued to become gradually milder, drier and less stormy, until by the time of the Roman conquest of England in the 1st century AD it was probably very similar to our own day; and from about 106 AD, when the Danube was bridged at the Iron Gates under the Emperor Trajan, there seems to have been a very long run of immunity from winters severe enough to produce ice that could break the many piers of the bridge. After one or two noteworthy winters (at least in central Europe) in or about AD 250-263, a fairly steady climate, somewhat warmer and drier than now, seems to have maintained itself until about the year 400. This was followed by a colder time with great advances of the Alpine glaciers and glaciers in Norway, comparable with the positions reached a thousand years earlier and rivalling those around the 16th to 19th centuries AD. Just how sudden the break was needs further investigation, but glaciological and tree-ring studies by Röthlisberger and Schneebeli (1976), supported by radiocarbon dating and archaeological finds, suggest that there may have been a sudden turn to wetter summers and colder winters close to AD 400 and more or less coinciding with the withdrawal of the Roman armies from territories north and west of the Alps.

Despite the evidence of more or less complete immunity from winters severe enough to produce heavy ice on the rivers of central Europe- particularly between AD 100 and 250 and between 360 and 400—reports have survived in the literature of about a dozen winters in England between AD 80 and 359 with notable frosts which sometimes froze the rivers and of four winters with great snowfalls.*

That the prevailing warmth and dryness in England and central Europe in the 3rd and 4th centuries AD was probably an aspect of a world-wide period of warmth is indicated by evidence of a high stand of world sea level around AD 400, with the sea flooding the English Fenland and a great estuary, or "Danish-type" fjord, reaching up to Norwich, as well as the drowning of earlier harbour works in the Mediterranean. By contrast, the land-locked Caspian Sea sank to a low level, possibly 4 metres below its average level in recent times, implying reduced rain- and snow-fall and slackening of the flow of the Volga and probably increased evaporation. Droughts do not figure much in the reports surviving from the British Isles and central Europe, but the frequency of reports of flooding declines sharply after AD 270. It seems likely that the condition of riverside land and marshes, as at the early Saxon village at West Stow in Suffolk, improved over the 130 years that followed and became wetter again in the 5th and 6th centuries, particularly after about AD 580.

Climate history on into the Dark Ages and after

Fig. 3:2 provides an at-a-glance survey of what is known of the relative frequencies of dry or wet summers, mild or cold winters, in the British Isles and central Europe and of floods and raininess in Italy, by half centuries over more than a thousand years. Variations in the completeness of reporting are allowed for by plotting the ratios of the frequencies of reports of dry and wet summers, and of mild and cold winters, rather than the actual numbers of reports. Even so the numbers of reports are often too low to provide reliable ratios, and in some half-centuries no estimates could be made. The general reliability of the sequence depicted is, however, supported by the rather close parallelism of the curves for the British Isles and central Europe. The warm dry climate of the time around AD 400, and perhaps for 150 years before that, is indicated alike in these countries and in Italy, as is also the sharp turn to-

* Frosts in or about AD 80, 134, 153, 173, 220, 230, 251 and 252, 291, 329 and 359.

Great snowfalls in AD 125, 242, 329 and 341.

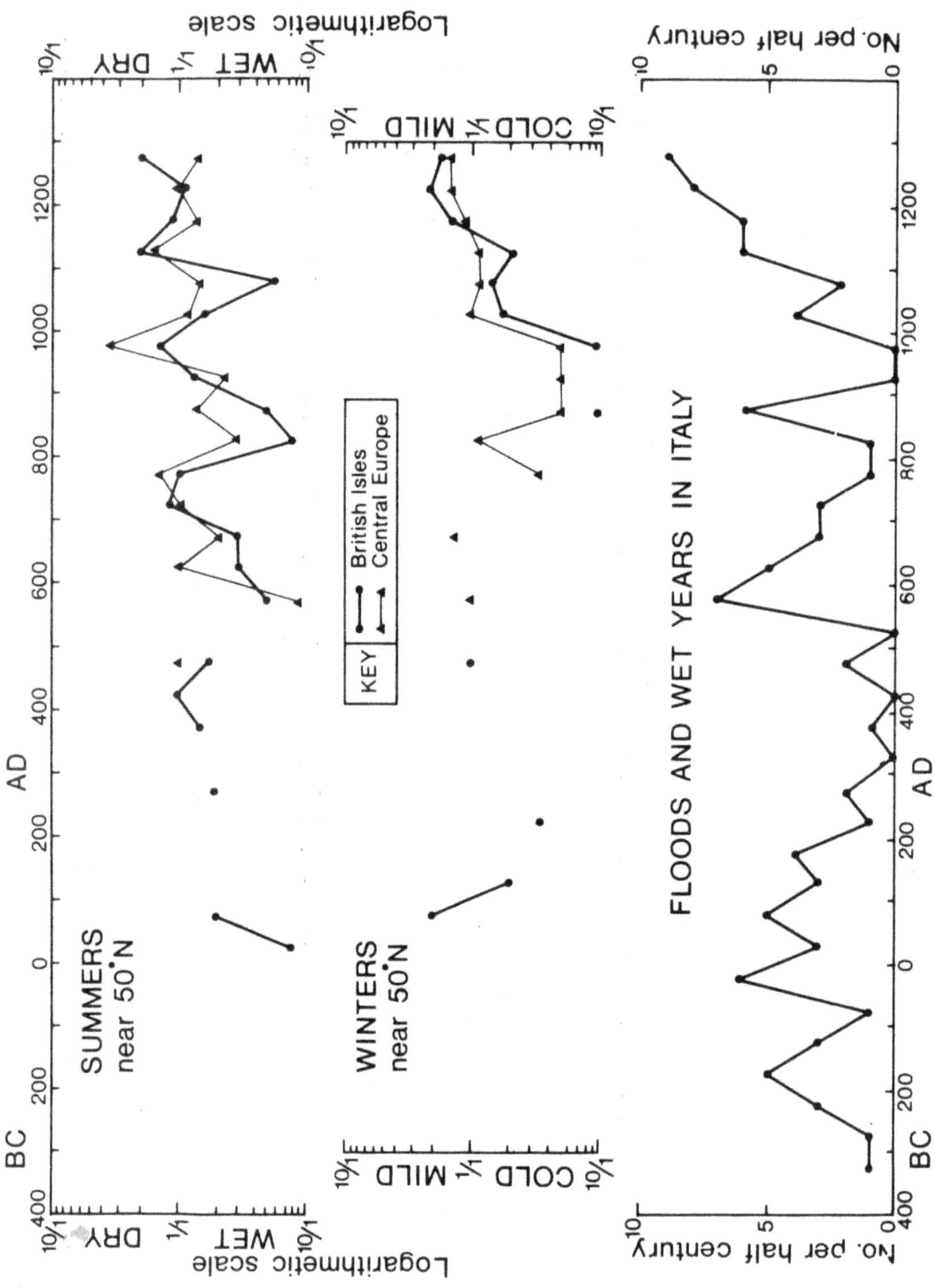
SUMMERS near 50°N
WINTERS near 50°N
FLOODS AND WET YEARS IN ITALY
KEY
British Isles
Central Europe
BC
AD
Logarithmetic scale
No. per half century
DRY
WET
MILD
COLD

Fig. 3.2 Prevailing character of the summers and winters in Europe near 50°N from AD1 to 1300 and numbers of wet years and years with flooding in Italy from 350 BC.

the character of the summers and winters is indicated by ratios of the numbers reported indicating dry/wet and mild/cold respectively by half-centuries. The ratios are plotted on a logarithmic scale.

wards colder summers and wetter conditions by the late 500s.* As far as Britain and Italy were concerned, there seems to have been little improvement before 650 to 700.

The 700s seem to have had a type of climate that is often described as "more continental", with drier and probably warmer summers and colder winters. Other fluctuations followed and are attested by more abundant reports. The 10th century shows another "continental" phase, with dry summers and cold winters, and sees the beginning of the build-up to the remarkable warmth of the early and high Middle Ages in Europe between about AD 980 and 1300, accompanied by an increasingly moist climate in the Mediterranean.

Corresponding to the dryness in central Europe, the Caspian Sea was again at a low level in the 8th century, but it began to rise before the end of the 10th century: climatic developments are not always in step over a span of sixty degrees of longitude. In Europe droughts and a great heat were reported in many of the summers of the 10th century, particularly in the 980s and 990s; and the moisture stress seems to be confirmed by remarkable sequences of narrow tree-rings in oaks that grew in western Germany in the early and late decades of that century. Moreover, Fletcher (1974) reports good agreement between the German oak series and the few tree-ring series so far available for southern England during that time. And by AD 1000 world sea level seems to have been once more as high as it had been in 400, presumably owing to the melting of glaciers. Fig. 3:2 indicates the same characteristics of a "continental" climatic regime—with dry summers (which are usually also warm) and cold winters—in the 10th century as in the 8th. Archaeological findings in both Norwich and York tend to support this: the positioning of the industrial (Potter(s) gate) area of Saxon-Danish times in Norwich to the west of the city points to provision for a pollution problem with light easterly winds in the winters, and in York an abundance of (bone) skates belongs to the same period. We may perhaps deduce frequent cold anticyclonic weather in the winters, while the other seasons of the year tended to be warm and sunny. The dryness of the climate of the 10th century also points to frequent anticyclones in the latitude of Britain, which on some grounds of meteorological theory is a likely accompaniment of a time when global mean temperature was rather higher than now.

No derivation of the probable temperature or rainfall values during the first millennium AD seems so far to have been worked out, but an assessment going back to AD 800 (Lamb 1965) suggests that the temperatures in England

---

* There are many references, e.g. in Hennig (1904), to a remarkable run of wet summers and autumns characterised by heavy rains and flooding in France, Switzerland and northern Italy, and a few (despite less reporting) in England, the southern Netherlands and Germany between AD 575 and 590—e.g. 575 in England 575 or 576 in central Europe, 579 in Britain and the Rhone valley; 580 in France, Switzerland and Italy; 583 in France, Switzerland and central Europe generally; 585, 587 and 590 similarly and in Italy as well; 586 exceptional rains in Italy; and 590 is also described as a very wet year in England: altogether 8 or 9 years out of 16, though there are only reports for three of them in England.

averaged over the period between 800 and 1000 were probably 0.5 to 0.7° C lower than between 1200 and 1300. A change almost twice as great as that might be suggested by the 200-metre rise of the upper limit of cultivation and human settlement in southern Norway over the same period (Holmsen 1961). A difference of that order between the change in central England in 52-53° N and southern Norway in latitude 60-62° N may, however, be real and was apparently repeated in the time of general cooling of world climate in the late Middle Ages between about AD 1300 and 1700.

We find also a very rough confirmation of the suggested sequence of changes during the first millennium, if comparisons be made with sequences of oxygen isotope measurements in tree rings near the upper tree-line in California and in the Greenland ice sheet.* Although the details and timing are manifestly different in those widely separated regions from what we discern in England, all three agree in indicating a warm period round about AD 300 or 400 followed by a sharp temperature minimum, which in Greenland occurs as early as 400-450 and in California around 500. In both these regions, however, the recovery of warmth to the early Middle Ages seems to have set in some centuries earlier than in Europe.

A comparison of the character of the climatic sequence from a point so close to East Anglia and other parts of eastern England as to be affected by nearly all the same weather systems is offered by the studies of coast changes carried out in the Netherlands (Jelgersma et al. 1970). Towards the end of the warmest postglacial times, some four to five thousand years ago, the melting of glaciers and of the former great ice-sheets and consequent rise of world sea level had extended the North Sea to a coastline about 10 km farther east than the present coast of the Netherlands. In the stormier and colder climatic phases which came at times during the last two millennia BC sand dunes began to form, and in the calmer intervals vegetation began to be established on the dunes. One such calmer period coincided more or less with Roman times and another with the early medieval warmth between about AD 900 and 1100. Storms and the formation of new sand dunes seem to have been increasingly active towards AD 700 and again between 800 and 900. Storminess setting in at a time when sea level was particularly high after some centuries of an increasingly warm regime obviously tends to breach any older dunes which have for a time protected a low-lying coast. So it was at the end of the relatively calm medieval period that great floods in 1251, 1287 and 1421** formed the Zuyder Zee, which has only been drained in our own times. Similarly, the Goodwin Sands in the English Channel were first formed in storms in or about AD 400 and 419, which submerged part of a former island, the process being completed by further storms in the Middle Ages.

The most valuable further light on the sequence in Europe is that shed by studies of the glaciers and related matters in the Alpine region. A study

---

* According to results kindly made available by Dr. Irving Friedman of the United States Geological Survey and Dr. W. Dansgaard of the University of Copenhagen.

** The same storms also produced sea floods in East Anglia.

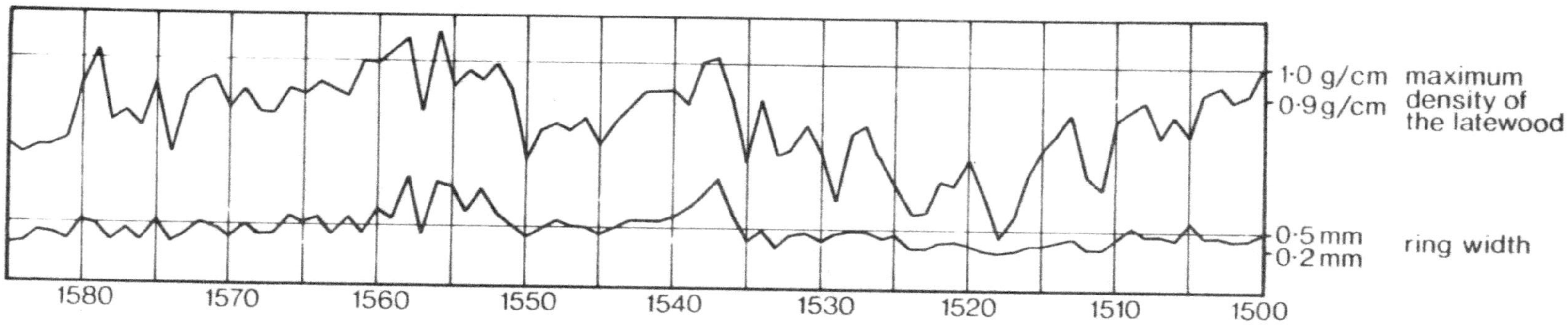

Fig. 3.3 The ring widths (lower curve) and maximum density of the late-season's wood (upper curve) in larches that grew near Zermatt, Switzerland from 1,585 to 1,500 years "before present" (1950)— i.e. from about AD 365 to 450.

(After Rothlisberger 1976).

(Delibrias et al., 1975), based on many radiocarbon datings of trees pushed over by advancing glaciers in the neighbourhood of Grindelwald in central Switzerland, and later evidence of renewed spread of the forest up the valley when the glaciers retreated again, indicates that there was a glacial advance between AD 400 and 750 following a long period of retreat from 300 BC. In the Val de Bagnes, in south-west Switzerland, survey and radiocarbon dating of old moraines (Schneebeli 1976) indicates that the Glacier de Brenay advanced from its source on the eastern side of the main valley right across the course of the river in the valley bottom to a position in AD 670 (± 85 years), close to the position of the glacier snout in 1820, its farthest forward position in recent centuries. As in the period 1820 to 1850 the snout of another glacier, advancing from the other side of the same valley, reached a position between AD 430 (± 60 years) and 790 (±65 years) within 100 to 300 metres of the Glacier de Brenay, almost closing off the upper part of the valley.

The closer dating provided by a tree-ring record from AD 200 to 450 (Röthlisberger 1976), in particular larches in the upper margin of the forest near Zermatt where the variations of ring-width and density of the latewood formed towards the end of the season may be presumed chiefly due to variations of summer warmth, shows (Fig. 3:3) the fluctuating build-up to periods of warmth in and around the 4th century and a sharp decline setting in in AD 414, preceded by initial shocks between 400 and 411. There are hints in archaeology and legend (Röthlisberger loc. cit.) of a very interesting pattern of the impact of this climatic change, which coincided with the withdrawal of the Roman armies from Britain (and by later in the same century from all northwestern and central Europe). It has been possible to trace ancient ways leading across the Alps, and between valley and valley on the Swiss side, earlier than the known medieval routes and still known locally in parts by the Latin name Via Antiqua: at one point a stone bridge of Roman architectural style fell only in recent years. Latin inscriptions have been found, as well as the remains of presumed Roman guard posts, along this route which leads to the high Col d'Herens over into Italy. It seems clear that the route was interrupted by advancing glaciers before the Middle Ages, possibly soon after AD 400. A local legend of a king whose kingdom was overwhelmed by ice is tentatively interpreted as referring to the experience of the local commander of one of the Roman guard posts.

Acknowledgement

The account here given owes much to the Historical Weather Mapping Project in the Climatic Research Unit, University of East Anglia, Norwich, generously funded by the Rockefeller Foundation, New York.

Bibliography

Bernabo, J. C. & Thompson Webb (1977). Changing patterns in the Holocene pollen record of northeastern North America: a mapped summary. Quaternary Research 8(1) 64-96. Seattle.

Birks, H. J. B., Deacon, J. & Peglar, S. (1975). Pollen maps for the British Isles 5000 years ago. Proceedings of the Royal Society of London, B 189 87-105.

Blytt, A. (1876). Essay on the Immigration of the Norwegian Flora. Christiania.

Brooks, C. E. P. (1949). Climate through the ages. London: Ernest Benn, Second Edition.

Delibrias, G., Ladurie, M. & Ladurie, E. L. (1975). Le forêt fossil de Grindelwald: nouvelles datations. Annales: économies, sociétés, civilisations 1 137-147. Paris: Armand Colin.

Fletcher, J. M. (1974). Annual rings in modern and medieval times. In: The British Oak, (eds. M. G. Morris and F. H. Perring) Farringdon, Berks: E. W. Classey for Botanical Society of the British Isles 80-87.

Gams, H. (1937). Aus der Geschichte der Alpenwälder. Zeitschrift des Deutschen und Österreichischen Alpenvereins 68 157-170. Stuttgart.

Godwin, J. (1975). History of the British flora. Cambridge University Press. Second Edition.

Hennig, R. (1904). Katalog bemerkenswerter Witterungsereignisse von den ältesten Zeiten bis zum Jahre 1800. Abhandlungen des Kgl. Preuss. Met. Inst., Bd. II, No. 4. Berlin.

Holmsen, A. (1961). Norges historie. Oslo: Universitets forlaget.

Ingram, M. J., Underhill, D. J. & Wigley, T. M. L. (1978). Historical climatology. Nature 276 329-334. London. (See related Lamb et al. 1979).

Jelgersma, S., De Jong, J., Zagwin, W. H. & van Regteren Altena, J. F. (1970). The coastal dunes of the western Netherlands: geology, vegetational history and archaeology. Meddedelingen, Rijks Geologische Dienst (nieuwe serie), No. 21. Maastricht.

Lamb, H. H. (1965). The early medieval warm epoch and its sequel. Paleogeography, Palaeoclimatology, Palaeoecology 1 13-37. Amsterdam: Elsevier.

Lamb, H. H. (1974). Climate, vegetation and forest limits in early civilised times. Philosophical Transactions of the Royal Society of London, A, 276 195-230.

Lamb, H. H. (1977). Climate: Present, Past and Future. Volume 2 - Climatic history and the future. London: Methuen.

Lamb, H. H. (1979). Climatic variation and changes in the wind and ocean circulation: the Little Ice Age in the northeast Atlantic. Quaternary Research 11 1-20. Seattle.

Lamb, H. H., Ingram, M. J., Underhill, D. J. & Wigley, T. M. L. (1979). Historical climatology. Nature 279 169.

Manley, G. (1974). Central England temperatures: monthly means 1659 to 1973. Quarterly Journal, Royal Meteorological Society, 100(425) 389-405. London.

Moore, P. D. & Chater, E. H. (1969). The changing vegetation of west-central Wales in the light of human history. Journal of Ecology 57 361-379.

Pears, N. V. (1972). Interpretation problems in the study of tree-line fluctuations. In: Research Papers in Forest Meteorology: an Aberystwyth Symposium. (ed. J. A. Taylor). Aberystwyth: Cambrian News pp. 31-45.

Röthlisberger, F. (1976). Gletscher und Klimaschwankungen im Raum Zermatt, Ferpècle und Arolla. Die Alpen 52(3-4) 59-150. Luzern: Schweizer Alpen-Club.

Schneebeli, W. (1976). Untersuchungen von Gletscherschwankungen im Val de Bagnes. Ibid., 52(3-4) 5-57.

Turner, J. (1965). A contribution to the history of forest clearance. Proceedings of the Royal Society of London, B 161(984) 343-354.

Wigley, T. M. L. (1977). Geographical patterns of climatic change 1000 BC - 1700 AD. Final Report (Contract No. 7-35207). Washington, D.C.: National Oceanic and Atmospheric Administration.

4. THE VEGETATION

Judith Turner

Introduction

There are two major sources of evidence, other than historic records, available for reconstructing former vegetation; plant material that has been found on archaeological sites and plant remains that occur in natural deposits. This chapter is concerned with the latter. It is widely recognised that pollen analysis of peat bogs and lake sediments especially when combined with radiocarbon dating, is one of the most powerful methods of studying the vegetational changes of prehistoric and historic times. Pollen analysis has its limitations, however, some of which are particularly relevant to the Iron Age and Anglo-Saxon periods, and it is therefore considered first.

The nature of the evidence

(a) Pollen diagrams

Most pollen diagrams from deposits of the last three thousand years show relative pollen frequencies, and indicate how the amount of tree pollen has changed with time in relation to the amount of herbaceous pollen. Decreases in the amount of tree pollen are usually interpreted as meaning that forest was cleared. Increases in the pollen of grass and of certain pasture-weed taxa indicate that pastures were created, and increases in the pollen of crop and arable-weed taxa that more land was being cultivated. Similarly increases in the pollen of heather, sedges and other heath and mire taxa mean that some of the former forested areas were going over to heath or blanket mire.

In many pollen diagrams from lowland areas the tree pollen frequency fluctuates more than once indicating that the forest was cleared and later allowed to regenerate, then cleared and allowed to regenerate again, several times. With such diagrams it is possible to use the tree pollen frequency as an indicator of the relative amounts of forest and farmed land within the pollen catchment and to compare it with known settlement patterns.

In other diagrams, however, including many from upland areas the tree pollen frequency decreases either steadily or stepwise from the Neolithic onwards and this is normally interpreted as meaning that the trees were unable to regenerate after clearance, partly because of continued grazing pressure, but also because the soil was becoming podzolised and less suitable for tree growth. Such areas may have been maintained as rough grazing, indicated by high values for the grass frequency, or have gone over to heath or blanket bog, as evidenced by high heather and sedge frequencies. In the English Lake District, for example, once the forest had been cleared soil erosion occurred, washing the mor humus layer into the lakes, and thus allowing grassland, heath and bog to spread (Pennington, 1978). It is now widely believed that much of Britain's moorland formed in this way and is thus largely anthropogenic in origin.

Clearance phases on pollen diagrams can be dated by the radiocarbon method, although not all published diagrams covering the last two and a half thousand years have been dated in this way. Thus, in recognising periods of intensive land use by man on their diagrams many authors have associated them with the Iron Age or later cultures on the basis of the local archaeological evidence. Such correlations are of course valuable and some have stood the test of subsequent radiocarbon-dating. But not all. It is clearly preferable to have pollen diagrams with a secure radiocarbon chronology, so that conclusions about the nature of the vegetation, whether it was previously forest, pasture, crop, heath or blanket bog etc. at any one time may be independent of the local archaeological evidence.

(b) The availability of deposits

It is perhaps stating the obvious to say that useful studies can only be made in areas with pollen bearing deposits. But this must be stressed because there are parts of Britain, for example, the drier more calcareous south-eastern lowlands, where there are few such deposits for the period and about which therefore little can be inferred from the palynological evidence. By contrast areas such as the north of England contain numerous suitable deposits and thus have more potential for such studies.

(c) The size of a deposit and its significance

Suitable deposits vary a good deal in size, and from what is known about pollen transport in forested areas (Tauber 1977) it may be assumed that some pollen diagrams refer to very small areas and others to large ones. Lakes or mires less than a hundred metres in diameter will yield diagrams indicating the local vegetational history only, i.e. that of the few hundred metres around the site, whereas pollen diagrams from the centre of lakes or mires over a kilometre in diameter may be recording the regional vegetational history, which could be of a pollen catchment up to several kilometres around the site. It is very important therefore in interpreting a particular diagram to know the size of its pollen catchment. It must also be remembered that one pollen catchment, whatever its size, is only a sample of the region as a whole and it is of course best to have a number of pollen diagrams before deducing a region's vegetational history.

(d) The ideal deposit

The ideal deposit for this purpose is an undisturbed peat bog which has been, and preferably still is, growing at a reasonable rate.

Peat bogs are preferred to lake sediments because of their more highly resolvable pollen stratigraphy. The pollen rain of successive years appears to preserve within them in a more exact chronological sequence than it does at the bottom of a lake where the current's year's pollen tends to become mixed with that of earlier years by burrowing organisms and the resuspension of sediment. Such mixing can obscure, in a way that does not happen with peat, the evidence for such short-term vegetational changes as episodes of forest clearance. That this is the case has been elegantly demonstrated by H. H. Birks (1972) in her studies of the vegetational history of southern Scotland, for she found in the pollen stratigraphy of Snibe Bog a series of small temporary forest clearances which were not present in the contemporaneous sediments of the neighbouring lake, Loch Dungeon, only 4.8 km from it.

It is also best to use bogs that are still growing. Unfortunately most bog surfaces are no longer in their natural state. They may have been burnt regularly in an attempt to maintain the vegetation for rough grazing or deliberately drained in land reclamation schemes. Local inhabitants may have been cutting peat from around their margins for centuries causing neighbouring intact surfaces to dry and become more liable to erosion. Too often bogs have lost the pollen record of some or even all of the last two millennia. Even when this has not happened there is still a serious risk that the surface layers will have become so dry that modern plant roots will have penetrated rather deeply and rendered the peat unsuitable for accurate radiocarbon dating. This may not always be apparent when examining the peat. Such was the case at Pow Hill Bog, where the surface appeared to be intact but where a radiocarbon-dated horizon from 26-28 cm depth gave a modern date (Turner & Hodgson, in press) when there were good grounds for expecting one of about 2000 radiocarbon years b.p.

(e) Dating the vegetational change

1. The nature of the sample

On some pollen diagrams the sample of peat dated represents the vegetational change quite accurately. Such is the case when the peat has been forming quickly, the pollen samples taken at close intervals and the radiocarbon sample is thin. The first extensive clearance of woodland at Tregaron dated to 2354 b.p. ± 110 (Q-596) is a good example (Turner 1964). The estimated rate of peat formation was fast, 1 cm in 3.8 years, the pollen samples were fairly closely spaced at 2 cm intervals and the radiocarbon sample was 2 cm thick. This meant that the sample of peat dated must have formed within at the most 15 years of the actual clearance and thus represents it reasonably accurately.

On other diagrams, however, such accuracy is not possible, as for example with the reversion to woodland at Steward Shield Meadow dated to 840 ± 100 b.p. (Gak-3032) (Roberts, Turner & Ward 1973). The estimated rate of peat formation was very slow, 1 cm in 150 years, the pollen samples at 2 cm intervals and the radiocarbon sample was 1 cm thick. In this particular case the dated slice could have formed up to 225 years after the regeneration had actually occurred. The possibility of such errors should not be overlooked.

2. The accuracy of the radiocarbon estimate

All radiocarbon dates are published in the journal, Radiocarbon, with a standard error based on the variation in the rate of emission of β-particles. This statistic, i.e. ± x years, corresponds to the 68% level of probability. The date is based on the Libby half-life for the radiocarbon isotope of 5568, rather than the present best estimate of 5730. It is recommended that dates based on the Libby half-life should be multiplied by 1.03 to accord with the new estimate. Dates are also uncalibrated. Discrepancies between the radiocarbon and dendro-chronologies are not large for the Anglo-Saxon period but may amount to centuries rather than decades during the early Iron Age. There are various tables available for calibrating dates (see Switsur, 1973 and Watkins (Ed.), 1975 for further details). Radiocarbon dates should be interpreted with care bearing these facts in mind.

The vegetation

(a) At the beginning of the Iron Age

The vegetation at the beginning of the Iron Age was quite different from today's and more closely resembled that of the mid-Flandrian or late Mèsolithic. In a few particularly favourable areas permanent fields for pasture and arable crops had long been established, as for example on the chalklands of the south-east and on some of the northern limestones and coastal areas. Also on many of the poorly drained upland soils blanket mire had already replaced the original trees. However, elsewhere natural or semi-natural woodland still persisted. This woodland was a temperature mixed oak forest over most of England, Wales and southern Scotland, although as today there were birch and hazel woods in the north and north-west and pine woods in the north-east of Scotland. For as much as two thousand years neolithic and Bronze Age peoples had practised a form of shifting agriculture within them but the amount of cleared land at any one time appears to have been comparatively small and in most lowland areas the forest regularly re-established itself. The biological evidence for this has been reviewed recently in Simmons and Tooley (in press) and will not therefore be detailed here.

(b) During the Iron Age and Roman period

This was a time of rapid change and by the end of the Roman occupation forest may well have been almost as scarce in some parts of Britain as it is today. Much of the palynological evidence for this has already been reviewed (Turner, in press) and the conclusions only are summarised here. This process of more extensive forest clearance began possibly as early as the 5th century bc in the south and occurred progressively later in the north. For example in the mid-Pennines it has been dated to the 4th century bc, in north-east England to as late as the 1st century ad and in the north-west England and the central lowlands of Scotland to the late 5th century ad. In general terms this increased amount of forest clearance during the Iron Age can readily be understood in terms of the greater efficiency of iron tools for felling trees and farming, although detailed correlations with prehistoric cultures are best made at the local level. There are of course exceptions to this general picture of intensive forest clearance during the Iron Age and Romano-British periods. In addition to the areas of upland heath and bog already mentioned there were also some lowland areas that appear to have retained their woodland nature. One such is part of Epping forest which has been shown by Baker *et al.*, (1978) to have been wooded throughout this and the succeeding Anglo-Saxon period. Another is Neasham Fen in south-east county Durham where there appears to have been a little clearance between 2850 ± 60 bp (SRR-98) and 2488 ± 75 bp (SRR-100) but no major clearance until Saxon times, 1213 ± 60 bp (SRR-96).

(c) <u>During the Anglo-Saxon Period</u>

By contrast with the Iron Age, the Anglo-Saxon period appears on balance to have been one with a lower intensity of human pressure on the land. The majority of pollen diagrams indicate a regenerated forest and a lower proportion of arable and pasture land, being maintained for several centuries, sometimes until the Viking invasions or later. Some show no change and only a very small proportion indicate a higher level of activity than in the Iron Age.

This variety is well illustrated by the pollen diagrams from the north-east of England where the following diagrams show a decreased level of activity. In that from Fellend Moss, near Hadrian's Wall (Davies and Turner, 1979), after its low Iron Age levels the tree pollen frequency begins to rise at a level dated to 1330 ± 40 bp (SRR-875) and it does not fall again until just below a level dated to 945 ± 40 bp (SRR-874). Similarly at Steng Moss in Northumberland it rises at a level dated to 1490 ±60 bp (Q-1519) and remains high until a level dated to 1085 ±35 bp (SRR-1041). Also at Hallowell Moss near Durham City (Donaldson and Turner 1977) it begins to rise at a level dated to 1355 ±50 bp (SRR-413) although at this site it does not fall again until much nearer the present day. There is also evidence from Fen Bog in the North York Moors for forest regeneration at a level radiocarbon-dated to 1530 ±130 ad (T-1086) (Atherden 1976).

Two diagrams from the North-East show the farming activity of the Iron Age continuing in their catchment areas throughout Anglo-Saxon times, the tree pollen not rising until later. At Thorpe Bulmer in south-east Durham (Bartley <u>et al</u>. 1976) it does not rise until a level dated to 852 ±60 bp (SRR-405) and at Steward Shield Meadow in Weardale (Roberts <u>et al</u>. 1973) not until 840 ±100 bp (GaK-3032).

At only one of the sites so far studied in the North-East does any new forest clearance actually occur during the Anglo-Saxon period and that is the site of Neasham Fen already mentioned (Bartley <u>et al</u>. 1976). At this site there appears to have been some clearance in the late Bronze Age, but the first major clearance, equivalent to that of the Iron Age elsewhere in the region, has been dated to 1213 ± 60 bp (SRR-96).

A generally lower level of land use during Anglo-Saxon times is similarly indicated in other regions. For example a diagram from Bolton Fell Moss (Barber in press) in Cumberland shows the tree pollen frequency rising steadily from a level dated to 1860 ±60 bp (Hv-3085) to a level dated to 1170 ±50 bp (Hv-3084) the wooded phase lasting until 920 ±70 bp (Hv-3082) (Barber in press). Pennington's diagrams from numerous tarns in the north and east of the Lake District show increased forest clearance during the Iron Age and a later episode of clearance, which has been dated at Blelham Tarn to between 910 ± 50 bp (SRR-255) and 930 ±50 bp (SRR-254) (Pennington <u>et al</u>. 1977). There was no clearance during Anglo-Saxon times. At Rishworth in W. Yorkshire the forest began to regenerate after a clearance which had reached its maximum at 1920 ± 80 bp (GaK-2825), and continued to do so throughout the Anglo-Saxon period. From Derbyshire there is evidence at Featherbed Moss (Tallis and Switsur 1973) that forest regeneration had occurred by 1400 ±50 bp (Q-852) and that renewed clearing did not begin until 1023 ±50 bp (Q-851).

There is similar evidence from Ayrshire in southern Scotland where at Bloak Moss the tree pollen frequency rises at a level dated to 1370 ±105 bp (Q-724) and remains high for some considerable time (Turner 1965). At Loch Pityoulish in the Eastern Highlands (O'Sullivan 1976) a pollen zone interpreted as one representing forest regeneration terminates at 1014 ± 55 bp (SRR-586) which almost certainly means it includes the Anglo-Saxon period.

From The Chains, Exmoor, there is a radiocarbon-dated level showing increased tree pollen frequencies at 1500 ± 60 bp (UG-816) (Merryfield and Moore 1974).

As for the north-east of England a few diagrams from elsewhere in Britain show no tree regeneration in Anglo-Saxon times. For example at Tregaron Bog W. Wales (Turner 1964) the high grass pollen frequencies which began during the Iron Age appear to have lasted until just below the level dated to 768 ± 90 bp (Q-390).

So far reference has been made to radiocarbon-dated pollen diagrams. There are many others with episodes of forest clearance and regeneration which have been dated by association with known prehistoric and historic events, and whilst such evidence is not as secure as that discussed it nevertheless gives a similar picture.

It may be concluded therefore that forest clearance was taking place on a hitherto unprecidented scale during the Iron Age and Romano-British period and that in contrast, during Anglo-Saxon times considerable tracts of land were falling waste and forest re-establishing itself with the Iron Age level of farming being continued in only a few places.

Bibliography

Atherden, M.A. (1976). The impact of late prehistoric cultures on the vegetation of the North York Moors. Transactions. Institute of British Geographers I 3 284-300.

Baker, C. A., Moxey, P. A. & Oxford, P. M. (1978). Woodland continuity and change in Epping Forest. Field Studies 4 645-669.

Barber, K. (in press). Peat stratigraphy and climatic change. A palaeoecological test of the theory of cyclic peat bog regeneration. Balkema, Rotterdam.

Bartley, D. D. (1975). Pollen analytical evidence for prehistoric forest clearance in the upland area west of Rishworth, W. Yorkshire. New Phytologist 74 375-381.

Bartley, D. D., Chambers, C. & Hart-Jones, B. (1976). The vegetational history of parts of south and east Durham. New Phytologist 77 437-468.

Birks, H. H. (1972). Studies in the vegetational history of Scotland II. Two pollen diagrams from the Galloway Hills, Kirkcudbrightshire. Journal of Ecology 60 183-217.

Davies, G. & Turner, J. (1979). Pollen diagrams from Northumberland. New Phytologist 82 783-804.

Donaldson, A. M. & Turner, J. (1977). A pollen diagram from Hallowell Moss, near Durham city, U.K. Journal of Biogeography 4 25-33.

Merryfield, D. L. & Moore, P. D. (1974). Prehistoric human activity and blanket peat initiation on Exmoor. Nature, 250, 5465, 439-441.

O'Sullivan, P. E. (1976). Pollen analysis and radiocarbon dating of a core from Loch Pityoulish, Eastern Highlands of Scotland. Journal of Biogeography 3 293-302.

Pennington, W. (1978). The impact of man on some English lakes: rates of change. Polski Archiwum Hydrobiologii (Warschau) 25 429-437.

Pennington, W., Haworth, E. Y., Bonny, A. P. & Lishman, J. P. (1977). Interpreting the environmental record in the sediments of Blelham Tarn. Report. Freshwater Biological Association 37-47.

Roberts, B. K., Turner, J. & Ward, P. F. (1973). Recent forest history and land use in Weardale, Northern England. In: Quaternary Plant Ecology (Eds. Birks, H. J. B. and West, R. G.) Oxford: Blackwells Scientific Publications 207-221.

Simmons, I. G. & Tooley, M. J. (in press). The Environment in British Prehistory London: Duckworth.

Switsur, R. (1973). The radiocarbon calendar recalibrated. Antiquity 47 131-137.

Tallis, J. H. & Switzur, V. R. (1973). Studies on southern Pennine peats VI. A radiocarbon-dated pollen diagram from Featherbed Moss, Derbyshire. Journal of Ecology 61 743-751.

Tauber, H. (1977). Investigations of aerial pollen transport in a forested area. Dansk Botanisk Arkiv 32 1-121.

Turner, J. (1964). The anthropogenic factor in vegetational history I. Tregaron and Whixall Mosses. New Phytologist 63 73-90.

Turner, J. (1965). A contribution to the history of forest clearance. Proceedings of the Royal Society of London B, 161 343-354.

Turner, J. & Hodgson, J. (1981). Studies in the vegetational history of the northern Pennines II. An atypical pollen diagram from Pow Hill. Journal of Ecology 69.

Turner, J. & Hodgson, J. (in press). Studies in the vegetational history of the northern Pennines II. An atypical pollen diagram from Pow Hill. Journal of Ecology.

Watkins, T. (Ed.) (1975). Radiocarbon calibration and prehistory. Edinburgh University Press.

5. VALLEY SEDIMENTS AND ENVIRONMENTAL CHANGE

Martin Bell

Charles Darwin (1881: 131) in his Formation of Vegetable Mould wrote as follows 'we should bear in mind how many branching valleys intersect most countries, the whole length of which is very great, and that earth is steadily travelling down both turf-covered sides of each valley... (on the valley floor)... a thick bed of alluvium will accumulate ready to be washed away in the course of centuries as the stream in the middle meanders from side to side'. In this succinct and perspicacious way Darwin described processes which we now suspect were responsible for fundamental change in many prehistoric and early historic landscapes. This paper will briefly review those changes and explore the role which valley sediments can play in elucidating the environmental background to later prehistory.

Although erosion and valley sedimentation is obviously the result of ever-present factors, changes in the rates at which they operate may result from climatic changes, natural seral changes in plant communities or anthropogenic effects. Clearance, in particular, is likely to lead to a significant increase in soil erosion (Dimbleby 1976) which will have an effect on the pattern of valley sedimentation.

The types of sediment with which we are dealing are basically two: alluvial and colluvial. Alluvial sediments are the result of deposition by running water and are liable to be sorted to some extent. Colluvial deposits are those which have moved downwards under the force of gravity as a result of processes of subaerial weathering which do not involve major components of wind and water transport. Colluviation occurs because any process causing the movement of particles on a slope is liable to incorporate a component of downslope movement. Agencies responsible for setting particles in motion include rainsplash, seasonal soil creep, the activities of biota and anthropogenic effects like tillage. The resulting sediments are characteristically unsorted and build up by gradual increments to their surface. The rate at which downslope movement occurs will obviously increase enormously on devegetated and cultivated slopes.

Colluvial valley sediments

Accumulations of colluvial sediment tend to occur at the boundaries of cultivation and points where there are pronounced breaks of slope, thus they consistently recur in the following situations (Fig. 5: 1):-

(i) Plateau edge deposits, substantial banks of colluvial soil at a point where a gently sloping hilltop of plateau surface dips abruptly to a valley or scarp face.

(ii) Lynchets.

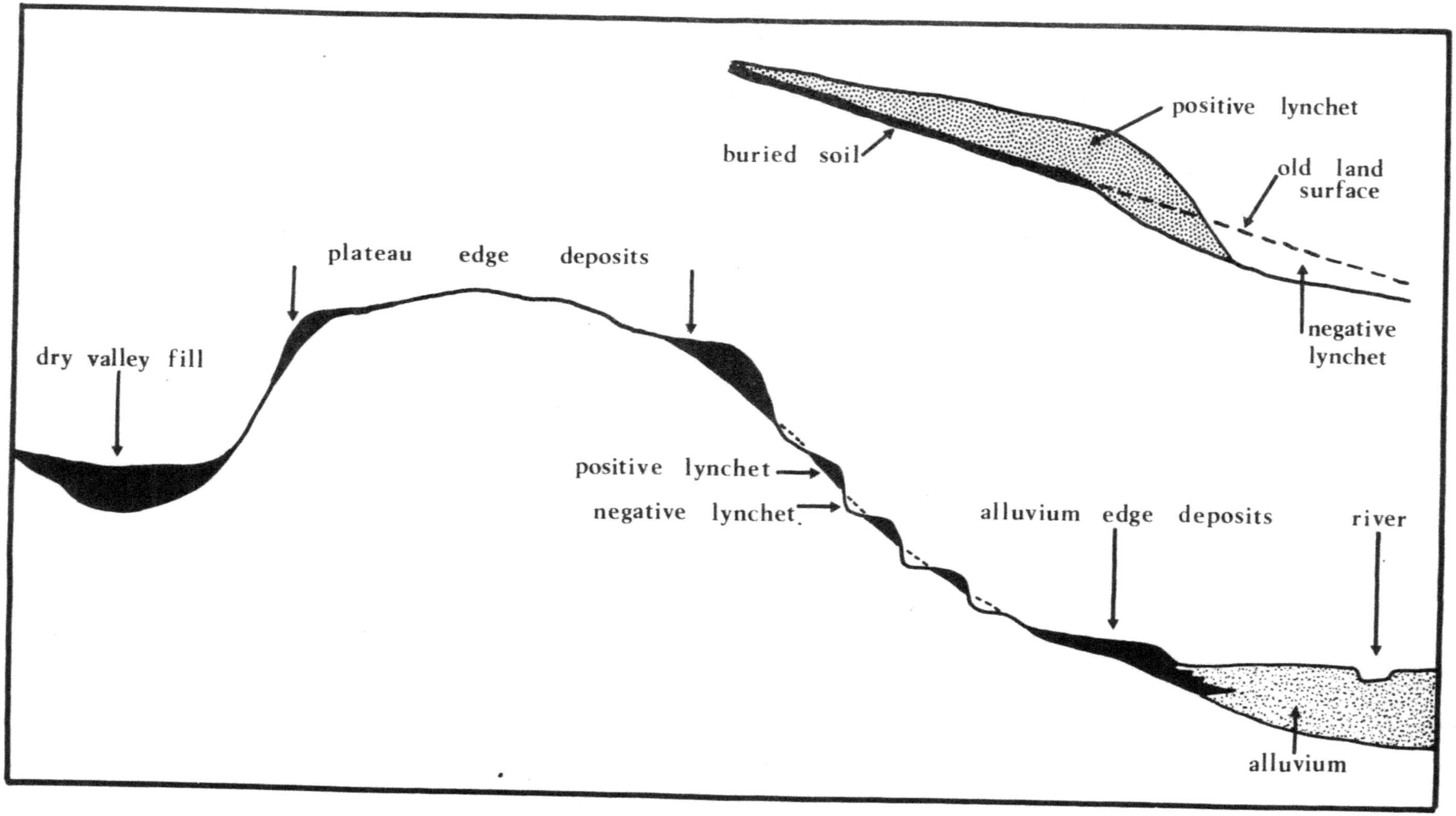

Fig. 5.1 Schematic diagram to show the main situations in which colluvial deposits consistently recur.

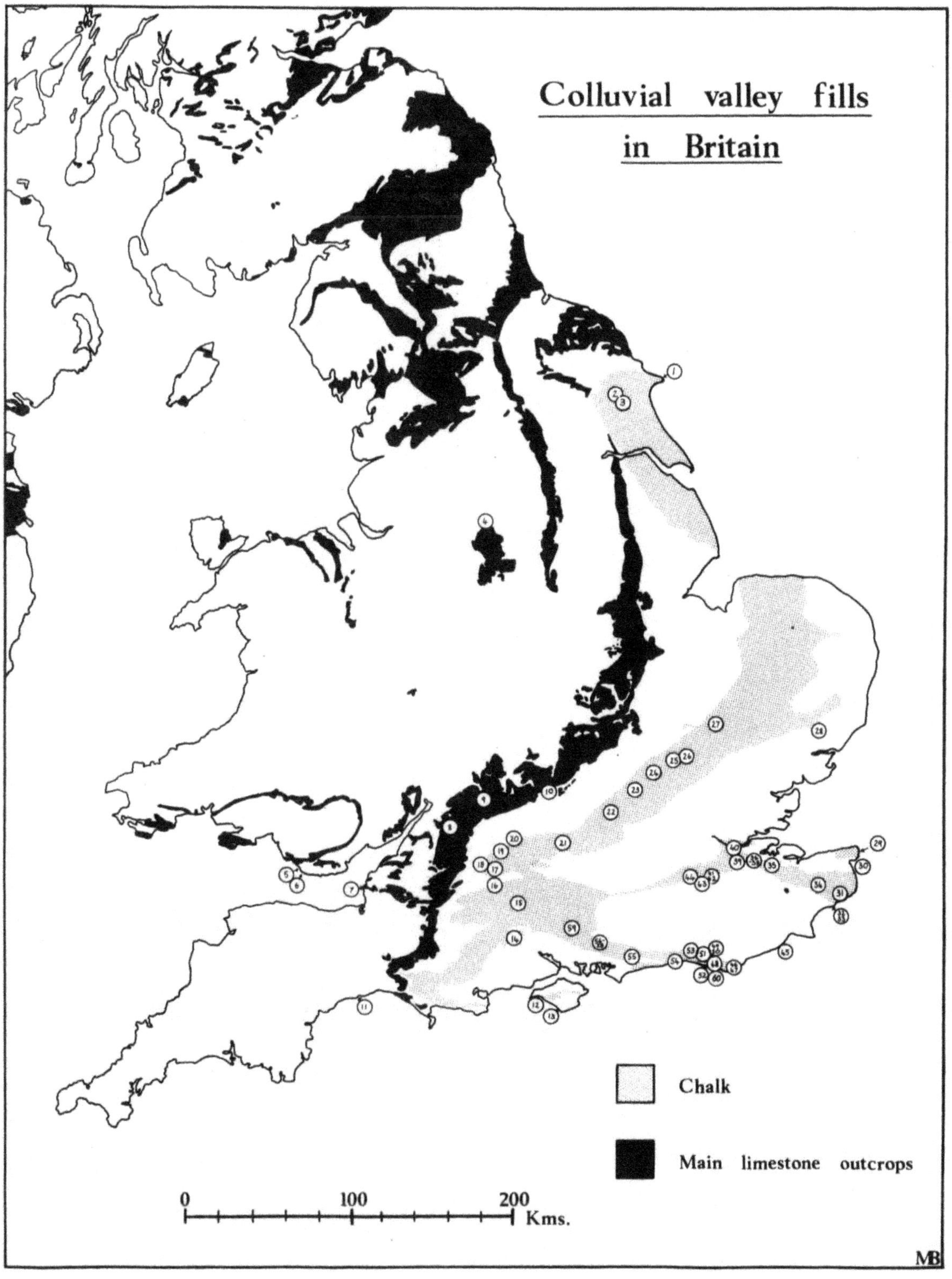

Fig. 5.2 Distribution of the main documented valley bottom colluvial deposits in Britain.

Plate 5.I

The excavation of dry valley sediments at Kiln Combe, East Sussex. Photo. Brenda Westley.

Plate 5.II

The excavation of dry valley sediments in Trench b, Chalton, Hampshire. Photo. Brenda Westley.

(iii) Alluvium edge terraces, accumulations on the margins of river valleys where cultivation has ceased. Often they accentuate existing geomorphological features such as river terraces and erosion cliffs.

(iv) Dry valley fills.

There has been a certain amount of controversy regarding the extent of dry valley colluvial deposits. Godwin (1967) was categorical that there was no evidence of them on the chalk but there is a growing number of documented exposures where valley bottom colluvium is associated with archaeological finds. The most important of these are at Brook, Kent (Kerney et al. 1964) and Pitstone, Buckinghamshire (Evans 1966) where it was possible to establish the ecological history of the area from studies of the mollusc faunas. The distribution of the main documented colluvial valley deposits (Fig. 5:2) shows that they are largely encountered in permeable chalk and limestone areas where colluvial processes predominate and where only limited movement of material occurs down the axis of the valley. Many of the examples are chance exposures in cliffs and quarries. Further indications of the extent of these deposits may be obtained from aerial photographs and a certain amount may be inferred from deposits marked as Valley Gravels by the Geological Survey and from the various colluvial soil series mapped by the Soil Survey. These sources do not, however, discriminate between pleistocene and postglacial deposits, which is obviously of crucial importance to the archaeologist, nor do they provide much indication of the thickness of deposits.

With the aim of confronting some of these problems and examining the potential of dry valley sediments for studies of early land-use, three study areas on the South Downs were selected on the basis that they were already fairly well known archaeologically. The areas were: Kiln Combe, Eastbourne, farmed by Mr. E. D. Williams and the subject of a recent exhaustive survey and excavation programme by Peter Drewett (1977); Itford Bottom, Lewes, farmed by the Brickell family and adjacent to the important second millennium Itford Hill complex (Burstow and Holleyman 1957, Holden 1972); and Chalton, Hampshire, farmed by John Budden who, with Professor Cunliffe, has conducted an intensive archaeological survey (Cunliffe 1973). In each area one or two machine cut trenches 30 m long by 3 or 4 m wide were made from the valley centre uphill. The trench provided a section of the postglacial stratigraphy (Fig. 5:3) which was recorded on the scale 1:10 and from which samples were obtained for molluscan and sedimentalogical analysis, the aim being to extablish the nature of the deposits and the environmental conditions under which they had been laid down.

In order to date the strata a strip was carefully hand excavated alongside the machine dug trench (plates 5:I and 5:II). In every case this revealed a surprising amount of artefactual material, mainly pottery sherds, flint flakes and tools, and charcoal, the three dimensional co-ordinates of which were recorded. From Kiln Combe there were 3278 artifacts, from Itford Bottom 2289, from Trench A at Chalton, only a small excavation, 374, and from Trench B at Chalton, a much larger excavation, 3239. The density of artifacts varied between 72 and 214 to the cubic metre of hand excavated soil. Most important for dating were pottery sherds which, in some cases, could be related by fabric, form, decoration, etc., to stratified assemblages on nearby

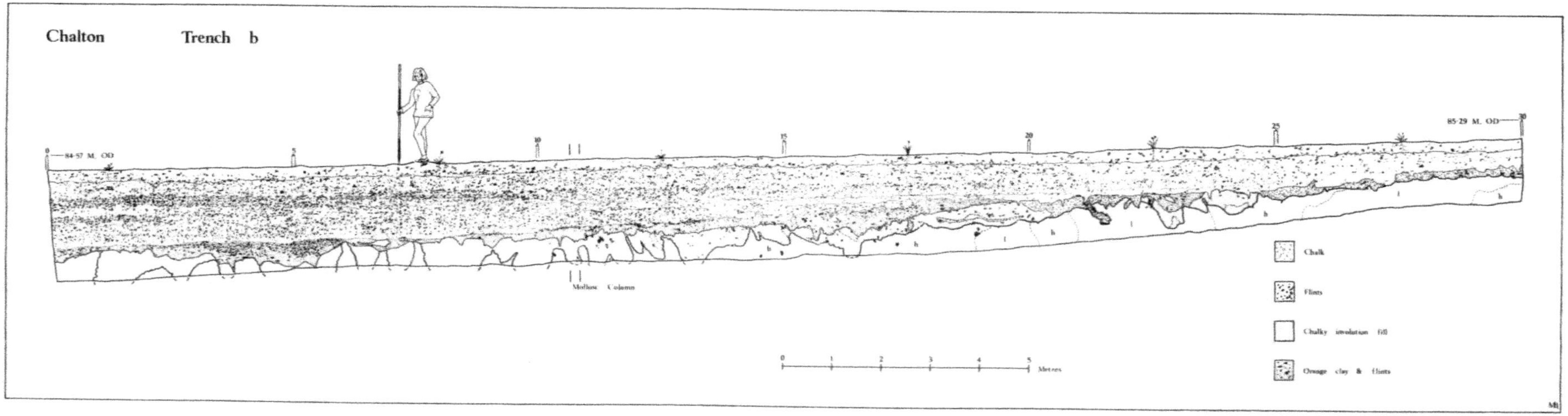

Fig. 5.3 A section through the sediments investigated by Chalton, Trench b. O represents the valley centre.

sites. When the distribution of the various types was plotted out it frequently showed a fairly clear horizonation indicating the approximate date of the various layers. For instance in the Kiln Combe trench there was a pronounced horizon of Beaker sherds (Fig. 5:4) in the stone accumulation horizon of a basal palaeosol, and no material which was demonstrably later, Here, indeed, the concentration of sherds and their condition was such as to suggest an actual Beaker occupation site buried below later colluvial deposits. The upper layers in the same valley were dated, by a heavy concentration of sherds, to the medieval period (fig. 5:5) indicating that this formed part of an arable area associated to a nearby, recently excavated, medieval farmstead (Drewett 1977: 18).

At the base of the Postglacial stratigraphy in the Itford Bottom trench there were subsoil hollow features, one of which produced a good closed woodland mollusc fauna. Nothing remained, however, of the contemporary woodland soil; it is presumed that this soil had subsequently been eroded down the axis of the valley. The earliest palaeosol that survived contained fragments of Beaker and early Bronze Age pottery and also charcoal for which a radiocarbon date of 1770 ±120 bc (BM-1545) was obtained. We can interpret this dated charcoal horizon as a clearance episode which was followed by colluviation. Comparable episodes have been reported at Brook, Kent, dated 2590 ± 105 bc (BM-254, Barker et al. 1971: 169) and Pitstone, Buckinghamshire, dated 1960 ±220 bc (HAR-327, Evans and Valentine 1974). In the case of Itford Bottom, however, we are dealing with secondary clearance since the molluscan evidence indicates that the burning involved isolated trees and shrubs in a fairly open landscape. The character of the pottery from the palaeosol and overlying colluvial deposits makes it highly probable that the cultivation episode relates to the second millennium complex of sites on the neighbouring spur of Itford Hill. The colluvium continued to accumulate into the Iron Age during which period a lynchet, integral to a surviving 'Celtic' field system, began to form. Previously it had been assumed that the field system was Romano-British because sherds of that date have been located within 0.5 km of the trench. In the event it transpired that Romano-British material simply formed a thin veneer over the earlier colluvial deposit. For this reason it is suggested that surface sherd scatters do not give an accurate indication of the date of field systems, nor do they enable us to detect settlement sites in valley bottoms, where they may be buried by blankets of colluvium as the Beaker site in Kiln Combe demonstrated (Fig. 5:4).

Although colluvial deposits seemed to form fairly continuous blankets on the valley floors of each of the study areas, as the trenches, soil pits and chance exposures demonstrated, some interesting variations in thickness were encountered. Chalton Trench A was located in a major dry valley which runs south from Butser Hill and in which is the demonstration area of the experimental farm. This trench encountered little more than 50 cm of colluvium, much of it Romano-British or later. Contrasting with this is one of its own, very minor, tributary valleys sectioned by Chalton Trench B and found to be choked with colluvial sediments up to 1.8 m thick (Fig. 5:3). What probably accounts for this contrast are seasonal streams (lavants) in the main valley; these are last attested about the turn of the century and may have been to some extent responsible for the removal, resorting and localization of the colluvial

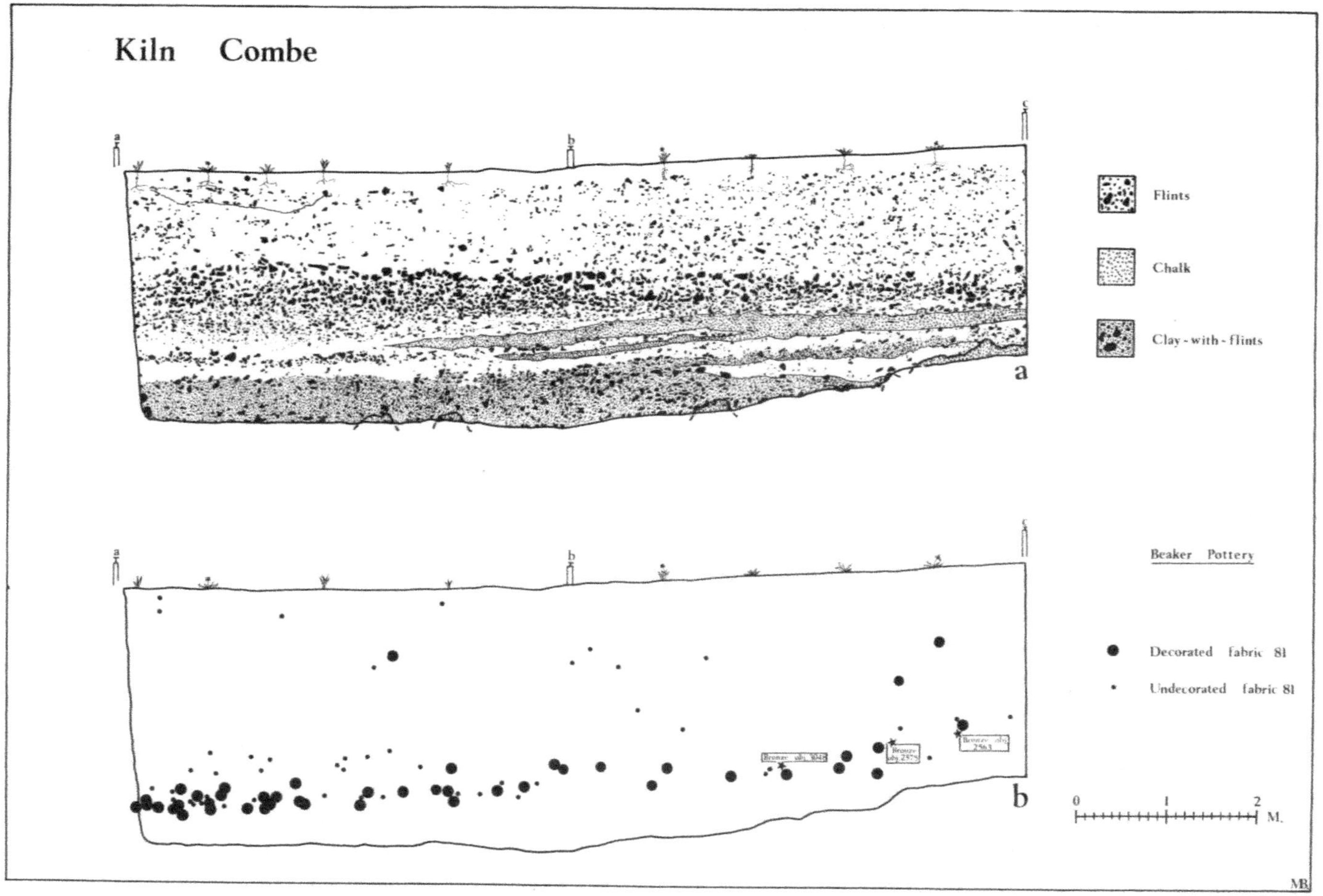

Fig. 5.4 Kiln Combe, Eastbourne: (a) section of the sediments (b) distribution of Beaker sherds.

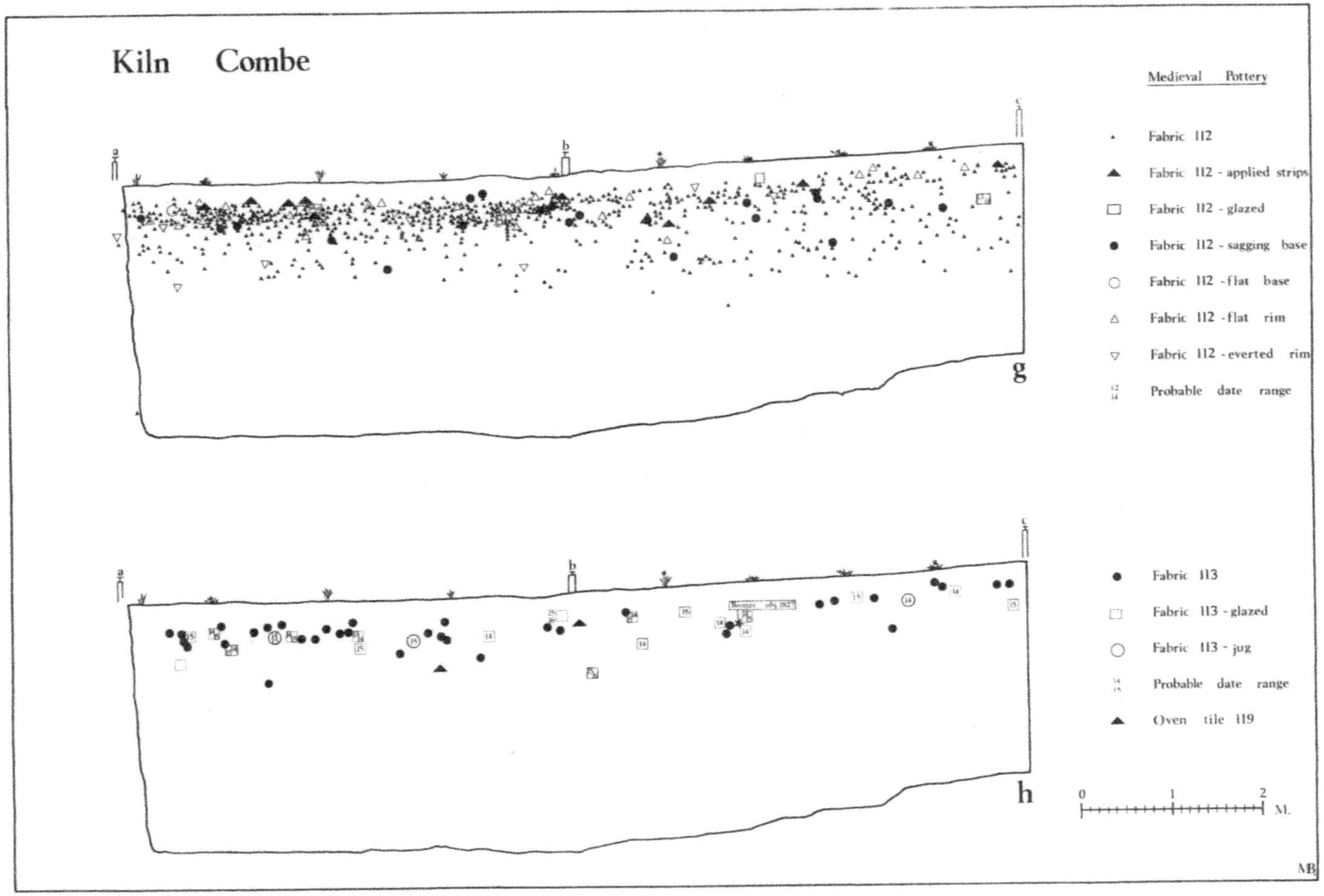

Fig. 5.5 Kiln Combe, Eastbourne: diagrams showing the distribution of various types of medieval pottery.

deposits. If that is so the former more widespread occurrence, and greater erosive competence, of seasonal streams in major 'dry' valleys on the chalk may go some way towards explaining the apparent paucity of colluvial deposits in some dip slope valleys.

Reference has already been made to the processes responsible for colluviation but it is appropriate to say something about its link with human activity. In each of the three valleys it was closely linked to evidence of cultivation. Not only do the deposits themselves closely resemble lynchet deposits but the mollusc fauna they contain is a restricted one of dry open conditions. The link with human activity is further strengthened by the large numbers of artifacts which the deposits contain. Sites recorded to date do not seem to support the notion that colluviation was particularly prevalent during certain episodes, correlating, for instance, with episodes of apparently higher rainfall, such as the Sub-Atlantic. Neither does the temporal range of colluvial deposits support the frequently stated idea (e.g. Bradley 1978: 123) that colluviation was predominantly a process of later prehistory, particularly the Iron Age. Several valley fills began to accumulate in the Neolithic and Bronze Age, others accumulated in various periods up to the Post-Medieval.

The evidence for fairly extensive colluvial deposits on the chalk must lead us to consider the nature of the eroded material and its implications in pedological terms. At the surface of the pleistocene strata in Kiln Combe was a stripe of predominantly coarse silt with a mineralogical composition comparable with that of pleistocene loess. Coarse silt was also the major particle size fraction in the overlying colluvial deposit. In all three study areas the basal postglacial soil horizons were non-calcareous but the colluvium became increasingly calcareous towards the surface because once the superficial deposits had been eroded a greater proportion of chalk would be brought down. There can be little doubt that, as Catt (1978) and Limbrey (1975) have already emphasised, prehistoric soils on the chalk were substantially different from the shallow rendsinas we see today. Not only prehistoric soils, but also hilltop and hillslope occupation horizons, have been eroded away. This was apparent from excavations in the Yorkshire Wolds (Manby 1975) and at Bishopstone, Sussex (Bell 1977) where neolithic occupation was represented only by large pits and hollows, the Iron Age settlement by ditches, pits and, generally large, postholes and the Anglo-Saxon settlement by buildings of very much smaller postholes. Circular huts, or other absolutely convincing domestic structures were absent from the Iron Age settlement, reminding us that at Danebury these more ephemeral structures only survive where they are protected both by extensions to the ramparts and a colluvial build-up against them (Cunliffe 1976).

If, as I am suggesting, significant erosion of chalkland soils has occurred during, and since, prehistory we must consider whether this helps to explain some of the broad trends of land-use on the chalk and adjoining areas. The concentration of prehistoric field monuments on the chalk downs is well known and in marked contrast to the pattern of surviving villages in river valleys and on the margins of the upland. Traditionally it has been suggested that the lighter chalk soils were more attractive to prehistoric communities but the introduction of heavier, wheeled ploughs in the Anglo-Saxon period facilitated cultivation of the clay vales.

Recently Taylor (1970, 1972 and 1973) has drawn attention to the accumulating evidence for prehistoric settlement on lowlands surrounding the Downs and argued that the chalk uplands were marginal land in prehistory as they are today. This idea has proved a most stimulating one prompting us to extend archaeological research beyond the traditional areas of field monument survival. The danger exists, however, that we will come to overemphasise the marginal nature of the chalklands in prehistory. In fact the valley sediment and lynchet accumulations do seem to imply fairly steady and continuous cultivation of many areas. Consequently we are still forced back to enquire how it is that areas intensively utilised in prehistory were very much less intensively utilised during historic times. Soil erosion and particularly the removal of fertile loess horizons may be a partial answer. Bradley (1979) has suggested that the only areas to sustain intensive occupation after the Bronze Age were those with brickearth deposits and certainly intensive land-use seems to have gone on longest in areas, like Kiln Combe and Chalton, where some superficial loess cover survives. It may be that throughout prehistory and the early historic periods there was a gradual but progressive reorientation of the focus of agricultural activity away from the chalk. That shift finds its first manifestation in the low ground distribution of Romano-British villas later to be reinforced during the Anglo-Saxon period by the pattern of settlements in the charters and Domesday Book, many of which survive today as villages.

As time went on the maintenance of a sufficient depth of fertile soil would increasingly have become a problem. Manuring and marling was a partial solution but we should also consider the possibility that the lynchets themselves represent attempts at soil conservation, rather than accidental artifacts of the erosion process. As early as 1946 the soil scientist E. J. Russell (1946: 62) wrote 'the old method of doing this (stopping erosion) was by a system of terracing such as the early prehistoric lynchets on the chalk downs of England.' In Holland and North Western Germany Groenman-van Waateringe (1979) has suggested that certain 'Celtic' fields represent attempts to counter excessive erosion of the fields. In this country investigations of the medieval strip lynchets in both Wiltshire (Wood and Whittington 1959: 336) and Wharfedale (Raistrick and Chapman 1929: 173) have indicated that some degree of deliberate construction augmented natural plough erosion. The same thing is suggested by the very flinty or stony nature of some lynchet banks which form 'Celtic' fields where it seems that stone clearance was being combined with an attempt to retard further downhill soil movement.

Alluvial valley deposits

The foregoing discussion of colluvial sediments has drawn attention to the existence of close links between colluvial and alluvial sedimentation. The presence of alluvium edge colluvial deposits has been mentioned and colluviation, along with sheet, rill, gully, bank and bed erosion, etc., contributes material for subsequent riverine transportation, sorting and deposition. The extent of erosion on the interfluves is conditioned to a large extent by the amount of runoff and the density of the ground cover which obviously relates to land-use. Studies of the recent geomorphological history of river systems (e.g. Wolman 1967; Leopold 1956) show that on many bedrocks little erosion occurs under woodland conditions but there is a dramatic increase consequent upon clearance and cultivation. All things being equal increased erosion and bed-load is liable to lead to increased sedimentation but we must remember that the pattern of deposition is also very much dependent on the velocity of the stream. This may be modified near the coast by sea level changes or even features like coastal bars; changes in salinity also play an important part. Upstream local changes in base level, produced by beaver dams, dams, mill-ponds, etc., might also lead to the onset of riverine sedimentation.

A number of recent studies of less complicated upstream courses have produced an essentially similar pattern of sedimentological change. At the base of the Flandrian sequence there are often fairly coarse gravels and sands which are overlain by reasonably organic rich, slowly accumulated deposits in which reducing conditions operated giving them a black or grey colour. Above these there are often mineral rich oxidised silts with a much smaller organic matter content. The best documented sequence of this kind is that described by Shotton (1978) from three exposures in the Severn and Avon valleys. The transition to mineral rich silts, which was probably the result of clearance, occurred c. 650 bc and in one case it was marked by a mass of drifted trees and branches, which might have derived from clearance. In the case of the exposure at Pilgrim Lock it was even possible, from the colour of the recent alluvium, to deduce that it had been eroded as the result of clearance of Keuper Marl some 6 km or more upstream. A similar transition from grey reduced silts to oxidised silts was observed by Samuels and Buckland (1978) in a strategically placed J.C.B. trench next to a Romano-British settlement alongside the course of the Idle in South Humberside. There the transition was associated with flooding of the archaeological levels and must have occurred in the late Roman period or shortly after. They suggested that the change of sedimentation could be correlated with palynological evidence, from elsewhere in the region, for extensive forest clearance in the Iron Age and Romano-British periods.

On the Thames at Farmoor, Oxfordshire, there is evidence of a considerable increase in the rate of alluvial deposition following middle Iron Age seasonal occupation of the floodplain, and the build-up of alluvium which seals the Iron Age horizons seems to have substantially accumulated by the 4th century AD (Lambrick and Robinson 1979: 118). Sediments in the valley of the Windrush, a tributary of the Thames, have been investigated near Witney, Oxfordshire, by Hazeldine and Jarvis (1979). They report sandy gravels overlain by clay alluvium with alder roots near the top of the gravel. These roots have a radiocarbon date of 710 ± 85 bc (I-9337) which it was suggested indicates deposition of the clay alluvium since that date. However, Mark Robinson (pers. comm.) emphasises the need for caution in interpreting this date because it is derived from root material which may, like other sections he has observed, have been preferentially preserved below the water table and have decayed the alluvial clay through which it grew. In Ireland Mitchell (1976: 150) has obtained dates for the onset of floodplain alluviation and suggested that it relates to clearance in the catchment. Alluviation at Ballinvegga, Co. Wexford commenced c. 1875 bc and at Scariff Bridge on the Boyne c. 270 ad.

Even in situations where the sediments are not very precisely dated some indication of the environmental conditions under which they were laid down may be possible. For instance alluvium of the Tame at Tamworth (Shillitoe

1961) contained pollen and Mollusca indicating that it had been laid down following clearance and cereal cultivation. Similarly the Bourne, a tributary of the Tame at Shustoke, revealed organic-rich deposits with a woodland flora and fauna overlain by floodplain alluvium with a very open country flora and fauna (Kelly and Osborne 1964). A rather different situation seems to have obtained in the Willow Brook valley at Apethorpe, Northamptonshire (Sparks and Lambert 1961). There was a typical sequence of grey/black silty detritus mud overlain by brown mottled clay, but the pollen and molluscan remains produced little sign of more open conditions associated with the sedimentological change. It was argued instead that the brown clay was the result of occasional flooding of the valley pasture after the stream had been confined by man in its present course.

In a number of valleys the presence of archaeological material underlying the recent mineral rich alluvium provides a terminus post quem for its deposition. At Earls Barton, on the banks of the River Nene, there was evidence that fertile loam soils on the floodplain were covered by alluvium which post-dated a 'Wessex Culture' barrow (Jackson 1972). Large numbers of Roman sites, in particular, are sealed below recent alluvium. Well documented instances include alluvium covering a Roman road and features at Thornborough, Buckinghamshire (Johnson 1975), evidence of flooding levels at Alchester and Beckford (Limbrey 1978), late Roman alluvium interstratified with colluvial deposits at Braughing, and evidence of post Roman alluviation adjacent to the Roman fort at Watercrook, Cumbria (Potter 1976, 1979). A number of similar sites have recently been described and listed by Gater (1979). In the light of all this evidence for late, or post-, Roman alluviation Potter argues that a connection may exist with the contemporary evidence for innundation of the Fens. As for the cause of late Roman flooding he suggests there may be a link with the broadly contemporary younger fill deposits in Mediterranean valleys for which a climatic cause has been argued by Vita-Finzi (1969). Clearly in view of the climatic disparity between the two areas, this link is a rather problematical one. Furthermore recent work is beginning to suggest that even in the Mediterranean the role of climatic change may have been overemphasised and insufficient consideration given to the role of erosion on the interfluves, particularly a colluvial contribution, and the relationship of the sediments to human activity.

Synthesis and future work

When we come to bring together the evidence from colluvial and alluvial sequences in terms of their implications for palaeoenvironmental studies in the Iron Age to Anglo-Saxon periods some interesting contrasts emerge. Colluviation has been referred to as a significant feeder of sediment for subsequent riverine transportation, yet we have evidence for large scale colluviation considerably preceeding the evidence for large scale deposition of recent mineral rich floodplain alluvium. This is partly explained by the fact that most of the colluvial sequencies are in quite different geographical areas and on different bedrocks from the well documented sequencies of floodplain alluvium. It may also be a reflection of the fact that colluvium basically reflects localised erosion whereas alluvium reflects much larger scale changes affecting a significant portion of the catchment. Even if the original clearance of some areas took place at a comparatively early date it may only have been much later, with the creation of continuous expanses of arable land, that mineral material found its way into streams in considerable quantities. Shotton (1978: 31) has argued this in the case of the Severn/Avon sediments and pointed out that the inception of winter sown crops with late autumnal ploughing might have led to increased soil erosion. However, Jones (this volume) has argued that winter sowing may already have been practised in the Bronze Age. A further factor is that pedological change and soil erosion, coupled perhaps with innovations and changes in agricultural practice, may have led to a progressive reorientation of the focus of agricultural activity. If this occurred, not only on the chalk but in other areas, it might help to account for the evidence of increased alluviation in the Roman and Post-Roman periods.

Much remains to be done to realise the full potential of valley sediments in palaeoenvironmental studies. The role of such sediments in contextual studies of the environmental or eroded sites is obvious. They also provide one way in which we can escape from the confines of individual sites and confront more directly the broader issues of land-use and landscape history. It is particularly interesting that we can do this by investigating an environment of deposition which itself will often have formed a key part of the exploitation zone of neighbouring archaeological sites. Many of the problems of chronology and causation require much refinement and there is an obvious need for studies of colluvial sediments to be extended to other geological areas, particularly the limestone. Floodplain margins would also repay further detailed study, here we might find a range of palaeoenvironmental evidence interstratified within colluvial and alluvial deposits which would enable us to assess the extent of the colluvial contribution. In fact, excavators on a variety of bedrocks might well find that a good deal of information could be obtained at very little cost from a few strategically placed soil pits in neighbouring valleys.

Acknowledgements

This paper represents an interim report on a study of valley sediments as evidence of prehistoric land-use which is being prepared as a London University thesis. I am pleased to acknowledge grants towards my investigation of dry valley sediments from the I. D. Margary Research Fund of the Sussex Archaeological Society and the Central Research Fund of London University. I should also like to thank the landowners (already mentioned) who made our field investigations both possible and enjoyable and all those who assisted in the excavation and recording of the trenches, particularly Jennifer Foster, Peggy Norman, Brian Oldham, the Stevens family, Ken and Marion Suckling and Brenda and Jim Westley, without whom little would have been achieved.

Bibliography

Barker, H. et al. (1971). 'B.M. natural radiocarbon measurements VII' Radiocarbon 13 169.

Bell, M. (1977). 'Excavations at Bishopstone' Sussex Archaeological Collections 115 1-299.

Bradley, R. (1978). The prehistoric settlement of Britain (London: Routledge).

Bradley, R. (1979). 'The implications of hillwash - the archaeological framework' unpublished lecture at symposium on hillwash and archaeology British Museum 5.3.79.

Burstow, G. P. and Holleyman, G. A. (1957). 'Late Bronze Age settlement on Itford Hill, Sussex' Proceedings of the Prehistoric Society 23 167-212.

Catt, J. A. (1978). 'The contribution of loess to soils in lowland Britain' In: The effect of man on the landscape: the lowland zone (eds. Limbrey, S. and Evans, J. G.) (CBA Res. Rep. 21) pp. 12-20.

Cunliffe, B. W. (1973). 'Chalton, Hants: the evolution of a landscape' Antiquaries Journal 53 173-190.

Cunliffe, B. W. (1976). 'Danebury, Hampshire: second interim report on the excavations 1971-5'. Antiquaries Journal 56 198-216.

Darwin, C. (1881). The formation of vegetable mould through the action of worms with observations of their habits (London: Faber and Faber edition published 1955).

Dimbleby, G. W. (1976). 'Climate, soils and man' Philosophical Transactions of the Royal Society B 275 197-208.

Drewett, P. L. (1977). 'Rescue archaeology in Sussex 1976' Bulletin of the Institute of Archaeology. London University 14 1-57.

Evans, J. G. (1966). 'Late glacial and post-glacial subaerial deposits at Pitstone, Bucks' Proceedings of the Geologists Association 77 347-364.

Evans, J. G. and Valentine, K. W. G. (1974). 'Ecological changes induced by prehistoric man at Pitstone, Buckinghamshire' Journal of Archaeological Science 1 343-351.

Gater, J. A. (1979). 'Flooding and flood deposits in Britain: some implications' Unpublished undergraduate dissertation Dept. of Archaeological Science, University of Bradford.

Godwin, H. (1967). 'Strip lynchets and soil erosion' Antiquity 41 66-7.

Groenman-van Waateringe, W. (1979). 'Nogle aspekter af jernalderens agerbrug i Holland og NV Tyskland' In Thrane, H. (ed.) Fra jernalder til Middelalder (Odense).

Hazelden, J. and Jarvis, M. G. (1979). 'Age and significance of alluvium in the Windrush valley, Oxfordshire' Nature 282 291-2.

Holden, E. W. (1972). 'A Bronze Age cemetery-barrow on Itford Hill, Beddingham, Sussex' Sussex Archaeological Collections 110 70-117.

Jackson, D. (1972). 'The Earls Barton barrow' Current Archaeology 3 238-241.

Johnson, A. E. (1975). 'Excavations at Bourton Grounds, Thronborough 1972-3' Records of Buckinghamshire 20 3-56.

Kelly, M. and Osborne, P. J. (1964). 'Two faunas and floras from the alluvium at Shustoke, Warwickshire' Proceedings of the Linnean Society of London 176: 1 37-65.

Kerney, M. P., Brown, E. H. and Chandler, T. J. (1964). 'The late glacial and post-glacial history of the chalk escarpment near Brook, Kent' Philosophical Transactions of the Royal Society B 248 135-204.

Lambrick, G. and Robinson, M. (1979). Iron Age and Roman riverside settlements at Farmoor, Oxfordshire (CBA Res Rep 32).

Leopold, L. B. (1956). 'Land-use and sediment yield' in Man's role in changing the face of the earth (ed. Thomas W. L.) Chicago pp. 639-647.

Limbrey, S. (1975). Soil science and archaeology London: Academic Press.

Limbrey, S. (1978). 'Changes in the quality and distribution of the soils of lowland Britain' In: The effect of man on the landscape: the lowland zone (eds. Limbrey S. and Evans J. G.) CBA Res Rep 21 pp. 21-27.

Manby, T. G. (1975). 'Neolithic occupation sites on the Yorkshire Wolds' Yorkshire Archaeological Journal 47 23-60.

Mitchell, F. (1976). The Irish Landscape London: Collins.

Potter, T. W. (1976). 'Valleys and settlement: some new evidence' World Archaeology 8 207-19.

Potter, T. W. (1979). Romans in North-West England Cumberland and Westmorland Antiquarian and Archaeological Society Research Series 1.

Raistrick, A. and Chapman, S. E. (1929). 'The lynchet groups of Upper Wharfedale, Yorks' Antiquity 3 165-181.

Russell, E. J. (1946). A student's book on soils and manures Cambridge.

Samuels, J. and Buckland, P. C. (1978). 'A Romano-British settlement at Sandtoft, South Humberside' Yorkshire Archaeological Journal 50 65-75.

Shillitoe, J. S. (1961). 'Borings in the first terrace and alluvium of the River Tame at Tamworth' Proceedings Coventry Natural History and Scientific Society 3 pt 4 133-8.

Shotton, F. W. (1978). 'Archaeological inferences from the study of alluvium in the lower Severn-Avon valleys' In: The effect of man on the landscape: the lowland zone (eds. Limbrey, S. and Evans J. G.) CBA Res Rep 21 pp. 27-32.

Sparks, B. W. and Lambert, C. A. (1961). 'The post-glacial deposits at Apethorpe, Northamptonshire' Proceedings of the Malacological Society of London 34 302-315.

Taylor, C. C. (1970). The Dorset landscape London: Hodder and Stoughton.

Taylor, C. C. (1972). 'The study of settlement patterns in pre-Saxon Britain' In: Man Settlement and Urbanism (eds. Ucko P. J., Tringham R. and Dimbleby G. W.) London: Duckworth 109-13.

Taylor, C. C. (1973). The Cambridgeshire landscape London: Hodder and Stoughton.

Vita-Finzi, C. (1969). The Mediterranean valleys: geological changes in historical times Cambridge.

Wolman, M. G. (1967). 'A cycle of sedimentation and erosion in urban river channels' Geografiska Annaler 49A 385-95.

Wood, P. and Whittington, G. (1959). 'Further examination of strip lynchets north of the Vale of Pewsey in 1958' Wiltshire Archaeological and Natural History Magazine 57 322-338.

THE EXPLOITATION OF ENVIRONMENTAL RESOURCES

6. THE DEVELOPMENT OF CROP HUSBANDRY

Martin Jones

At the beginning of the Iron Age, a range of crops that are now virtually unknown was being grown in fields that were tilled with a light ard, an implement no longer seen in this country. By the end of the Anglo-Saxon period, heavy wheeled ploughs were being drawn across the lowlying clay loams, and the species of crop that were being sown remain in use on those same soils today. The period under study was one of considerable change in crop husbandry and one in which the development can be followed of methods of agriculture that prevailed throughout the Middle Ages.

These changes have been recognised in general terms by a number of authors, and have been linked with a concurrent 'shift of emphasis' in the settlement evidence from the lighter soils on higher ground to the intervening valleys (e.g. Fox 1932, Ellison and Harriss 1972, Bell, this volume). Various attempts to define these developments more closely in time have been both made and refuted. Current authors are generally reticent about the nature of agricultural change in this period. Cunliffe (1978), while recognising the stimulating effect of late Iron Age trade, comments:-

> "Once the economy had re-adjusted to the new regime [in the first half of the first millennium BC], there was little scope for significantly improved productivity".

A. H. M. Jones' assertion that:-

> "no changes in agricultural methods are recorded under the Roman Empire" (Jones 1964).

is considered a little extreme by some. Applebaum (1972) states that:-

> "it is difficult to hold that Roman Agriculture was not an improvement on what preceded it".

though it is questionable how many of his listed improvements are really associated with the Conquest or indeed the Roman period at all. Percival (1976), taking a moderate line argues against the introduction of a radically different system in the Roman period. Finally, Finberg (1972) suggests:-

> "There is no evidence whatsoever that the Anglo-Saxon immigrants brought with them better ploughs or a more advanced agricultural technique than those they found when they arrived".

It is difficult to isolate mechanisms for the changes outlined in the opening sentences of this paper against the overview provided by these authors. A more rigorous analysis of crop husbandry in the period is necessary, and is now possible as a result of various recent developments in environmental and archaeological research.

Knowledge of the associated artefactual and structural evidence has improved, with for example the very useful synopses of agricultural implements (Rees 1979) farm buildings (Morris 1979) and field systems (Bowen and Fowler 1978). Developments have also occurred in the analysis of a key form of environmental evidence, the preserved remains of crop plants and crop impurities. These are here considered in detail.

Records of Crop Plants and Crop Impurities

In the forties and fifties, the Danish workers, Jessen and Helbaek produced two surveys of British crop plant records derived from the charred remains of seeds and chaff, and their impressions in pottery (Jessen and Helbaek 1944, Helbaek 1952). Their compilations have since formed the basis for all discussions of prehistoric and early historic crop husbandry. Since their work it has become apparent, as the result of the routine use of flotation techniques on site, that charred remains in particular are considerably more commonplace than had previously been supposed (Jones, M. 1978). As a result, the number of sites for which crop records exist has more than trebled. In addition the occurrence of assemblages of weeds associated with carbonised crops, richer than any examined by Jessen and Helbaek in Britain, has allowed new types of inference to be drawn (Jones, M. 1978). The wide range of species present has produced hitherto unavailable evidence on the precise ecological state of early fields.

The interpretation of the crop plant record itself has been approached more critically by Dennell (1974) and Hubbard (1975) both of whom raise doubts about certain of Jessen and Helbaek's conclusions, and also indicating new possible forms of inference.

A less obvious development of equal importance is the more refined chronological framework attainable today. The records compiled by Jessen and Helbaek were organised according to traditional cultural groupings, such as 'Early Iron Age', 'Roman' and 'Anglo-Saxon'. As the appearance of new crops was associated with the arrival of new cultural groups, this method of organisation was appropriate to the current research questions. With today's more critical approach to continuity and change (e.g. Burnham and Johnson 1979), the date and context of a particular record often needs to be far more precisely defined than it was in the forties and fifties.

In the light of these developments, an attempt has here been made at a synthesis of the information that now exists for the nature and development of crop husbandry from the Iron Age to the Anglo-Saxon period.

The distribution of records

Figures 6:1, 6:2 and 6:3 show the distribution of sites for which, to the best of the author's knowledge, records of particular crop species have been made. The number of such records is constantly increasing, and while such figures will always be incomplete, they serve to show the general distributional biases in the extant data. It is immediately apparent from the figures that such biases exist. Virtually nothing is known of highland zone crops in the Iron Age. In the Roman period, knowledge of Wales and the South West is minimal, and Anglo-Saxon records are biased markedly towards the East. These biases are to some extent inevitable, when researchers are looking for secure cultural contexts for their botanical material. Nonetheless, a range of records from contrasting cultural zones would allow a far greater insight into the relative contribution of cultural and non-cultural factors in the development of crop husbandry.

In addition to the gaps, there are areas where records are unusually common. These too, if carelessly interpreted, can lead to spurious inferences. The domination of Anglo-Saxon records by barley may be due to a large number of records of this crop from a restricted area in East Anglia. A number of earliest crop records occur around an axis running between Gloucester and Southampton. This may also be a spurious consequence of the concentration of records in this area. A species that is widely and evenly distributed at a very low concentration is more likely to show up where the greatest density of records occurs.

Having made these points, it should be mentioned that certain of these distributional biases are being rectified in particular by the work of Hillman in Wales, and Murphy in East Anglia, and their results may well modify current ideas on development in crop husbandry.

The economic plants

In the period under study, a very large number of economic plants appear in the botanical record. The last decade has seen an enormous increase in the records of food plants for the Roman period, largely in conjunction with rescue archaeology conducted at London, York, Winchester and Colchester (Willcox 1977, Greig 1976, Kenward and Williams 1979, Murphy 1977). The list contains such fruits as grape (*Vitis vinifera*), fig (*Ficus carica*), date, (*Phoenix dactylifera*), mulberry (*Morus* spp.) and olive (*olea europea*), the nuts walnut and pine-nut (*Pinus pinea*), the vegetables cucumber (*Cucumis sativus*), cabbage (*Brassica oleracea*), celery (*Apium graveolens*) and fennel (*Foeniculum vulgare*), the grain crops lentil (*Lens culinaris*) and millet (*Panicum miliaceum*) and the herbs, coriander (*Coriandrum sativum*), dill (*Anethum graveolens*) and anise (*Pimpinella anisum*). The general view is that many of the new additions are less likely to represent crop introductions than luxury imports. The vast majority of these records are from urban sites, although coriander, dill and one grape pip are known from rural Romano-British sites (Lambrick and Robinson 1979, Jones, A. 1977, Murphy 1977). Certain of the seeds have clearly been imported. Williams has pointed out that since figs grown in Britain do not set seed, the fruits from York with well developed seeds must be of foreign origin (Kenward and Williams 1979) and Murphy (pers. comm.) has made the same point about figs from Colchester. Helbaek (1964) romantically suggested that the lentils and chickpeas from a deposit in the legionary fortress at Isca was imported by the Roman troops:-

> "in an attempt to produce beer and polenta of the same attractive properties as at home".

The grape seeds from York are probably beyond the northern limit for the successful growth of the vine (Addyman *et al*. 1976) and the evidence for Romano-British viticulture in general is now considered slight (Williams 1977). A supposed vineyard at North Thoresby in Lincoln could just as easily be a market garden for a number of other foodstuffs (Webster *et al*., 1967).

50
52
51
27
28
29
25
40
4
23
2
3
32
1
24
15
33
34
30
31
16
53
26
46
39
45
37
36
38
49
35
21
22
42
17
19
18
47
44
20
48
43
10
7
9
13
11
12
8
14
5
6
41

ENGLAND

Bucks.: 1. Bierton, MJ

Cambs.: 2. Abington Piggots, HG
3. Aldwick Barley, HG
4. Wandlebury, Renfrew 1965

Cornwall: 5. Killibury, GH
6. Trebarvath, J&H

Devon: 7. Hembury, HH

Dorset: 8. Corfe Castle, FG
9. Corfe Mullen, Percival 1934
10. Gussage-All-Saints, Evans and Jones 1979
11. Maiden Castle, HH
12. Portland, HH
13. Poundbury, FG
14. Worth Matravers, HH

Essex: 15. Little Waltham, Wilson 1979

Glos.: 16. Uley Bury, GH

Hants.: 17. Danebury, MJ
18. Micheldever Wood, PM
19. Old Down Farm, PM
20. Owselbury, PM
21. Portway, PM
22. Winnall Down, PM

Herts.: 23. Ashwell, J&H
24. Prae Wood, J&H

Hunts.: 25. Godmanchester, HG

Kent: 26. Wilmington, GH

Norfolk: 27. Hunstanton, PM
28. Micklemoor, HG

Northants. 29. Twywell, Arthur 1975

Oxon: 30. Abingdon, MJ
31. Berinsfield, MJ
32. Chastleton, J&H
33. Hardwick, MJ
34. Radley, J&H

Somerset: 35. Casterley, HH
36. Glastonbury, HH
37. Meare, HH
38. Small Down, HH
39. Worlebury, HH

Suffolk: 40. West Stow, PM

Sussex: 41. Bishopstone, Arthur 1977

Surrey: 42. Hascombe, PM

Wilts.: 43. Fifield Bavant, HH
44. Highfield, HH
45. Lidbury, HG
46. Little Solsbury, HH
47. Swallowcliffe Down, HG
48. Tollard Royal, Evans and Bowman 1968
49. Winklebury, HH

SCOTLAND 50. Dun Mor Vaul, Renfrew 1974

WALES 51. Breidden, GH
52. Dinorben, GH
53. Wenvoe, GH

Fig. 6.1 Distribution of Iron Age sites with specific crop records. Abbreviations used: FG, cited in Green 1979 and this volume; GH, Hillman 1978, 1981a&b and pers. comm.; HG, cited in Godwin 1975; HH, Helbaek 1952; J&H, cited in Jessen and Helbaek 1944; MJ, Jones, M. 1978, in Lambrick and Robinson 1979, and forthcoming; MM, cited in Monk 1978; PM, Murphy 1977 and pers. comm.

46
49
48
47
45
43
44
20
3
2
27
54
23
21
22
24
25
37
26
38
33
39
7
8
52
10
9
31
30
1
18
32
29
28
55
50
51
53
42
35
19
34
5
12
17
11
16
13
33
41
4
14
15
40
6

ENGLAND

Bucks.: 1. Bierton, MJ
Cheshire: 2. Lindow Moss, HG
3. Wilderspool, GH
Dorset: 4. Iwerne, JH
5. Stockton, HG
6. Studland, Field 1965
Essex: 7. Colchester, PM
8. Rivenhall, HH, HG
Glos.: 9. Frocester, Clarke 1970
10. Upton-St-Leonards, Clarke 1971
Hants.: 11. Micheldever Wood, PM
12. Neatham, PM
13. Owselbury, PM
14. Portchester, PM
15. Purbrook, PM
16. Winchester, PM, FG
17. Winnall Down, FG
Herts.: 18. St. Albans, HH, HG
Kent: 19. Keston, GH
Lancs.: 20. Holcroft Moss, HG
Norfolk: 21. Ashill, Jones, A. 1977
22. Caister-by-Norwich, HG
23. Caster-on-Sea, PM
24. Southery, PM
25. Thetford, PM
26. Weeting, PM
Notts.: 27. Bunny, HG
Oxon: 28. Abingdon, MJ
29. Berinsfield, MJ
30. Farmoor, Lambrick and Robinson 1979
31. North Leigh, HG
32. Radley, MJ
Somerset: 33. Catsgore, GH
34. Meare, HH
35. Star Villa, HG
Suffolk: 36. Icklingham, PM
37. Lakenheath, PM
38. Somersham, PM
39. Stowmarket, PM
Sussex: 40. Fishbourne, FG
Wilts.: 41. Rotherley, HH, HG
42. Windmill Hill, HH, HG
Yorks.: 43. Malton, J&H, HG
44. York, Greig 1976, Kenward and Williams 1979

SCOTLAND

45. Birrens, J&H, HG, Wilson 1975
46. Castle Cary, J&H, HG
47. Edinburgh, J&H, HG
48. Glasgow, J&H, HG
49. Old Kilpatrick, J&H, HG

WALES

50. Caerleon, HG
51. Caerwent, HG
52. Camarthen, GH
53. Llandough, GH
54. Newtown, GH
55. Usk, GH

Fig. 6.2 Distribution of Romano-British sites with specific crop records. Abbreviations as in Fig. 6.1.

9
39
38
20
23
22
21
24
32
17
8
7
29
31
33
30
4
1
5
6
2
3
37
10
27
25
26
28
12
35
34
19
18
14
16
15
11
36
13

Beds.: 1. Sandy, MM
2. Stratton Park, FG

Cambs.: 3. Barrington, J&H, HG
4. Cambridge, J&H, HG
5. Haslingfield, J&H, HG
6. Little Wilbraham, J&H, HG
7. Mildenhall, J&H, HG
8. Tuddenham, J&H, HG

Durham: 9. Durham, MM

Glos.: 10. Gloucester, FG

Hants.: 11. Chalton, MM, FG
12. Cowdery's Down, FG
13. Portsmouth, MM
14. Portway, FG
15. Southampton, MM
16. Winchester, MM, FG

Hunts.: 17. Buckden Down, MM

Kent: 18. Canterbury, MM
19. Graveney, HG, Wilson and Conolly 1978

Norfolk: 20. Downham Market, J&H
21. Great Yarmouth, MM
22. Old Buckenham, HG
23. Stanford, J&H, HG

Northants.: 24. Rothwell, J&H, HG

Oxon: 25. Abingdon, J&H, HG, MJ
26. Berinsfield, MJ
27. Oxford, MJ
28. Sutton Courteney, J&H, HG

Suffolk: 29. Icklingham, J&H, HG
30. Lackford, J&H, HG
31. Redgrave, J&H, HG
32. Somersham, J&H, HG
33. West Stow Heath, J&H, HG

Surrey: 34. Hawkshill, HG
35. Leatherhead, MM

Sussex: 36. Walton, MM

Worcs.: 37. Fladbury, MM

Yorks.: 38. Newmark, J&H, HG
39. York, MM

Fig. 6.3 Distribution of Anglo-Saxon sites with specific crop records. Abbreviations as in Fig. 6.1.

These introductions are then of more relevance to long distance trade than to British crop husbandry. It is of course a point of interest to consider how far such perishable foodstuffs as fig or cucumber could move within classical trade networks. Evidence for comparable movement of foodstuffs does not exist for the Iron Age, but one wonders whether rigorous environmental analysis of late Iron Age oppida might not produce similar evidence.

Turning to the probable home-grown crops, the major period of innovation remains as Helbaek (1952) noted, in the first half of the first millennium BC. Helbaek's compilation suggested that six new species of crop appear at this stage, spelt wheat (Triticum spelta), bread wheat (T. aestivum/compactum), rye (Secale cereale), celtic bean (Vicia faba minor), common oat (Avena sativa) and bristle oat (Avena strigosa). A number of these crops are now known in earlier contexts; the neolithic date of the spelt wheat from Hembury has been confirmed (Field et al., 1964); Bronze Age bread wheat is now known from Abingdon, Oxon (Jones, M. 1978) in addition to the neolithic records known to Helbaek; a late neolithic seed impression has been identified as Vicia faba (Hillman, pers. comm.). It still appears however that this period saw a considerable influx of new crop types. The introduction of a new species of staple is a rare occurrence, and in these terms this period is the main period of change between the early Neolithic and the 16th century AD. Although this innovation is normally attributed to the onset of the Iron Age, its actual timing is difficult to determine with any confidence. The crop remains described as 'Early Iron Age' by Helbaek are dated by associated pottery assemblages, which under modern scrutiny can be placed anywhere between the 9th and 5th centuries BC (David Knight, pers. comm.). Two early occurrences of these species are associated with radiocarbon dates. An Early Iron Age pit at Godmanchester, Cambridge contained a grain deposit that included a trace of spelt, and which gave a radiocarbon date of 930 ± 80 bc (Green, H. J. M., unpublished typescript). Spelt wheat and cultivated oats were recovered by Hillman (pers. comm.) from a deposit sealed by the inner rampart of Dinorben fort North Wales, with radiocarbon dates ranging between 440 ± 45 bc and 545 ±65 bc. In the light of the caution expressed by Hubbard (1975) as to the significance of Bronze Age grain impressions, it would be useful to examine more carbonised samples from this period before suggesting too precise a date for these innovations.

A point that the improved records suggest is that these innovations are associated with Southern Britain. An agriculture involving emmer wheat as a principal component continues in Northern Britain. However, the new species do penetrate; oats are being grown by the Roman period in the north, and there are isolated records of other species. Both emmer and oats are typically spring sown in northern Europe and a system without winter sowing may be envisaged, such as continued to exist into the 17th century in some parts of the highlands (Applebaum 1954). In this later period, autumn sowing was avoided in order to maximise winter grazing land. The same may well have been true in earlier periods.

The records are still too patchy to compare these developments in different regions of southern Britain, but it is interesting to note the absence of any time lag between innovations in Southern England and in the more remote parts of Wales (Helbaek 1964, Hillman 1979 and pers. comm.).

In attempts to explain the innovations discussed above, attention has generally been focused on the species that comes to dominate the crop record, spelt wheat (e.g. Applebaum 1954). This cereal is well suited to autumn sowing, the Applebaum suggests its introduction may be linked with the development of the autumn sowing practice. The main advantage of mixed autumn and spring sowing is that it spreads the labour load at sowing and harvest time, the peak seasons of the year.

The remaining species have received less consideration; rye and oats have often been dismissed as arable weeds. However there seems little reason to suppose they occur merely as weeds; the carbonised assemblages in which they are recorded are virtually free of other weeds and would correspond to Dennell's system to a fully processed crop, (Dennell 1974). The introduction of the whole range of crop plants is more difficult to explain than the introduction of spelt alone, whose crop characters are not shared by all the other incoming species. In the light of a certain amount of confusion that exists about the ecological preference of different cereals, a few points are here made about the crops in cultivation at this period.

Barley (Hordeum sp.)

Various classical authors refer to barley being grown for human and animal consumption and for brewing (Applebaum 1972). A range of six-row barleys occur throughout the crop record, both naked and hulled, and dense and lax eared forms. Caution is needed in making the latter distinction, as examination from the large body of material from the Ashville site, Abingdon, Oxon has suggested that early crop varieties may not have been as genetically discrete as their modern counterparts, (Jones, M. 1978). This probably applies to many crop species which are inherently more likely than wild species to have undergone micro-evolutionary change. The common assumption that all crop plants encountered in archaeological contexts fall within the genetic range of the motley collection of varieties that have reached modern herbaria and plant breeding stations is an unsound basis for examination, and may often lead to spurious results. This caution also applies to crop ecology. Current ideas about what is good land for barley are dominated by the requirements of the brewing industry, which requires a very specific quality of barley. The range of conditions in which barley can grow is extremely broad; from the Arctic Circle to the subtropics, and to heights of 12,000 ft in S.E. Asia. It grows well on both heavy and light soils, and can be sown in either winter or spring (Bell 1948, Bland 1971). The supposed correlation between different forms of barley and best time of sowing; inferred by Applebaum (1954) and subsequently quoted by Cunliffe (1974) and Bradley (1978), is based on a misunderstanding of Beaven (1947) and there is no evidence that it is generally applicable.

Barley may potentially be grown universally in Britain, except where the drainage is very poor, or where the pH is below 6 (Bland 1971). Throughout the period under study, barley occurs with greater frequency in Southern Britain and Wales, and the paucity of Romano-British samples from this region in Helbaek's list may explain the apparent rarity of barley in his Romano-British record (Helbaek 1952). Present evidence in fact suggests that barley could be the dominant cereal in the Somerset area at this stage. In contrast, as mentioned earlier, barley may be over-represented in the Anglo-Saxon period.

Wheat (*Triticum* spp.)

In the same way that barley is not necessarily a crop of light soils, nor are all the wheats necessarily associated with heavy soils. The various species of wheat have varying and sometimes contrasting ecological preferences and should be treated individually.

(a) Emmer (*Triticum dicoccum*)

The ecological preferences of emmer wheat are very different from those of modern bread wheat. Whereas the latter grows well on heavy loams, emmer is well-suited to the light and dry extremes. Columella (II 9.3) suggests that heavy clay wet soil is suitable for *far adoreum*, which is normally translated as emmer. This is puzzling in the light of the modern ecology of this cereal, and it is suggested here that Columella may be referring to a wheat more similar to spelt.

Emmer is relatively frost susceptible and for this reason, although winter sown varieties are known, it is probably better suited to spring sowing (Percival 1921).

Emmer is the principal wheat in neolithic and Bronze Age records from Britain. Within the limits of inference from sparse records, its importance seems to continue in the highland zone into later periods. For the Iron Age, records are absent north of Hertfordshire and Oxfordshire, and crop husbandry is an unknown in the bulk of the highland zone. However, as one moves from east to west in Southern Britain, an increase in the occurrence of emmer is clearly discernable.

A number of records of Roman date in the Edinburgh and Glasgow region, and one early Christian record at Edinburgh, suggest that, as mentioned above, emmer remained the commonest wheat in this part of the highland zone. In the lowland zone, it is not uncommon in either period, but generally secondary in importance to spelt wheat. It was still grown in scattered localities throughout the world in the 19th century and is still an important crop in the Ethiopian Highlands.

Emmer wheat produces flour of exceptionally good quality, both in nutritional and gastronomic terms. Its protein content is substantially higher than that of modern bread wheat, and it was favoured in 19th century Germany for its fine white flour and its baking quality. In soup it was considered superior to pearl barley or spelt (Percival 1921).

(b) Spelt (*Triticum spelta*)

The records suggest that this crop enjoyed great success in South Britain throughout the Iron Age and Roman periods, but diminished in prominence thereafter, with the growing occurrence of bread wheat.

Spelt will grow on the heavier soils associated with bread wheat, and also on drier and lighter soils. Its main characteristic is its hardiness to cold, wind, diseases and pests. This hardiness allows it to be grown today in the Schwabian Alps near Stuttgart to altitudes of 3000 ft. It also makes it ideal for winter sowing in NW Europe (Percival 1921).

Spelt has excellent baking and milling properties and for this reason, its cultivation in Germany was encouraged during the 3rd Reich, as Hitler was concerned that his foreign hard wheat supplies would be cut off. Today people commute for miles to buy dinkelbrot from the one baker in Stuttgart who still uses spelt wheat.

(c) Bread Wheat (Triticum aestivum/compactum)

This cereal grows well on 'strong land' with a significant proportion of silt or clay, and also responds well to high nitrogen levels (Bland 1971). It has become typically associated with deep clay loams in Britain. A dense eared form normally referred to as 'club wheat' (Triticum compactum is known, and may have been important in the Roman period. However it is genetically very close to bread wheat sensu stricto, T. aestivum, and the distinction may not be significant.

One thing that emerges from an examination of the so-called primitive wheats is that they do have positive characteristics not shared by bread wheat. There has been a tendency to regard bread wheat, which has since become the most extensively used biological resource of all time, as inherently superior to primitive cereals. This is clearly refuted by the crop record, in which bread wheat remains no more than a minor component for at least four millennia following its early neolithic introduction. Indeed, it does have the advantages of winter hardiness, a high yield potential and a loosely packed ear that is easily threshed. But that same character of loose packing in the ear renders the seed far more vulnerable to attack, especially from birds and fungi. Bread wheat is also a poorer competitor with weeds and needs a greater soil fertility than other wheats. It is typical of artificially bred species in that during the course of selection for single characters, it has lost many of its natural survival characteristics, and therefore demands far more care and attention for its higher potential yields to be realised.

By the Anglo-Saxon period, bread wheat has become the most frequently occurring wheat. Monk (1978) after closer analysis of samples from the Wessex area, puts bread wheat's rise to dominance in the late Roman period. He suggests that this rise is due to bread wheat's greater suitability for heavy soil. A link with heavy soil is substantiated by grain finds from late Iron Age contexts at Bierton, Bucks. a site whose catchment is dominated by clay soils. Although bread wheat is normally only a minor component of assemblages of this date, it is easily the most abundant cereal at Bierton (Jones, M. forthcoming). However the wheat that precedes it, spelt, is also suited to heavy soils. The relationship with heavy soils is only a partial explanation. It is here suggested that bread wheat was favoured when, and only when it became both possible and desirable to invest the greater amount of fertiliser, and man hours in the form of cultivation and weeding, that would be necessary to obtain the high yield potential of bread wheat, and that this was only one stage in the gradual shift from lighter to heavier soils.

Green (1979) has pointed out that the ease of threshing and therefore removal of waste, would substantially reduce transport costs, and this could be a contributory factor in the choice of bread wheat at a stage when trade in foodstuffs was expanding.

Oats (Avena spp.)

Cultivated oats are unfortunately only distinguishable from wild oats when certain fragments of chaff, the floret bases, are present. These are relatively infrequent, but enough occur to isolate two species of cultivated oat in at least two Iron Age sites in Dorset, Fifield Bavant and Maiden Castle. Records of oats for the Iron Age in general cluster around the Gloucester Southampton axis, but as outlined earlier, this may be the result of uneven sampling. The same is true of Anglo-Saxon records of oats, which concentrate in the south and east, as do the sites for which records exist. In the intervening Roman period, oats are more extensive, and occur in virtually all samples recovered from the highland zone. Applebaum (1972) has suggested that this highland cultivation of oats may have been stimulated by the military demand for horse fodder. The tolerance of oats to acid and infertile soils is high, and the increasing development of these soils in the highland zone may be a sufficient explanation. It would be useful to establish whether oats were as common in the highland zone in the Iron Age, but sufficient data have not been collected.

Oats prefer a milder moister growing season than wheat or barley and are less frost hardy. They are therefore generally better suited to spring sowing than winter sowing, although winter-sown varieties do occur. Because of their high water requirement, they grow best on deep-retentive loams and clay loams (Bell 1948).

Rye (Secale cereale)

This very hardy crop can be grown on all soils except the very heavy clays. It prefers good soil but like oats, can tolerate acidity and low fertility, and can also tolerate very dry soils better than most cereals in Britain. Rye is suitable for both autumn and spring sowing (Bell 1948, Bland 1971).

Rye appears in four Iron Age sites in Dorset, Wiltshire and Hampshire. Eleven occurrences are known to the author from Roman Britain spanning from Owslebury in Hampshire to the south (Murphy 1977). Carmarthen to the west (Hillman 1978) and Edinburgh to the north (Jessen and Helbaek 1944), and including records scattered around the midlands and north England. In the Anglo-Saxon period, six records are known, again widely scattered, though it is not recorded in the few records in Northern England.

Chess/Brome (Bromus spp.)

In addition to the conventional species use may also have been made of plants that are normally considered to be wild. The most frequenly referred to is the genus of grass Bromus, and in particular, Bromus secalinus. Helbaek noticed that seeds of Bromus occur frequently among Iron Age grain samples. Chess can occasionally dominate carbonised assemblages (Jones 1978: 101), and Hubbard (1975) has argued that it may have been a crop in its own right. It would certainly be of interest to know whether the Bromos referred to by Pliny (N.H. XVIII: xx: 93) and normally translated as 'oats' in fact refers to this species.

To ask whether or not plants such as Bromus were crops may be an invalid question. Until recently, and in many parts of the world still today, the division drawn between crop and wild plant was less sharp by far than it is in the modern western world. A heavy reliance on species other than what was sown in the fields was common. In Denmark to the present century, the seeds of chess were collected and ground to flour, as a famine food in years the rye crop failed. Chess would be the weed that dominated these failed fields. In many parts of the Near East today, the cultivated ground is not expected to produce the sown crop in isolation. In some cases the harvest includes a greater yield of what we would call the weeds, than of the sown crop itself (e.g. Zohary 1967: 226). Outside the modern Western world, the binary distinction between crop and wild plant is too coarse; a whole range of possible man-plant relationships must be considered (cf. Higgs et al. 1972).

Celtic bean (Vicia faba var. minor)

This small form of bean is common throughout prehistoric Europe. It appears in Britain in three Iron Age sites around the Somerset Levels, and two other Iron Age sites, in Dorset and Hampshire. Two late Bronze Age records are cited by Green (this volume), and they also occur sporadically in later deposits.

Like all legumes, Celtic bean has the ability to restore nitrogen to the soil, wherein lies its importance to agriculture. They may be sown in spring or winter, but the latter only in mild winters followed by warm dry springs (Bland 1971). In the light of climatic developments early in the first millennium BC (Lamb, this volume) it seems likely that Celtic bean was used at this stage as a spring-sown crop.

The introduction of the autumn sowing practice is insufficient as an explanation of the influx of crops in the first millennium BC. Whilst it is appropriate for spelt, and bread wheat, rye is equally well-suited and oats and beans are better suited to spring sowing. In addition, the improved knowledge of contemporary weed floras allows an independent estimate of sowing time. Brenchley and Warington (1930) emphasised that time of sowing is one of the key determinants of the weed flora, which consequently becomes an ideal indicator of sowing time (van Zeist 1974, Jones, M. 1978).

Autumn germinating weeds such as Galium aparine and Veronica hederaefolia are now known as weed impurities in earlier deposits of carbonised grain, suggesting that the autumn sowing practice can be back dated at least to the Bronze Age. One wonders whether better evidence will link its introduction with the appearance of organised field systems, but this is outside the scope of this paper.

Whilst their ecological preferences vary, the common feature of the first millennium crop introductions is their ability to cope with unfavourable conditions in one form or another. Spelt wheat and bread wheat are more frost-hardy than emmer, and could produce a crop on soils too damp or heavy for either emmer or barley. Rye and oats could be successfully grown on the acid and infertile soils that had become increasingly common in later prehistory, and celtic bean would replenish nitrogen to sour soils. The introduction of this group of crops can perhaps best be explained as a diversification of the crop repertoire in order to allow both the sustenance of cultivation on deteriorating land the the expansion onto hitherto marginal lands, in the face of the increasingly frequent winds and storms suggested by Lamb (this volume). It is at this period also that the rate of alluvial soil movement seems to increase (Bell, this volume) and that the rate of new woodland clearance gains fresh momentum (Turner, this volume).

This explanation is supported by the potential crop that did not form part of this influx, millet (Panicum miliaceum). This cereal occurs frequently in the continental records, and its absence from Britain may be explained in terms of the lack of demand for a new cereal that was not tolerant of unfavourable situations in the same way as the other incoming species.

Few further introductions occur in the crop record from rural sites over the next two millennia. The field pea (Pisum sativum) appears at Iron Age Bishopstone (Arthur 1977) and Roman Owslebury (Murphy 1977) and flax (Linum usitatissimum) in Roman deposits at Meare, Somerset and two Upper Thames Valley sites (Jessen and Helbaek 1944, Robinson 1979 and forthcoming). Flax is also known from neolithic and Bronze Age contexts (Helbaek 1952), and its absence from Iron Age deposits may be associated with its poor tolerance of frosts and heavy rains (Kirby 1965). Hops (Humulus lupulus), woad (Isatis tinctoria) and hemp (Cannabis sativa) appear in Anglo-Saxon contexts, and the latter makes an appreciable contribution to the pollen from Old Buckenham Mere, Norfolk (Godwin 1968).

So far, only qualitative change in the crop record has been discussed. There is clearly quantitative change too, although this is more difficult to assess satisfactorily. A number of statistical approaches have been suggested (e.g. Renfrew 1967, Hubbard 1975, Green 1979, Jones, G. 1979) though none has succeeded in meeting universal acceptance. Dennell (1974) has argued for the importance of understanding more fully the crop-processing activities to which archaeological remains had been subjected, before constructing such models. The distribution of sites for which records occur is at present perhaps too haphazard to allow detailed quantitative comment. Hubbard's presence value for rye reaching zero at a time when the Saxons referred to August as Ruggern, the month of the Rye harvest, is perhaps an example of the type of difficulty that can arise.

Nonetheless, distinct trends can be inferred from the qualitative changes alone. Certain of the crops that appear in the later periods, flax, hops and hemp in particular, show a preference for the deep retentive loams that in Britain occur on heavy lowlying ground (Kirby 1965). They are a reflection of the shift of emphasis, referred to at the beginning of this paper, from the lighter upland soils to the heavier soils in the valleys. The innovations of the first half of the first millennium BC may be directly related to this shift. If these innovations represent the ability to expand onto hitherto marginal land, the following developments represent the increasing concentration on the one type of marginal land that had long term potential, the heavy intractable soils. Once sufficient means of drainage and cultivation were acquired, these soils ceased to be marginal, and their greater depth made them the optimal soils for agriculture. It is significant that of the crop species that appear in the first half of the first millennium BC, species suited to heavier soils immediately predominate and continue to do so throughout further developments. The various lines of evidence that may reflect both the changes in emphasis on heavy and damp soils, and agricultural intensification in general, are considered below.

The Cultivation of heavy and damp ground

One consequence of improved fieldwork in recent years has been the realisation that heavy and damp soils were being exploited in this period to a far greater extent than had previously been envisaged. Areas of predominantly clay soils in the Weald, Essex and Northamptonshire were being settled prior to the 1st century BC (Cunliffe 1976) and at least one instance of pre-Roman ard-marks on clay is known (Everton and Fowler 1978). Bradley (1978) has pointed out that 22 per cent of early or middle Iron Age sites known to the Ordnance Survey in 1967 were situated in areas of heavy soil. The expanding cultivation of clay soils is reflected in the appearance in the carbonised weed flora of a species of mayweed. Anthemis cotula. This weed of cultivated ground is most frequent on heavy clay and clay loam soils (Kay 1971). Iron Age records are known from Abingdon and Dorchester-on-Thames in Oxon, and Bierton in Bucks, though only as individual seeds. It occurs abundantly in 1st century AD deposits from South Wales (Hillman, pers. comm.) and in many parts of the country is quite common in Romano-British deposits, and very common in medieval deposits. In York, it is typical of Romano-British and medieval assemblages (Addyman et al. 1976) and in Wessex it becomes a typical component by the late Saxon period (Green 1979). It would clearly be useful to follow the occurrence of this species in greater detail.

In a similar way, the exploitation of soils, either light or heavy, that are unfavourably damp is reflected in the behaviour of species of damp ground occurring as impurities in carbonised grain assemblages. Such species, and in particular, Eleocharis palustris are particularly common in carbonised assemblages of a wide range of dates, in various parts of Northern Europe (Helbaek 1954, van Zeist 1968) and on sites from a variety of locations, from river valley farmsteads to the well-drained site of Danebury hill fort (Jones, M. 1978 and in Cunliffe, forthcoming). Their presence is presumably associated with the cultivation of low-lying ground; in the Upper Thames Valley they are most abundant in sites close to the river. It has also been suggested that the dampness of soils in any position would have been enhanced by long periods of shallow cultivation, which would tend to create impermeable pans and subsoil streams (R. J. Chancellor, pers. comm.) which may in part explain the frequency of damp ground species in carbonised assemblages from such a variety of contexts.

The environmental evidence may be seen in the context of the contemporary developments in tillage technology as surveyed by Rees (1979). A wide range of tillage equipment is discussed by Rees in great detail. Without wishing to underrate the contribution of subsidiary tools of tillage, attention here will be confined to developments in the ard or plough, that are considered to be of most relevance to the tillage of heavy or damp ground.

It is likely that various types of ard were used throughout the period under study. Wooden parts of ards are known from as late as the 3rd century AD in lowland Britain (Fowler 1978) and later on the continent. The Dønnerupland-type share from Abingdon, Oxon shows the trace of a metal capping, and equivalent metal caps are far more common than the wooden parts themselves. The capping of shares with iron, which would greatly increase the durability of the implement, appears early in the Iron Age. Iron share tips are known from the first phase at Gussage-All-Saints, Dorset (Wainwright 1979), and from a promontory hilltop enclosure with a timber-laced rampart at Budbury (Wainwright 1970). Such a share was probably responsible for the criss-cross ard-marks on clay at Falfield, Avon (Everton and Fowler 1978). A series of one way ard marks of Romano-British date have been recovered, and one group on clay at Rudchester fort, shows evidence of slight ridging between the ard furrows. This can perhaps be associated with a further development in tillage technology, subsequently described by Palladius (O.A. I: XLIII: 1).

> "Ploughs; simple, or, if the open nature of the terrain permits, eared, so that with the latter it may be possible to raise up the plants on a deeper furrow to counteract the high water table in winter"—K. D. White's translation.

Tillage with "eared" ploughs would also have been useful for breaking subsoil pans that may have developed in the manner outlined above. The existence of eared ploughs in Britain is known from a series of bronze models of possible 2nd century or 3rd century AD date. This need not be their earliest occurrence; the Rudchester ard-marks are pre-Hadrianic, and a series of wooden shares are known, similar in form to the share of the Sussex model, one of which bears a radiocarbon date of 400 ± 90 bc.

A series of very significant developments has occurred by the 3rd or 4th century AD, when amongst the innovations are coulters and assymetric shares. Evidence of the mould-board itself is still lacking before the 12th century, but Rees is prepared to infer the existence of a mould board plough in the later Roman period, an adaptation clearly aimed at the more efficient utilisation of deep heavy loams. Two further developments may be observed in 10th century AD manuscript illustrations, the occurrence of wheels and of swing-shares (Payne 1947), signifying the existence of ever larger and heavier ploughs.

A further activity connected with the cultivation of heavy damp ground is the widespread cutting of drainage ditches. Robinson (this volume) has observed that a landscape involving ditch-digging on a large scale appeared in the Upper Thames Valley in the late Iron Age, and Miles (this volume) alludes to the same trend in the country as a whole. This may be related to the temporary disappearance of Eleocharis palustris from Iron Age carbonised weed floras in various parts of the Upper Thames Valley at this time (e.g. Jones, M. 1978). The reappearance of this species in the late Roman period perhaps suggests a further advance onto even damper ground.

Further evidence for intensification

(a) The maintenance of soil quality

An important aspect of land preparation is the maintenance of soil fertility by some means of marling, manuring or crop rotation. Changes in fertility can again be monitored by the occurrence of particular species of weeds. Leguminous weeds are known to flourish in conditions of decreasing soil nitrogen (Warington 1924). The gradual increase in the occurrence of such species in carbonised assemblages from the Ashville site, Abingdon, between the mid first millennium bc and the 3rd century AD was interpreted as evidence that overcropping was gradually lowering the level of soil nitrogen (Jones, M. 1978). A similar increase in leguminous species, probably mostly weeds, was observed in samples from the second half of the first millennium bc at Gussage-All-Saints (Evans and Jones 1979).

The practice of muck-spreading, as indicated by thin scatters of small sherds of pottery, goes back well beyond the Iron Age. Bell (forthcoming) has produced evidence for the use of seaweed as a fertiliser on an Iron Age site 10 km inland from the Sussex Coast, in the form of marine molluscs within a field lynchet. In the late 3rd century AD, a more intensive form of fertilising is known from North Thoresby, Lincoln; a rectilinear series of trenches cut in clay, were packed with household refuse and manure, perhaps serving as the basis for a market garden (Webster et al. 1967).

A further precaution against impoverishing the land is the use of crop rotation or various forms of fallow system, which also have the effect of cleaning the land of weeds and pests. Some form of rotation has often been argued in conjunction with villa farming (e.g. Applebaum 1972, Branigan 1977). The occasional seeds of non-cereal crops that occur in among carbonised cereal assemblages may have resulted from 'volunteer' plants surviving from previous courses in the rotation. The traces of flax and celtic bean in the cereal assemblage from a corn drying oven at Barton Court Farm, Abingdon is particularly suggestive, especially as flax is unlikely to be mixed purposely with a grain crop, nor is it a plant capable of surviving as a weed for many seasons. (Jones in Miles, forthcoming). Records of cultivated legumes, a key component of many rotation systems, are still rare for all periods, though they may be less likely to enter the archaeological record for reasons related to archaeological deposition rather than agricultural technique.

There is less indication of rotation systems in earlier periods. The use of legumes may in itself imply rotation, and flax, grown as early as the Neolithic, is too demanding a plant to be grown continuously on the same ground. However, other possible systems exist. Bradley (1967) has argued for the existence of long fallowing in Hampshire fields which were probably in use prior to the 1st century AD, on the basis of the molluscan assemblages within a series of associated lynchets.

It seems possible that various systems were in use in the Iron Age, and only by the later Roman period is there a strong indication of full crop rotation. It may well be that as well as experiencing soil deterioration at an early stage North European communities also developed methods of restoring quality to the soil in advance of contemporary communities in the Mediterranean. In classical Italy, a fairly simple system with cropping years alternating with fallow years was the standard practice (White 1970). Also of relevance is Pliny's comment (N.H. XVII: iv: 42) that marling was first practised in Britain and Gaul and subsequently developed in the Mediterranean.

(b) Harvesting and crop yields.

The harvest is one of the peak seasons in the agricultural calendar, when both manpower availability and harvesting technology are likely to be limiting factors.

The onset of the Iron Age sees the appearance of short crescent-shaped sickles which differ from their Bronze Age predecessors only in the metal of which they are constructed. The absence of any change in form suggests that little change in harvesting technique occurred at this stage. Such a change may be detected in the late Iron Age, when balanced sickles appear (Rees 1979). Such sickles are far less fatiguing and allow a smoother cutting motion (White 1967). These balanced sickles also occur in the Roman period and from the 4th century larger examples are known, up to 30 cm in length. Long scythes also appear in the Roman period, but most of the evidence suggests that these were used for cutting hay rather than arable crops.

It has suggested on the basis of the type of sickle in use, that the Iron Age harvest was taken just below the ear (Applebaum 1954). The presence of seeds of certain low-growing weeds among carbonised assemblages of harvested drain suggest that the harvest was cut further down the straw (Jones, M. 1978) Murphy (1977) has suggested that the presence of root tubers of the weed Arrhenatherum elatius in carbonised deposits, may imply that uprooting was used. The Roman authors, Cato, Varro and Columella describe a wide range of harvesting methods including uprooting, combing and cutting the straw at various positions (White 1967). Each author points out that the method varies a great deal from region to region, and there may have been a variety of methods in use in Britain.

The practice of combined winter and spring sowing has advantages which have been mentioned earlier. It was also suggested that this practice originated before the period under study.

Documentary sources give figures for a wide range of early yields, from Classical Egypt and Italy (Jones, A. H. M. 1964) Carolingian France (Slicher van Bath 1963) and medieval England (Titow 1972). The classical yields vary between 11 and 20 bushels to the acre, and the medieval yields vary considerably, although the majority fall below 20 bushels to the acre.

Modern yields given by Percival (1921) for primitive strains of wheat range between 4 and 30 bushels to the acre. Reynolds (this volume), using simple agricultural techniques has obtained yields far in excess of these figures using primitive strains.

In attempts to reconstruct early farming systems lower rather than higher yields, in the order of 8 bushels to the acre, have been assumed (e.g. Branigan 1977, Radford 1936). Applebaum (1972) has reached a similar figure by extrapolating from the floor areas of two structures on the villa at Pitney, which he interprets as a granary and byre.

The danger of assuming any particular yield, is that the yield itself varies according to the farming strategy adopted. With certain strategies it is expedient to invest the effort required to realise high yields on a specific plot of land, whereas with others it may be preferable to realise lower yields over larger areas. For this reason, modern cereal yields vary by a factor of over three from place to place in Europe alone (van Valkenburg and Held 1952).

The sustained low level of productivity in medieval England is seen as an unusual situation that in itself requires explanation (Postan 1977) and may have been far below prehistoric and classical yields. Substantial drops in productivity are known; successive records of Italian cereal yields by Cicero in about 70 BC and Columella in the mid-1st century AD suggest a drop in productivity by a factor of $2\frac{1}{2}$ over about a century (White 1970).

Reynold's results do not necessarily show that prehistoric and early historic productivity was high, but they do demonstrate that contemporary agricultural techniques and cereal strains were in principle capable of producing high yields.

(c) Processing and subsidiary activities

A development in archaeological studies since Helbaek's work that has aroused a certain amount of interest is the detection of crop-processing activities from the nature and context of carbonised deposits (Hillman 1973, Dennell 1974). Dennell has thus proposed a model to explain variations in the nature of carbonised assemblages on the basis of such activities as threshing, winnowing and food-preparation, and various refinements to the model have been proposed by Hillman (pers. comm.) and Jones, G. (1979). With certain limitations, his model has proved useful on rural sites in Britain (Jones, M. 1978: 107) and it is for example possible to detect the occurrence of threshing and winnowing within the hill fort at Danebury (Jones, M. in Cunliffe forthcoming). There is a certain amount of corroborative information from artefacts and documents on threshing and winnowing. Well-worn flints believed to be from a tribulum or threshing sledge have been recovered from Roman sites at Atworth and Angmering, Sussex (Rees 1979). Reference to flails appears in The Life of St. Jerome (4th century), The Exeter Riddle (8th century) and illustrated together with winnowing baskets and weed seives in The Old English Calendar (11th century) (Monk 1978).

The preparation of flour is manifest in fragments of querns and millstones. The earliest rotary querns are from early Iron Age deposits on the farmstead at Gussage-All-Saints. Parts of mechanical mills may be found in various Romano-British sites (Wacher 1966), and are referred to throughout the 1st millennium. The considerable increase in their use in the late Saxon period appears not to have a directly agricultural cause, but rather to be associated with the decline in slavery (Finsberg 1972).

A crop-processing structure that has recently come up for reassessment is the corn-drying oven. The traditional idea of the function of these ovens is for the drying of cereal crops in preparation for storage, threshing or milling. This idea has been criticised by Reynolds (this volume) largely in the basis of a continous floor separating the corn from the hot air produced. His alternative explanation, as malting floors, concurs with the find of germinated cereal grain in the corn drier from Abingdon, Oxon (Jones in Miles, forthcoming), and Catsgore, Somerset (Hillman 1981a). If this second hypothesis is correct, it lessens the direct relevance of these structures to total cereal productivity. However they may have implications related to changing economic patterns. The growth of a disembedded monetary economy in various Third World countries is known to have presented settled arable farmers with a seasonal cash flow problem; their saleable product is produced only once a year, and at the same time as every one else, encouraging its market value to drop. In Tanganyika, where this development has been closely studied (Beidelman 1961) one of the responses of the arable farming Kaguru to this

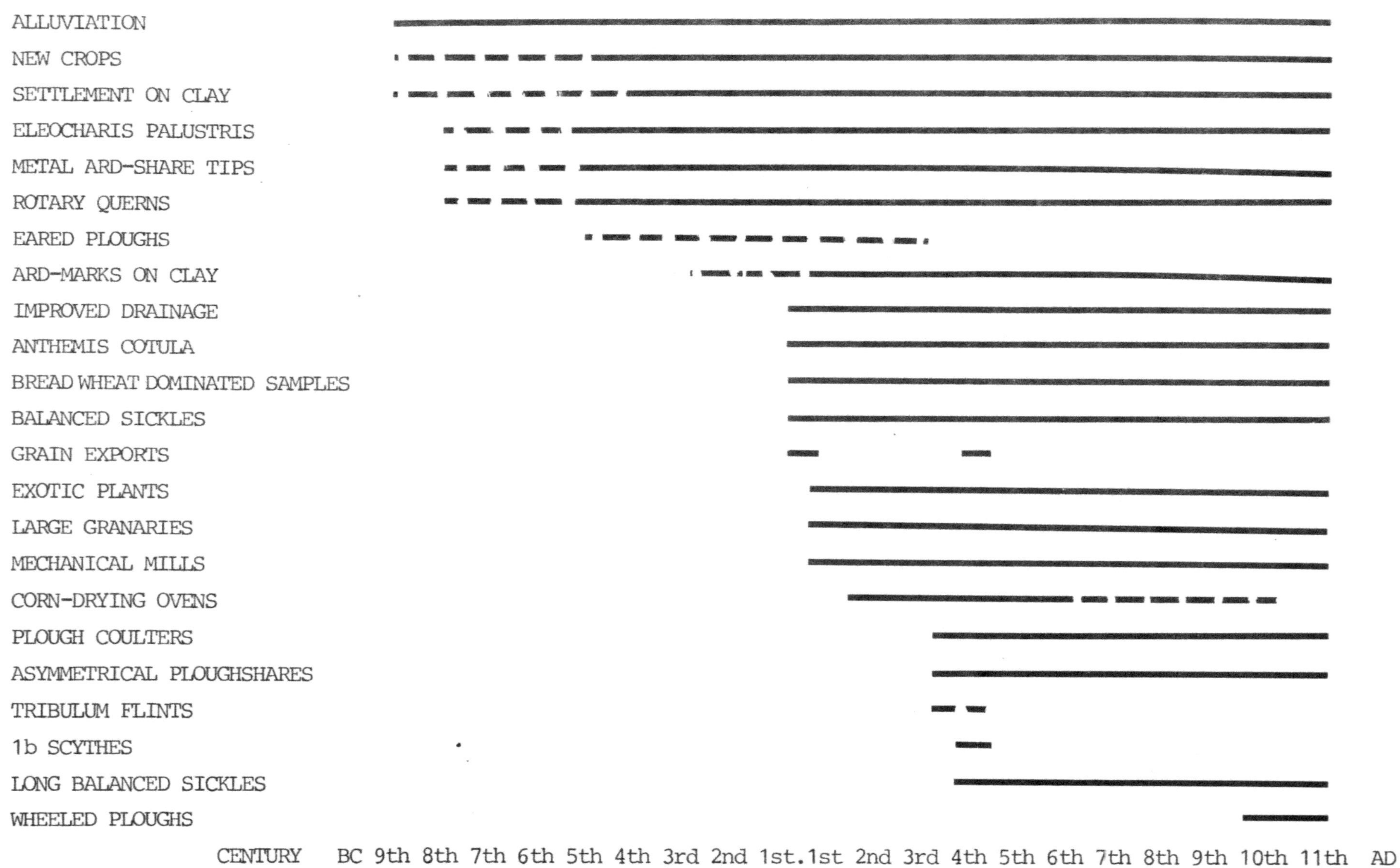
ALLUVIATION
NEW CROPS
SETTLEMENT ON CLAY
ELEOCHARIS PALUSTRIS
METAL ARD-SHARE TIPS
ROTARY QUERNS
EARED PLOUGHS
ARD-MARKS ON CLAY
IMPROVED DRAINAGE
ANTHEMIS COTULA
BREAD WHEAT DOMINATED SAMPLES
BALANCED SICKLES
GRAIN EXPORTS
EXOTIC PLANTS
LARGE GRANARIES
MECHANICAL MILLS
CORN-DRYING OVENS
PLOUGH COULTERS
ASYMMETRICAL PLOUGHSHARES
TRIBULUM FLINTS
1b SCYTHES
LONG BALANCED SICKLES
WHEELED PLOUGHS
CENTURY
BC 9th 8th 7th 6th 5th 4th 3rd 2nd 1st. 1st 2nd 3rd 4th 5th 6th 7th 8th 9th 10th 11th AD

Fig. 6.4 Evidence relating to inovations in crop husbandry between 1000 BC and 1000 AD. Continuous line denotes the period by which the innovation had definitely occurred, the broken line denotes the time range within which the innovation is though to have occurred.

problem, is the retention of a portion of the harvest for the brewing of beer all year round, which they can easily sell at the markets. Perhaps the appearance of these ovens as malting floors represents the response of British farmers to a similar seasonal cash flow problem, brought on by the expanding monetary economy. It may be relevant that prior to the 4th century AD, as many such ovens are known from the presumably poorer native-type settlements as are from villas (Morris 1979). One might also expect cash crops to figure more prominently in the palaeobotanical record for this period, but as mentioned above, such records so far tend to be from urban rather than rural sites.

The question of storage facilities for grain in the Iron Age has been considered in detail by Reynolds (1974) and a number of pits and four-posters are known from contexts in lowland Britain as late as 2nd century in date (Morris 1979). Soon after the conquest, various forms of larger granary appear, indicating that harvests are being dealt with the redistributed in larger quantities.

Analysis of change

Having considered the various forms of environmental and archaeological information available, the question of how and when the shift of emphasis towards a system of crop husbandry associated when intensification on the heavy loams may be approached. Each piece of evidence that has been considered above has been plotted in Fig. 6:4, after the manner of Brongers and Waltering (1973) later adopted by Bradley (1978). Some of these pieces of evidence have a more direct bearing on methods of crop husbandry than others, but all have been included owing to the difficulty of drawing a realistic line between them.

The 'agricultural revolution' that occurred early in the period under study is clearly evident from a broad front of innovations. The diagram suggests that further innovations were not evenly spread, but instead cluster in two further periods of innovation in the late Iron Age and the late Roman period.

The three items that appear soon after the Roman conquest, exotic food plants, mechanical mills and large granaries are more concerned with distribution and trade than with crop husbandry itself.

The Piercebridge model showing the appearance of groundwrests on the plough may be of 2nd century date, but as argued above, this introduction could well be earlier and perhaps be traced back to the early Iron Age.

In a traditional approach, periods of innovation are related to periods of increased international contact. Indeed both periods of innovation have been associated though not assuredly, with population movements. In this vein, Applebaum has as recently as 1972 suggested that Romano-British refugees, before fleeing to Ireland, remembered to pack their large plough coulters, thereby revolutionising Irish agriculture. However, what is now known about contemporary networks of trade and information exchange would suggest that something as significant as knowledge of improved agricultural technique could travel freely throughout the period, and that the stretches of sea cannot be regarded as a barrier to information transfer, only surmounted by the influx of a new cultural group.

CENTRAL PLACE	CROP HUSBANDRY
10th - 5th C. BC GROWTH of hilltop enclosures.	INNOVATION: Diversification of crops and soil types in use.
5th - 1st C. BC FULLY DEVELOPED hillforts.	STAGNATION: Soil exhaustion evident from weeds.
1st C. BC - 1st C. AD GROWTH of low-lying oppida.	INNOVATION: Cultivation of damp and heavy ground intensified. Export of grain.
1st - 3rd C. AD FULLY DEVELOPED Roman urban centres.	STAGNATION: Innovation confined to the redistributional sphere.
3rd - 4th C. AD GROWTH of small towns.	INNOVATION: Cultivation of heavy loams fully developed. Mouldboard plough in use. Export of grain.

Fig. 6.5 Chronological relationship between suggested phases in crop husbandry and suggested phases in the contemporary central places.

In fact, considering the density of early Iron Age settlement on heavy soils, it is perhaps surprising that the shifts in technique towards a system focused on these soils, were delayed until these later dates. The occurrence of periods of innovation is accounted for by this early expansion onto heavy soils. It is less easy to account for the intervening periods of stagnation. That these were periods of stagnation is emphasised for the first period by the behaviour of the weed flora from contemporary sites. As outlined above, the behaviour of leguminous weeds and weeds of damp ground during this period have been associated with deteriorating field conditions. In the later period of stagnation, the palaeobotanical record suggests that agriculture stimulated by Roman activity in the north was not based on the more developed hexaploid wheats (i.e. spelt wheat and bread wheat) but instead on the more traditional emmer, also suggesting a lack of innovation in this area.

A correlation may be observed between these periods of innovation and stagnation, and contemporary developments in society. Each period coincides with a stage in the development of the associated central place (Fig. 6:5). The three periods of innovation occur during the growth periods of defended hilltop enclosures, low-lying oppida, and small towns of Roman Britain, respectively. The two periods of stagnation coincide neatly with Cunliffe's phase of fully developed hillforts (Cunliffe 1978), and the period of dominance of large Roman towns based on the earlier oppida.

This association of developments in crop husbandry with the contemporary structure of society may be accounted for in the following way. None of the innovations considered in Fig. 6:4 involve merely a new technique; they each require a greater investment of some limited resource such as manpower, animal power, iron, or manure, to back up the new technique. Late Saxon documentary evidence concerning such themes as ploughing, clearly show that these factors were limiting in the later period. The survival of simpler techniques, such as the uncapped 1st century BC share from Waysland Rath (Wainwright 1971) and the 3rd century AD Dønnerupland-type share from the Upper Thames Valley (Fowler 1978) suggest that they were also limiting in earlier periods.

The ability of a farmer to invest these scarce resources would be closely related to what happened to any surplus produced. In each period of stagnation, there is evidence for the non-agrarian investment of surplus. Cunliffe (1978) suggests that:-

> "developed hillforts imply the coercive control of society's surplus in the interests of defense".

In the second period of stagnation, surplus was directed towards public building and administration. It would be wrong to suggest that during the growth phases of the central place there was any lesser tendency to extract surplus from the agrarian base. On the contrary, Cunliffe (1976) suggests that the situation in the second half of the first millennium BC in which:-

> "Much of the surplus produced by society was absorbed into the development of a class structure which consumed labour wealth to supply the increasingly extravagant demands of the ruling class".

reached a peak in the 1st century BC when:-

> "the aristocracy of the South East had become fully divorced from a productive role in society".

Again, the later period of innovation occurs in the context of ever-increasing imperial taxation (Jones, A. H. M. 1964). However in the growth phase of a new central place, when prospective sites are competing for viability, the productivity of their hinterland would be key to their survival, and it would be to their advantage to encourage innovation in crop husbandry, at least in certain quarters, by allowing surplus to remain within the agrarian sphere. The mechanism for encouraging agricultural growth in the face of heavy fiscal dues is well-documented in the later period, in the form of tax-concessions and various types of corruption (Jones, A. H. M. 1964). It may be relevant that the later innovations occur in the context of the newly embellished villas, presumably occupied by those members of society best able to take advantage of corruption.

In the mature phase of the central place, whilst the productivity of the hinterland remains of key importance, it may not be necessary to stimulate agricultural innovation in the same way, and the surplus necessary for innovation may be removed more completely from the agrarian sphere without immediately endangering the survival of the system.

Conclusion

The range of evidence for developments in crop husbandry have been reviewed, and the changes have been seen as relating to first, a period of arable expansion into a broader range of ecological situations, and secondly the gradual focusing of the emphasis on the heavy loams.

A model has been proposed in which the ecological conditions for these developments existed from the onset of the period under study, when settlement on heavy soil was already substantial, but that the social conditions for these developments occurred only sporadically in the following centuries.

In places, the evidence used in support of this model could be refined, greater detail and more precise dating would be desirable. The advantage of the interactions between man and his environmental resources in later prehistory and early history is that the potential archaeological and environmental data base for these periods is of such a quality that further research may provide these refinements. Whether or not the proposed model stands, it is hoped that this paper may encourage this further research.

Acknowledgements

I should like to thank my colleagues, Frank Green, Gordon Hillman, Mick Monk and Peter Murphy for allowing me to use much unpublished information and thesis material, and Martin Millett and Dennis Britton for reading through and commenting on the text. The author is currently engaged on research into carbonised assemblages, funded by the Science Research Council, and under the supervision of Dr. S. R. J. Woodell and Mr. Dennis Britton.

Bibliography

Addyman, P. V., Hood, J. S. R., Kenward, H. K., MacGregor, A. G. and Williams, D. (1976). Palaeoclimate in urban environmental archaeology at York, England; problems and potential. World Archaeology 8 220-233.

Applebaum, S. (1954). The Agriculture of the British Early Iron Age, as exemplified at Figheldean Down, Wiltshire'. Proceedings of the Prehistoric Society 20 103-114.

Applebaum, S. (1972). Roman Britain. In: The Agrarian History of England and Wales. 1(ii) (ed. Finberg, H. P. R.).

Arthur, J. R. B. (1975). The Carbonised Grain. In: An Iron Age Site at Twywell Northamptonshire. (ed. Jackson, D. A.) Northampton Archaeology 10 91-92.

Arthur, J. R. B. (1977). Plant Remains. In: Excavations at Bishopstone Sussex. (ed. Bell, M.) Sussex Archaeological Collections 115 273-275.

Beaven, E. S. (1947). Barley: fifty years of observation and experiment. London: Duckworth.

Beidelman, T. O. (1961). Beer Drinking and Cattle Theft in Ukaguru, Intertribal Relations in a Tanganyika Chiefdom. American Anthropology 63 534-549.

Bell, G. D. H. (1948). Cultivated Plants of the Farm. Cambridge: University Press.

Bland, B. L. (1971). Crop Production: Cereals and Legumes. London: Academic Press.

Bowen, H. C. and Fowler, P. J. (1978). Early Land Allotment in the British Isles. Oxford: BAR 48.

Bradley, R. (1967). Excavation on Portsdown Hill 1963-5. Proceedings of the Hampshire Field Club 24 42-58.

Bradley, R. (1978). The Prehistoric Settlement of Britain. London: Routledge and Kegan Paul.

Branigan, K. (1977). Gatcombe: the excavation and study of a Romano-British Villa estate. Oxford: BAR 44.

Brenchley, W. E. and Warington, K. (1930). Numerical estimation of viable seeds and observations on their natural dormancy. Journal of Ecology 18 235-272.

Brongers, J. A. and Waltering, P. J. (1973). Prehistory in the Netherlands: An Economic-Technological Approach. Berichten van de Rijksdienst voor het Oudheidkundige Bodemondergoek 7-47.

Burnham, B. C. and Johnson, H. B. (1979). Invasion and Response: The Case of Roman Britain. Oxford: BAR 73.

Clarke, H. H. (1971). The Plant Remains. In: Archaeology and the M5 Motorway, 1st Report, 1969. (eds. Fowler, P. J. and Walthew, C. V.) Transactions of the Bristol and Gloucestershire Archaeological Society 90 22-63.

Clarke, H. H. (1970). Appendix III. Grain. In: Frocester Court Roman Villa, Gloucestershire. (ed. Gracie, H. S.) Transactions of the Bristol and Gloucestershire Archaeological Society 89 81-82.

Cunliffe, B. W. (1974). Iron Age Communities in Britain. London: Routledge and Kegan Paul.

Cunliffe, B. W. (1976). The Origins of Urbanisation in Britain. In: Oppida: The Beginnings of Urbanisation in Barbarian Europe. (eds. Cunliffe, B. W. and Rowley, R. T.) Oxford: BAR S11 pp. 135-161.

Cunliffe, B. W. (1978). Settlement and Population in the British Iron Age: some facts, figures and fantasies. In: Lowland Iron Age Communities in Europe (ed. Cunliffe, B. W. and Rowley, R. T.) Oxford: BAR S48 pp. 3-24.

Cunliffe, B. W. forthcoming. Danebury Hillfort, Hampshire: Excavation Report.

Dennell, R. W. (1974). Botanical Evidence for Prehistoric Crop Processing Activities. Journal of Archaeological Science 1 275-284.

Ellison, A. and Harriss, J. (1972). Settlement and land use in the prehistory and early history of Southern England. In: Models in Archaeology (ed. Clarke, D. L.) Cambridge: University Press pp. 911-962.

Evans, A. M. and Bowman, A. (1968). Report on the Carbonised Grains. In: The Excavation of a Durotrigian Farmstead near Tollard Royal in Cranbourne Chase, Southern England. (ed Wainwright, G. J.) Proceedings of the Prehistoric Society 34 146.

Evans, A. M. and Jones, M. K. (1979). The Plant Remains. In: Gussage-All-Saints: An Iron Age Settlement in Dorset (ed. G. Wainwright) London. HMSO pp. 172-175.

Everton, A. and Fowler, P. J. (1978). Pre-Roman Ard-Marks at Lodge Farm, Falfield, Avon: a method of analysis'. In: Early Land Allotment (eds. Bowen, H. C. and Fowler, P. J.) Oxford: BAR 48.

Field, N. H. (1965). Romano-British Settlement at Studland, Dorset. Proceedings of the Dorset Archaeological Society 87 142-207.

Field, N. H., Matthews, C. L., Smith, I. F. (1964). New Neolithic Sites in Dorset and Bedfordshire. Proceedings of the Prehistoric Society 30 352-381.

Finberg, H. P. R. (1972). Anglo-Saxon England to 1042. In: The Agrarian History of England and Wales I ii (ed. Finberg, H. P. R.) Cambridge: University Press pp. 385-525.

Fox, C. F. (1932). The Personality of Britain. Cardiff: National Museum of Wales.

Godwin, H. (1968). Organic deposits at Old Buckenham Mere, Norfolk. New Phytologist 67 95-107.

Godwin, J. (1975). History of the British Flora. 2nd Edn. Cambridge: University Press.

Green, F. J. (1979). Mediaeval Plant Remains from Wessex. M.Phil. Thesis University of Southampton.

Greig, J. (1976). The Plant Remains. In: The Environmental Evidence from the Church Street Roman Sewer System. (ed. Buckland, P. C.) York: CBA.

Helbaek, H. (1952). Early Crops in Southern England. Proceedings of the Prehistoric Society 18 194-233.

Helbaek, H. (1964). The Isca Grain. A Roman Plant Introduction in Britain. New Phytologist 63 158-164.

Higgs, E. S. (1972). Papers in Economic Prehistory. Cambridge: University Press.

Hillman, G. (1973). Crop husbandry and Food Production: A modern basis for the interpretation of plant remains.' Anatolian Studies 23, pp. 241-244.

Hillman, G. (1978). Remains of Crops and Other Plants from Carmarthen (Church St.)'. Cambrian Arch. Assoc. Monographs and Collections Vol. I pp. 1 7-112.

Hillman, G. (1981a). Spelt Kilning at Catsgore, Somerset. In: Excavations at Catsgore (ed. Leech, R.) London: HMSO.

Hillman, G. (1981b). Grain Processing at 3rd century Wilderspool. In: Excavations at Wilderspool (ed. Hinchliffe, J. and Williams, J. H.) London: HMSO.

Hubbard, R. N. L. B. (1975). Assessing the botanical component of human palaeo-economies. Bulletin of the Institute of Archaeology. London University 12 197-205.

Jessen, K. and Helbaek, H. (1944). Cereals in Great Britain and Ireland in Prehistoric and early historic times. Wet Kongelige Danske Videnskaernes Selskab Biologiske Skrifter 3 2.

Jones, A. (1977). The Plant Remains. In: The Enclosure at Ashill. (ed. Gregory, T.) East Anglian Archaeology Report 5: Norfolk Gressenhall: Norfolk Archaeological Unit.

Jones, A. H. M. (1964). The Later Roman Empire. 284-602: A social, economic and administrative study. Oxford: Blackwell.

Jones, G. E. M. (1979). An Analysis of the Plant Remains from Assiros Toumba. M.Phil. Thesis Cambridge University.

Jones, M. (1978). The Plant Remains. In: The excavation of an Iron Age Settlement, Bronze Age ring ditches and Roman features at Ashville Trading Estate Abingdon (Oxon). London: CBA Research Report 28 pp. 93-110.

Kay, Q. O. N. (1971). Anthemis cotula. Journal of Ecology 59 623-636.

Kenward, H. K. and Williams, D. (1979). Biological Evidence from the Roman Warehouses in Coney Street. York: CBA.

Kirby, R. H. (1963). Vegetable Fibres: Botany, Cultivation and Utilization. London: Leonard Hill.

Lambrick, G. and Robinson, M. (1979). Iron Age and Roman riverside settlements at Farmoor, Oxfordshire. London: CBA Research Report 32.

Miles, D. forthcoming. Archaeology at Barton Court Farm.

Monk, M. A. (1978). The Plant Economy and Agriculture of the Anglo-Saxon periods in Southern Britain, with particular reference to the mart settlements at Southampton and Winchester. M.Phil. Thesis University of Southampton.

Morris, P. (1979). Agricultural buildings in Roman Britain. Oxford: BAR 70.

Murphy, P. L. (1977). Early Agriculture and Environment on the Hampshire Chalklands, circa 800 BC - 400 AD. M.Phil Thesis, University of Southampton.

Payne, F. G. (1947). The Plough in Ancient Britain. Archaeological Journal 104 82-111.

Percival, J. (1921). The Wheat Plant. London: Duckworth.

Percival, J. (1934). Wheat in Great Britain. Reading.

Percival, J. (1976). The Roman Villa: an historical introduction. London: Batsford.

Poston, M. M. (1975). The Medieval Economy and Society. Buckingham: Pelican.

Radford, C. A. Radlegh. (1936). The Roman Villa at Ditchley, Oxon. Oxoniensia 1 24-69.

Rees, S. (1979). Agricultural Implements in Prehistoric and Roman Britain. Oxford: BAR 69.

Renfrew, J. M. (1965). Grain Impressions from the Iron Age Sites of Wandlebury and Barley. Proceedings of the Cambridge Antiquarian Society 58 10-11.

Renfrew, J. M. (1967). Palaeoethobotany and the Neolithic period in Greece and Bulgaria. Ph.D. thesis Cambridge University.

Renfrew, J. M. (1974). Report on the Cereal Grains. In: Dun Mor Vaul: An Iron Age Broch on Tiree (ed. Mackie, E.) Glasgow: University Press pp. 210-213.

Reynolds, P. J. (1974). Experimental Iron Age Storage Pits: An interim report. Proceedings of the Prehistoric Society 40 118-131.

Slicher van Bath, B. H. (1963). Agrarian history of Western Europe. London: Edward Arnold.

Taylor, C. C. (1972). The study of settlement patterns in pre-Saxon Britain. Man Settlement and Urbanism. (eds. Ucko, P. J., Tringham, R. and Dimbleby, G. W.) London: Duckworth pp. 109-113.

Titow, J. Z. (1972). Winchester Yields a study in mediaeval agricultural productivity. Cambridge: University Press.

van Valkenburg, S. and Held, C. C. (1952). Europe. New York.

Wacher, J. S. (ed.) (1966). The Civitas Capitals of Roman Britain. Leicester: University Press.

Wainwright, G. J. (1970). An Iron Promontory Hillforst at Budbury, Bradford-on-Avon, Wilts. Wiltshire Archaeology and Natural History Magazine 65 108-166.

Wainwright, G. J. (1971). The Excavation of a Fortified Settlement at Walesland Rath, Pembs. Britannia 2 48-108.

Wainwright, G. J. (1979). Gussage-All-Saints. An Iron Age Settlement in Dorset. London: HMSO.

Warington, K. (1924). The influence ofmanuring on the weed flora of arable land. Journal of Ecology 12 111-126.

Webster, D., Webster, H. and Petch, D. F. (1967). A possible vineyard of the Romano-British period at North Thoresby, Lincolnshire. Lincolnshire History and Archaeology 2 55-61.

White, K. D. (1970). Roman Farming. London: Thames and Hudson.

White, K. D. (1967). Agricultural Implements of the Roman World. Cambridge: University Press.

Williams, D. (1977). A consideration of the Sub-fossil Remains of Vitis vinifera L. as Evidence for viticulture in Roman Britain. Britannia 8 327-338.

Willcox, G. H. (1977). Exotic Plants from Roman Waterlogged Sites in London. Journal of Archaeological Science 4 269-282.

Wilson, D. G. (1975). The Botanical Setting of the Roman Fort of Birrens, Dumfriesshire. In: Birrens (Blatobulginum) ed. Robertson, A. S.) Edinburgh: Constable pp. 259-268.

Wilson, D. G. (1979). Plant Remains. In: Excavations at Little Waltham 1970-71 (ed. Drury, P. J.) London: CBA pp. 142-144.

Wilson, D. G. and Conolly, A. O. (1978). Plant Remains including the Evidence of Hops. In: The Graveney Boat: a 10th Century Find from Kent. (ed. Fenwick, V.) BAR 53 133-150.

Van Zeist, W. (1970). Prehistoric and Early Historic Food Plants in the Netherlands. Palaeohistoria 14 41-173.

van Zeist, W. (1974). Palaeobotanical Studies of Settlement Sites in the Coastal Area of the Netherlands. Palaeohistoria 16 223-371.

Zohary, M. (1967). Plant Life of Palestine. New York: Ronald Press.

7. IRON AGE, ROMAN AND SAXON CROPS: THE ARCHAEOLOGICAL EVIDENCE FROM WESSEX

F. J. Green

The aim of this paper is to provide a review of recent work undertaken on plant remains from the Wessex region, in order to indicate the major sources of new information about crop plants since the pioneering work of Jesson and Helbaek (1944) and Helbaek (1940: 176-178, 1952: 194-233). For the purposes of this paper Wessex has been defined using historical criteria and consists of the counties of Hampshire, Dorset, Wiltshire, Devonshire, Berkshire and parts of the counties of Somerset, Gloucestershire, and Oxfordshire south of the river Thames.

Since most of the new data have been recovered very recently much of it remains unpublished. This paper aims to provide a synthesis as a basis for future lines of research and interpretation.

Harding (1974: 81) points out that Helbaek's basic conclusions are fundamental to our interpretation of Iron Age economy in Britain. Applebaum (1972: 108-121) relied entirely on Helbaek's work and Godwin's (1956) adaptation of it for the Iron Age and Roman periods. Documentary sources have been extensively relied upon by agricultural historians for an understanding of early to late Saxon agriculture and pollen evidence has been used for a considerable period to supplement this information. However, it is only very recently that the examination of plant macrofossils from archaeological sites in a specific region has been undertaken in an attempt to evaluate Iron Age to Saxon agricultural activities (Murphy 1977, Monk 1978, Green 1979).

A recent re-assessment of Helbaek's work has been undertaken by Dennell (1976: 11-23) in which apart from re-examining the pottery from Windmill Hill he has essentially just re-used the available information in a different way. The work has various limitations, Dennell only dealt with plant impressions from neolithic, Bronze Age and Iron Age pottery and the overall sample that he used was the same as that used by Helbaek. The lumping together of all the available information from sites of the Iron Age by both Helbaek and Dennell has produced a misleading impression of nearly a millennium of plant utilisation.

In the following examination of new evidence, the Iron Age has been split into three divisions which do not relate to current terminology for early, middle and late Iron Age based on ceramic chronology. Instead purely artificial divisions have been chosen and consist of sites prior to the 5th century BC, sites of the 5th to the end of the 3rd century BC and sites from the 2nd century BC to approximately 50 AD. By dividing the period in this way, specific trends can be seen which in many ways alter the impression given in the earlier work.

Preservation of the evidence

Until recently most of the evidence for cereal crops came from the impressions found in fired clay materials, in particular from pottery. Helbaek (1951: 231) and Godwin (1975: 405) argued that plant impressions are more representative of production and utilisation than caches of carbonised evidence, since plant impressions represent the chance sampling of material from sites. Dennell (1976: 12) has argued that in fact this is not the case, for impressions depend on where the pottery was manufactured, either locally on the site or imported to it from elsewhere. The inclusion of plant fragments into the fabric of the pottery depends as much on the availability of materials from crop processing as it does on the cultural activities and the season of the year during which these activities may have been undertaken. Different cultural groups in different geographical regions may have pursued these activities (even using stored crops) at different times as a result of which the plant information recovered may give a totally misleading impression of the importance of certain crops. Dennell has argued (1976: 18) that large seeded species such as legumes and fruit stones are less likely to be accidentally incorporated into the fabric of pottery and other fictiles (fired clay objects), and as such would distort the picture of their relative importance to cereal and other smaller seeded crops.

Even though plant impressions in fired clay objects may not be representative of the prevailing economy, they at least might tell us something about associated technology. For example, when a specific range of species regularly occurs in a particular type of fictile, this evidence could perhaps indicate a range of crop and crop products that were available, but were reserved for specific technological processes rather than being exclusively used for fodder, bedding and thatching. Valuable commodities such as straw and cereal chaff would surely not have been wasted in a mixed farming economy, unless there was an overwhelming surplus or some advantage to be gained in using specific plant remains in certain domestic or industrial processes.

It is interesting that during the medieval period daub was predominantly tempered with wheat straw (Green 1979b: 258) whereas on sites of earlier periods other materials have more commonly been recorded. Bracken was used in Iron Age cob at little Woodbury (Brailsford 1949: 159), and daub at Old Down Farm from the same period was also tempered with bracken. The examination of daub from sites such as Danebury (Poole, pers. comm.) indicates the use of a wide range of non-cereal products.

The examination of briquetage may well provide another class of information (Bradley 1975: 22-23). The available evidence does suggest that spelt (*Triticum spelta*) was used almost exclusively in its manufacture. The fired clay bell moulds of 10th century date from Cathedral Green, Winchester have produced a very specific range of cereal remains (oat grains, wheat chaff and weed seeds) only paralleled by evidence from other bell moulds and cauldron moulds (Green 1979b: 240-252 and forthcoming c). The fired clay loom weights so commonly recovered from Iron Age and Saxon sites may provide yet another class of data. The limited evidence recovered does begin to suggest that specific waste products were used for particular technological and domestic purposes and that such products may have been collected and stored during different stages in the crop processing cycle for future use. This may be compared with the intricate systems of collecting specific crop wastes for specific purposes, that have been observed in modern Turkey by Hillman (pers. comm.)

In contrast to the work of Helbaek and others this paper confines its attention to the actual plant macrofossil evidence. Apart from anaerobically preserved plant remains from some Roman and Saxon waterlogged sites, all the information discussed consists of carbonised plant material. No mineralized (calcium phosphate replaced) evidence as described by Green (1979c) having been recovered from pre-Saxon sites. It is important to note that, as with any study of plant remains, carbonised evidence clearly has limitations (Dennell 1978: 15-31, Begler and Keatinge 1979: 208 and 209). It is difficult to establish the range of crops grown and the part such crops played in the economy and diet of past populations. Carbonised plant remains are essentially indicative of waste disposal and chance burning. However, on archaeological sites where extensive sampling of all available features has been undertaken as advocated by Dennell (1978: 21, Halstead et al. 1978: 118), it may now be possible to assess reliably the range of crops available for particular periods, though Reynolds (1979: 58) holds more pessimistic views.

Carbonised evidence clearly does indicate potential crops. Recent work undertaken by Hillman (pers. comm.) has begun to show that the carbonised cereals, chaff and weed seeds so commonly recovered from Iron Age sites can be very closely compared with crop processing wastes from ethnographic parallels in modern Turkey, where the range of crops grown is similar to those recovered from British Iron Age sites and the climatic conditions under which they are grown is comparable. Hillman (forthcoming: a and b) has been able to show that at least twelve crop processing sequences can be detected from carbonised evidence on Iron Age sites. From this and related studies it is possible to evaluate which crops may be under-represented and which are likely to be over-represented. A higher degree of confidence as to which cereals constituted the main crops can be attained.

For the Saxon and medieval periods, Green (1979d: 50) has indicated that carbonised cereal evidence is the most consistently recovered form of plant material, and distinct patterning between sites can be seen. Whatever the cause of carbonisation, be it chance or deliberate activity, certain patterns emerge which indicate similar actions and attitudes to waste disposal, and therefore possibly a similar economy.

Since it is difficult to eradicate many of the problems associated with the examination of plant remains from archaeological sites the evidence cannot and possibly should not be used in a highly sophisticated way at the present time. The general lack of well-recovered seed samples from archaeological sites makes any such analysis difficult, since it is impossible to begin to cluster sites on the basis of their botanical component. However, with increased work in this subject more complex forms of analysis should be possible in the not too distant future. Apart from observing new species introductions through time, the only method of analysis available, albeit fraught with problems as indicated by Dennell (1978: 24-26), is that of presence used by Godwin (1956: 1975) and strongly advocated by Hubbard (1975: 198). This is the only method of analysis which allows the re-examination of data which have so often been published with no attention to providing details of methods of recovery, sampling and the methods of statistical manipulation employed. This problem is aggravated since many available reports only consist of identifications to genus and not to species. On many sites where preservation is poor this is understandable. However, it is crucial to any interpretation of prehistoric agriculture that identifications are made to as fine a level as is possible since different varieties of a cereal clearly require different techniques of harvesting, processing, and may have been used for widely differing purposes. Because of the limitations imposed on this work by the quality of the published data, the information presented in the bar-diagrams (Figures 7:1 to 7:5) has been calculated using presence analysis (Hubbard 1975: 197-205; and Green, forthcoming d).

In Fig. 7:1 the evidence recovered from pottery impressions by Helbaek and Jesson (1944) and Helbaek (1952), although as indicated of limited value, has been compared with the results obtained from carbonised grain from more recent excavations. Even considering the very small sample that Helbaek used there are some immediate points of comparison between results from the two independent sources of data. For example, Triticum dicoccum (Emmer wheat), Triticum species (wheat), Hordeum vulgare (hulled and naked six-row barley) show similar increases and decreases between the periods represented.

In Fig. 7:2 all the available information on cereal crops has been presented. It can be seen that wheat was present on more sites than barley, oats and rye. However, it can also be seen that wheat and barley are nearly equally represented possibly supporting Dennell's model (Dennell 1976: 11-23). The evidence in Fig. 7:2 for oats includes both the positive identifications of cultivated forms as well as grains that were not specifically identified as wild varieties. An increase in the presence of oats through time can be seen whereas rye (Secale cereale) fluctuates in occurrence not only from Iron Age and Roman deposits but from Saxon and later medieval ones (Green 1979b: 126 and 295). Examination of the relationships of the major cereal crops, although important, is clearly complicated due to the wide range of wheats and barleys that were cultivated. It is probably more important to monitor the fluctuations between the different types of wheat or barley and to see how individual sites relate to the general trends.

Cereal Crops

(a) Wheats (Triticum sp.)

There is no evid ence of Triticum monococcum (einkorn) from any sites for the period Iron Age - Saxon. This crop was clearly not being used and does not even occur as a weed of other cereal crops. The limited nature of the cereal evidence from Bronze Age sites does not make any discussion of this species any easier. It may well be that einkorn was of little importance during much of the Bronze Age.

The evidence for Triticum dicoccum (emmer) including both grain and the more diagnostic parts, such as rachis, spikelet forks and glume bases (designated in the text figures as chaff), clearly show that this type of wheat was second in importance to spelt (Triticum spelta) although it nearly always occurs with spelt (Fig. 7:3). Emmer, although never more important than spelt, occurs more frequently during the 5th to 3rd centuries BC than at any other time (Fig. 7:3). Such an apparent increase may in fact be due to the limitations of the existing sample rather than actual importance. Clearly more sites of this period will have to be examined to see if this pattern holds good.

Triticum spelta (spelt), although recovered from a larger number of sites of Iron Age and Roman date in Wessex than emmer, seems to decline in occurrence possibly as early as the 2nd century BC. There are, however, regional variations, spelt still formed a major component of cereal assemblages from post-Roman and Saxon Gloucester (Green 1979a, 188). It has been pointed out by Harding, 1974: 81 and 85, Wacher 1978: 109, and Cunliffe 1978: 178, albeit with reservations, that spelt was probably introduced during the Iron Age into southern England, possibly by invaders from the spelt growing regions of the middle Rhineland. This was originally suggested by Helbaek (1952: 210). The evidence from Wessex from the earliest Iron Age and latest Bronze Age sites is suggestive of continuity than abrupt introduction, possibly similar to continental evidence of the same period. It is tempting to suggest that the neolithic occurrence of this cereal at Hembury (Field, et al. 1964: 373) is not the isolated find that it is so often considered and that more detailed work on late Bronze Age and early Iron Age sites in Wessex may indicate that spelt was present at a much earlier period.

If the assumption is made that the occurrence of a single species, for example a single type of wheat, on a site or sites indicates that it had been taken into cultivation, then it can be suggested that bread wheat and club wheat (Triticum aestivum and Triticum compactum) were possibly grown as crops (or as a mixed crop) from at least the Roman period onwards. Information from two as yet unpublished 2nd-3rd century sites in Winchester, one suburban, the other intramural, indicate that free threshing wheats may have been grown for some specific purpose possibly a specialist crop to serve a particular urban requirement. These species occur somewhat later on rural sites, for example Poundbury (Monk pers. comm.). In many cases in other parts of the region, it is not until the early Saxon period that bread and club wheat occur, for example from the rural site at Stratton Park (Murphy pers. comm.). Free threshing bread type wheats have been recovered from an increasing number of sites of the middle and late Saxon periods. The indication is that by the late Saxon period, with the exception of the Gloucester evidence (Green 1979a: 188), club and bread wheats were the main types cultivated in the Wessex area.

Several hypotheses have been put forward to account for the appearance of bread wheat in the Iron Age. The absence of this species from the few Bronze Age samples examined to date does not of itself mean it was absent. Free threshing wheats have been recovered from neolithic sites and it may be reasonable to assume their presence all along. Continuity cannot be established at the present time and Dimbleby's suggestion that the neolithic varieties became extinct and another variety more suited to the British climate was introduced (Hubbard 1975: 200) remains open to question. However, as Hillman has pointed out there is a considerable risk that free threshing cereals are grossly under-represented at all periods (Hillman 1978: 169). For this reason the suggestion above, (that the occurrence of a single species indicates that it was specifically cultivated), although perhaps true for spelt and emmer, may not hold true for the bread and club wheats.

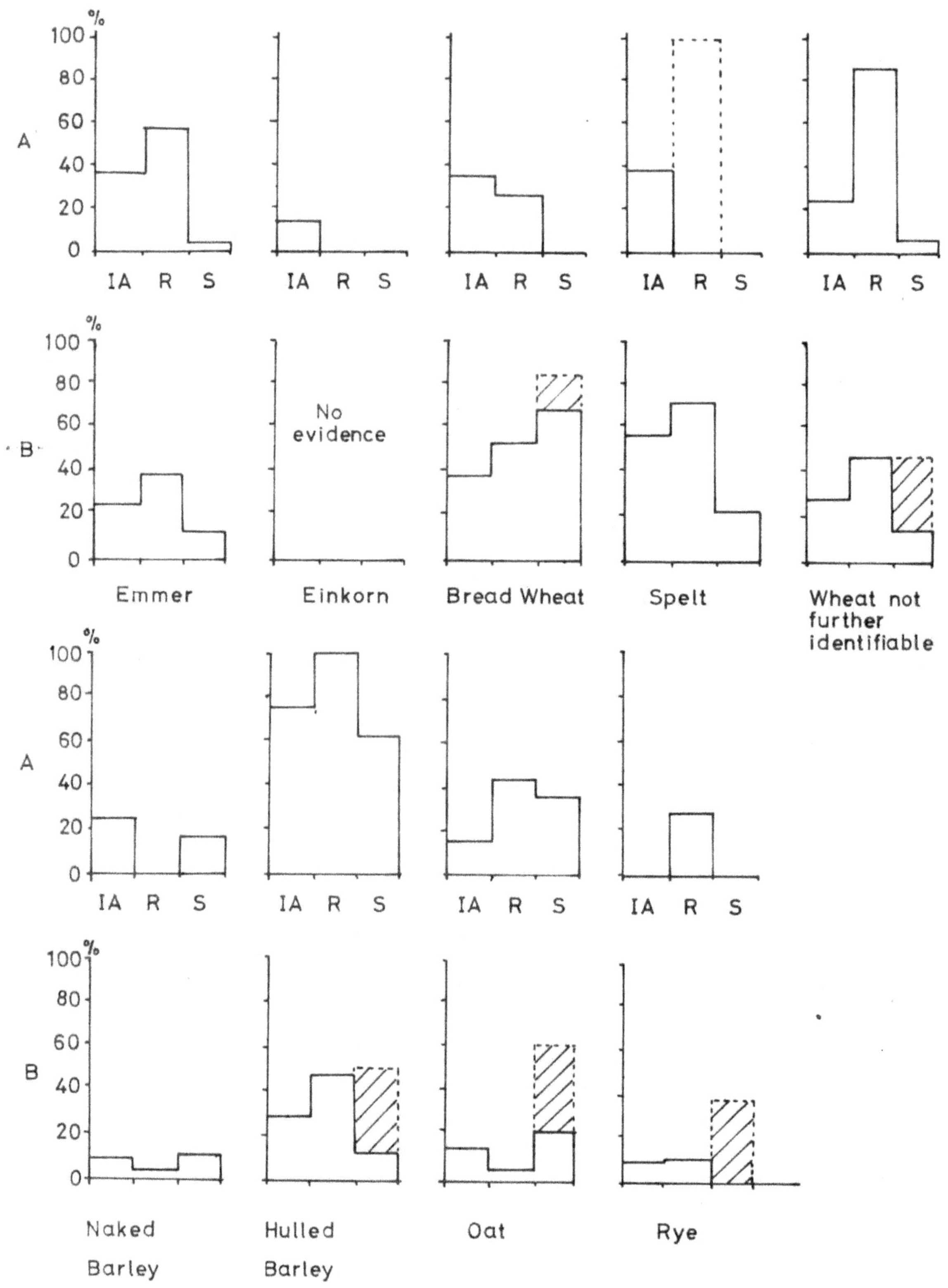

Fig. 7.1 Presence bar-diagrams to show the differences between the evidence of plant impressions recovered by Jesson and Helbaek (A) and the recent evidence of carbonised plant remains (B). Note, the hatched areas consist of additional late Saxon data and the dotted line for Spelt represents additional carbonised evidence recorded by Helbaek.

The Vertical axis represents the percentage of sites examined from which the different species were recorded.

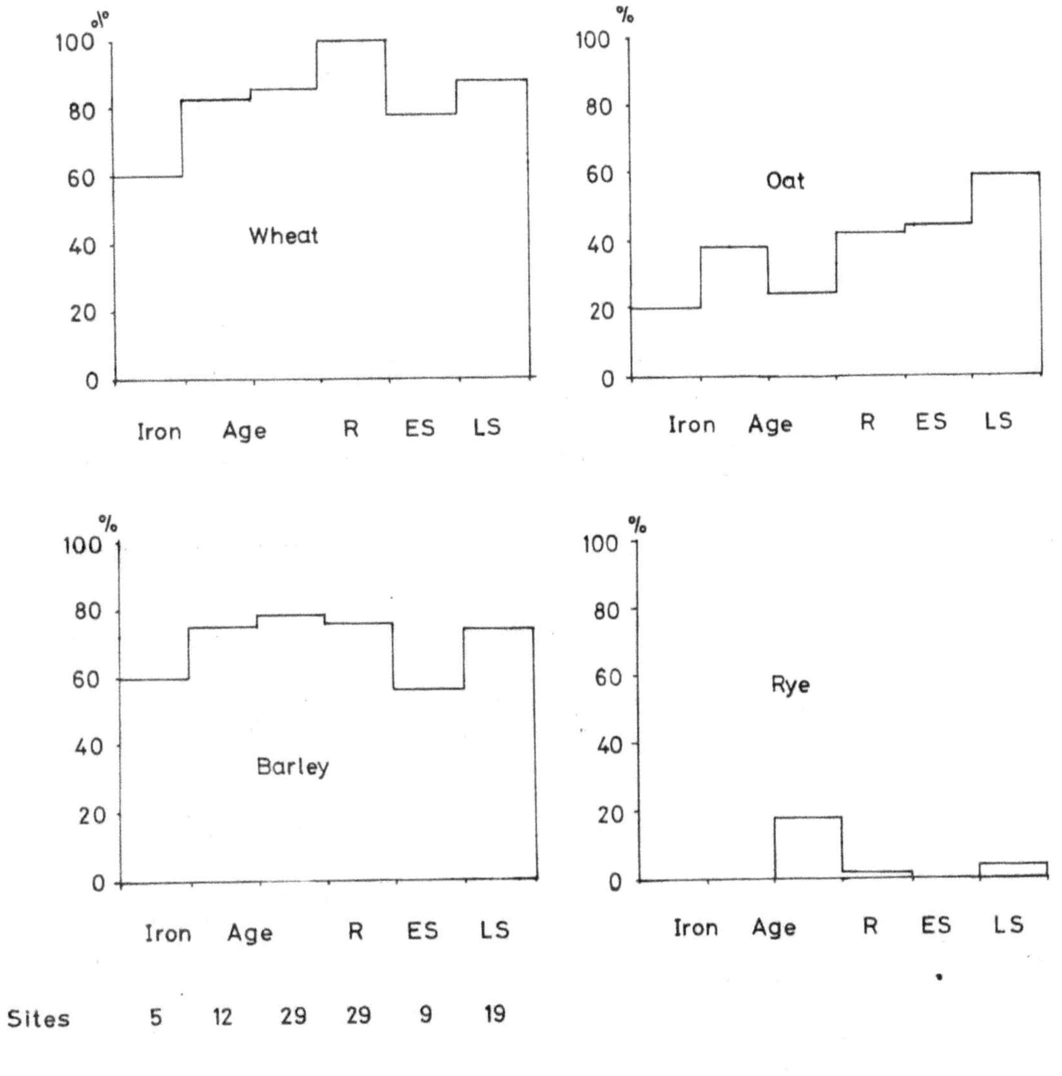

Fig. 7.2 Presence bar-diagrams of the main cereals, wheat, barley, oat and rye.

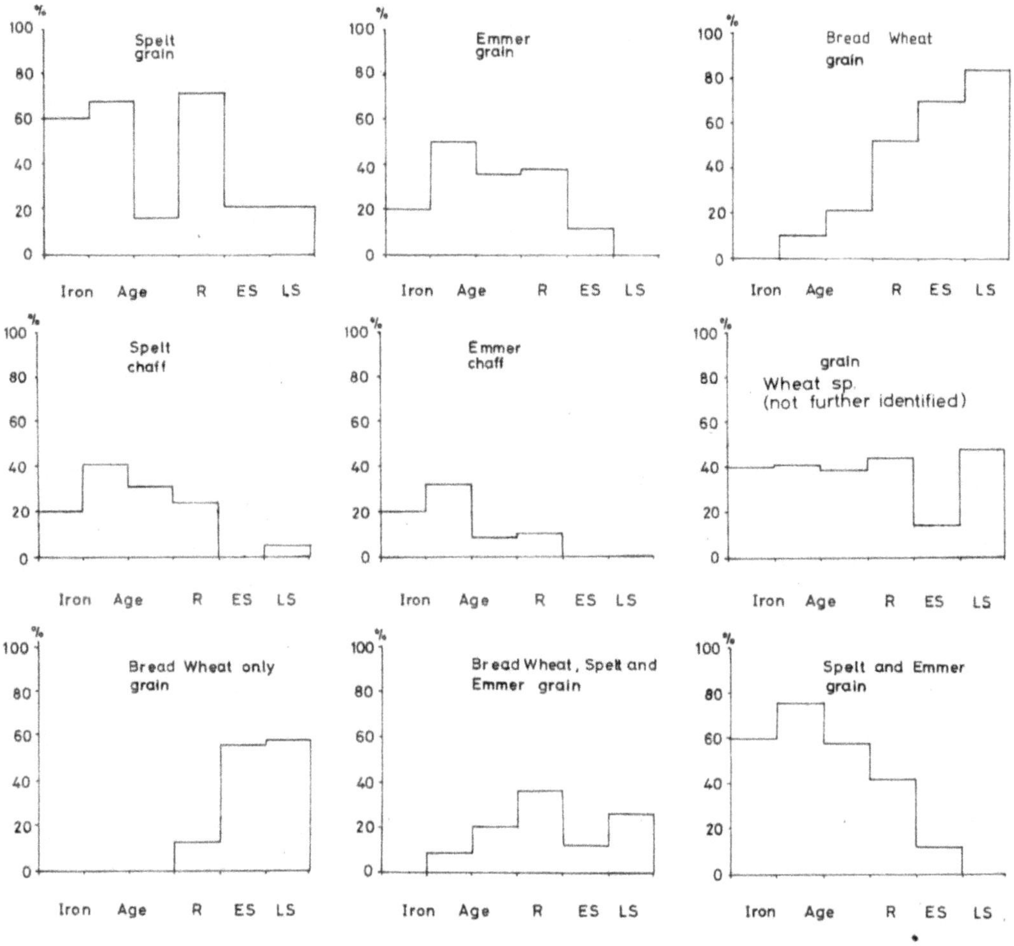

Fig. 7.3 Presence bar-diagrams of wheat species, showing the percentage occurrence of each species (grain and chaff) and also showing the percentage of sites where combinations of the different species occurred.

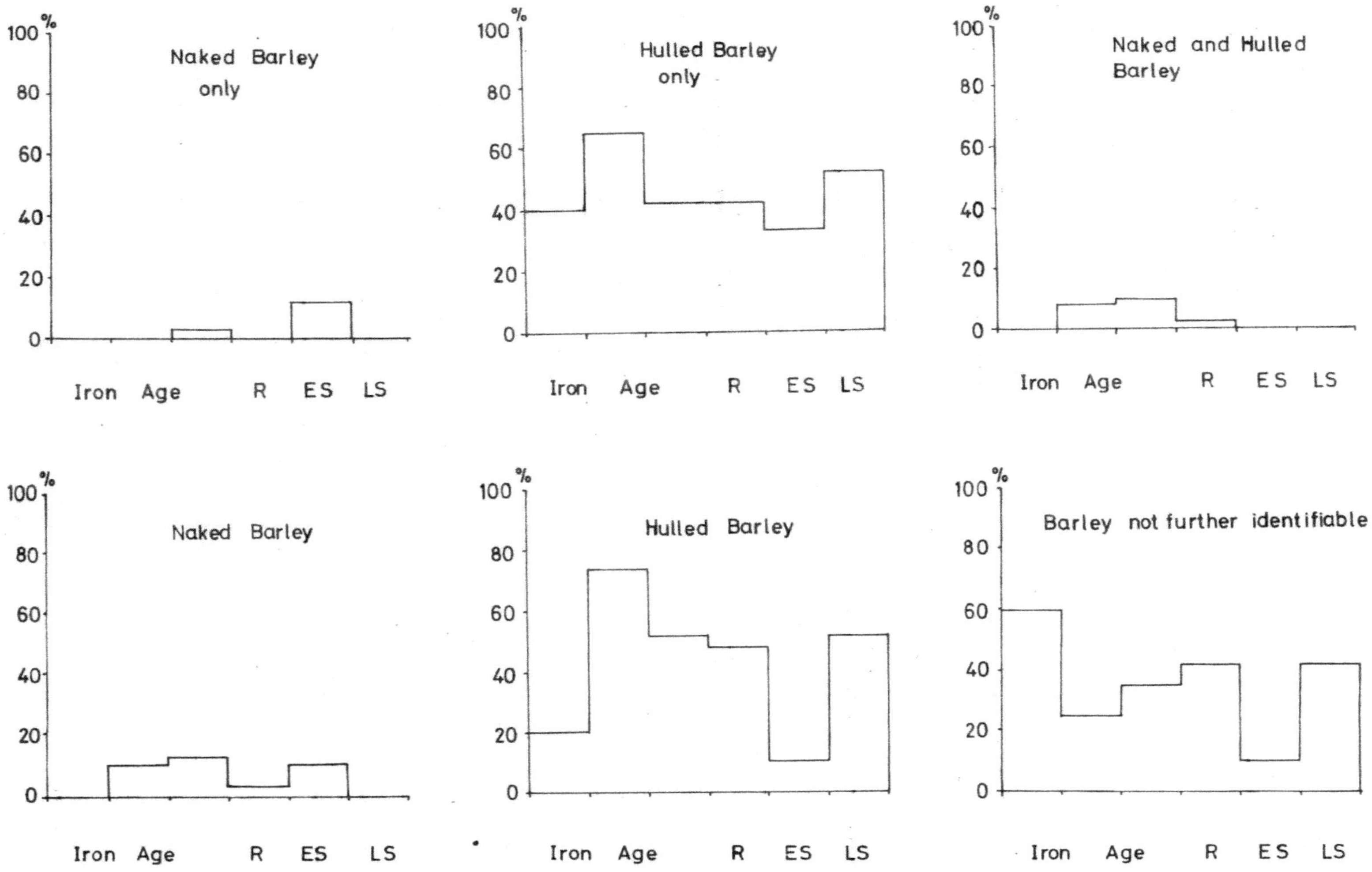

Fig. 7.4 Presence bar-diagrams of barley species, showing the percentage occurrence of each species and also showing the percentage of sites where combinations of these different species occurred.

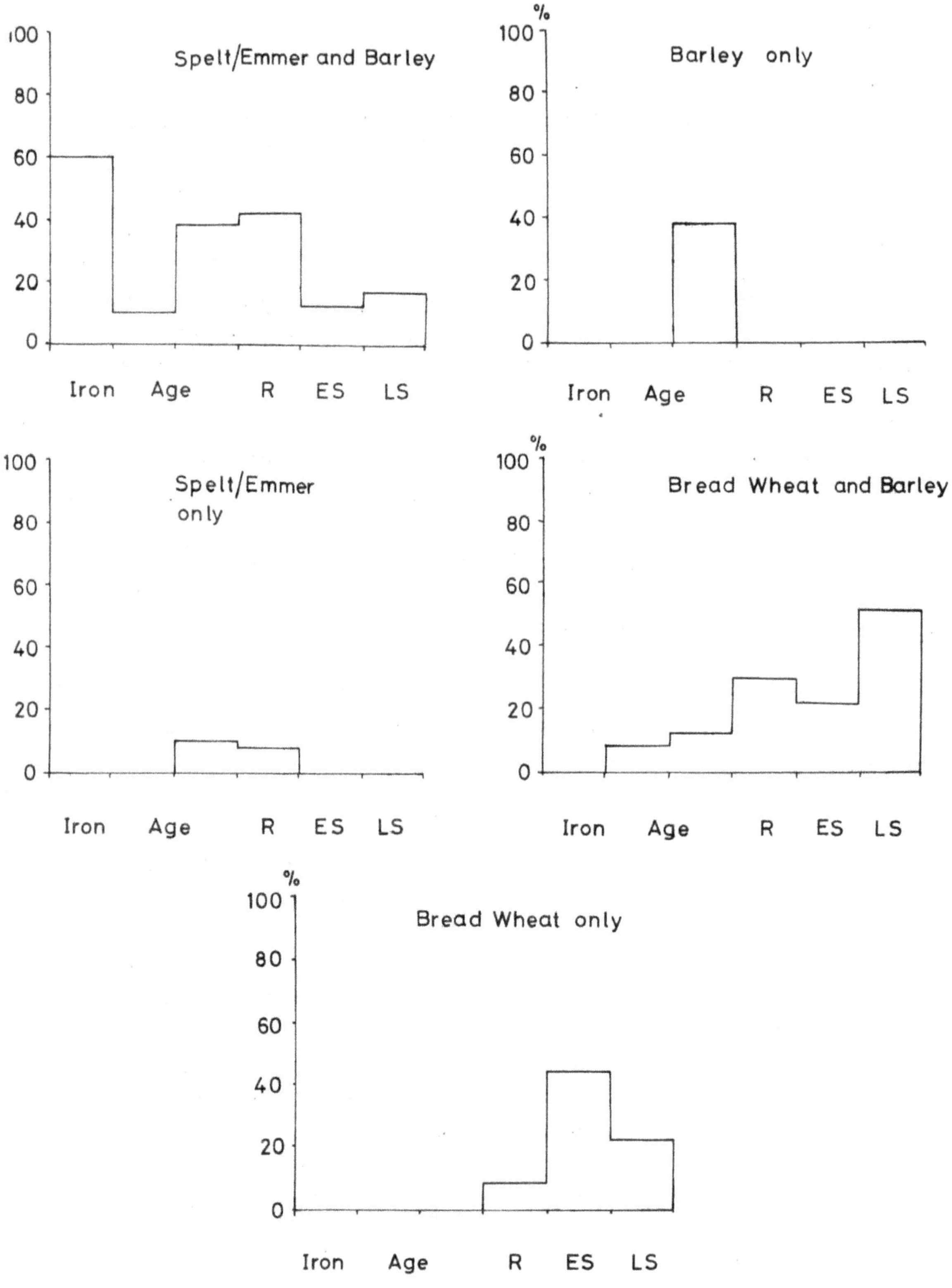

Fig. 7.5 Presence bar-diagrams to show the percentage of sites where different species of wheat and barley were recorded and also the percentage of sites where combinations of the different cereals were recovered.

The evidence for club and bread wheats from many of the sites on the Hampshire chalkland, for example Portway and Old Down Farm, indicates that free threshing cereals did not appear in this particular area, whereas the evidence from Gussage (Evans and Jones 1979: 172-175) shows that both varieties were present. It therefore seems that considerable variation in the introduction and cultivation of these species existed in the Wessex region during the Iron Age.

The types of bread wheat and club wheat cultivated seem to be morphologically diverse. It is common to find both dense eared and lax eared forms together in a way that indicates selection of specific strains clearly had not taken place until at least the late Saxon period.

(b) Barley (Hordeum sp.)

The percentage presence of barley species can be seen in Figure 7:4. Hulled six-row forms account for most of the evidence whereas naked six-row barley never occurs on more than 15% of the sites at any one period and even accounting for possible under-representation is probably of much less importance than Cunliffe suggests (1979: 178), (due to its being used for different purposes). From the archaeological information available it is difficult to establish whether barley really was a more important crop than wheat or vice-versa. Certainly there seems to be no great change in the occurrence of these two species through time (Fig. 7:2) and this is all the more strange when the change from brittle rachis glume wheats to the free threshing forms is taken into consideration, with supposed under-representation of bread and club wheats.

The lack of barley chaff fragments on sites does indicate that it was processed differently from the wheat crops. This is to be expected, especially during the Iron Age and was suggested by Liversidge (1968: 213). However, the balance between barley and wheat chaff during the late Saxon period suggests larger quantities of wheat. Barley chaff never seems to be the dominant chaff element even where large barley deposits have been recovered. It is therefore difficult at the present stage in the study of cereal crops to be as categoric as Applebaum (1972: 108) who states that hulled barley was the predominant crop of the Iron Age, a fact reiterated by others. At best hulled barley does seem to increase in frequency of occurrence from the 3rd century BC onwards as witnessed by evidence from Old Down Farm (Green forthcoming, a) and at Portway (Murphy 1977: 62). It has also been suggested that barley was the staple crop during the Saxon period. Once again the archaeological evidence does not support this hypothesis (Green 1979b: 148 and 295).

When examining the relationship between the barley and the wheat crops (Fig. 7:5) it can be seen that in the 1st and 2nd centuries BC barley appears to be more frequently found on its own. This may be coupled with the fact that bread wheat seems to occur on more sites where barley, emmer and spelt are absent. Such information certainly argues in favour of highly diversified crop cultivation/utilisation in the late Iron Age.

It is important to note that there is no evidence from Wessex that the barley crop was specifically grown for use in malting (Liversidge 1968: 215). Rarely have sprouted grains been recovered from pre-Saxon sites and yet sprouted grain has been commonly recovered from late Saxon and medieval ones. The implication is that barley may not have been exclusively grown for malting and that other crops may well have been grown for this purpose at earlier periods.

There seems to be no specific selection or preference for varieties of six-row hulled barley grown in this region from the Iron Age to the Saxon periods. Jones (1978: 104-105) has pointed out that the Iron Age barley at Abingdon cannot be divided into genetically meaningful varieties in terms of ear laxity. This certainly seems true on other sites, that is when sufficient evidence (rachis internodes) has been recovered for analysis. The information from the Roman and Saxon periods suggests genetically highly mixed crops of six-row barley. Barley crops seem to show less diversity within individual large grain deposits from the late Saxon period, whereas between such deposits differences can be closely compared with modern lax and dense eared crops.

(c) Oats (Avena sativa and Avena spp.)

The main problem when discussing oats is that a wide variety of wild oats exist as well as several cultivated forms. Without diagnostic parts of these plants (lemma bases) it is very difficult to separate poorly preserved grains of wild and cultivated species. As pointed out above, a general increase in the occurrence of oats can be seen through time although unlike many of the other cereals discussed, it is even more difficult to define at what period oats were grown as a crop in their own right. The available information from the Wessex region indicates that it is not until the very late Saxon period that sites begin to produce a dominance of this species over other crops and it is not until the early medieval period that sites produce large caches of this cereal.

It is difficult to dismiss oats as not being an important crop in Wessex prior to the Saxon period. It is conceivable that original utilisation of this crop may be biasing the archaeological picture. It is possible that future work will reveal a pattern similar to that noted by Jones for sites outside the Wessex region (Jones 1975: 104). However, it is also conceivable that the entire evidence for oats from Wessex in the pre-Saxon period may simply represent weeds of other crops or imports from other regions where this cereal is known to have been cultivated.

(d) Rye (Secale cereale)

This species certainly occurs in the 1st and 2nd centuries BC but always as a very minor component of cereal deposits. Hillman has indicated that this species is likely to be grossly under-represented in relation to other crops, especially non free-threshing cereals (Hillman 1978: 169). It has been stated by Godwin (1975: 415) and others that rye was a major crop in the Saxon and medieval periods. The evidence from sites in the Wessex region has clearly shown this not to be the case (Green 1979b, and Monk 1978). Certainly in the later medieval period rye is rarely mentioned in the historical documentation and at the same time rarely occurs on archaeological sites.

This species was clearly grown in the medieval period in this region, but like oats it may well have been grown on certain soil types for short periods. It is possible that rye was specifically grown for non-human consumption. Rye is ideal as an early bite (fodder crop) for sheep (Vancouver 1813: 157-158), and is also highly prized for its long straw for thatching (Stanes 1969: 53). The work on pollen by Godwin (1975: 415) has probably resulted in an over-emphasis of the importance of this crop, since the species produces an abundance of airborne pollen, the pollen being carried further than other cereal pollen. Also, the pollen of rye is distinctive and is therefore more easily separated from other large Gramineae pollen, and it is therefore not surprising that this species appears to be more important as a result of such evidence.

Legumes (Papilionaceae)

Legumes were probably of equal importance to the cereals for providing protein and vitamins for past populations. Unlike the cereals it is not possible to envisage that legumes were widely grown as specific field crops. There are several good reasons for this; the different varieties of *Vicia faba*, including modern broad beans and celtic beans are particularly prone to weevil and aphid attack. The celtic bean due to its extreme height is also easily damaged by strong wind and rain. It is therefore more logical to assume that these crops were grown in small cultivated garden areas where considerable control could be exercised over them. Evidence for celtic beans exists from the late Bronze Age both in carbonised form from outside Wessex (Murphy, pers. comm.) and as plant impressions from the Isle of Wight (Scaife, pers. comm.) and not just from the 2nd century BC as stated by Cunliffe (1978: 178). The evidence for this species recovered from the early excavations at Glastonbury (Reid 1917: 625-630) and other sites of the same period can now be seen as part of a longer continuum.

The evidence for peas (*Pisum sativum*) is less precise. Peas have only rarely been recovered from late Iron Age sites in the region for example at Bishopstone (Arthur 1977: 273-275). The evidence suggests that they may have been imported from the continent either as a specific crop import or as a weed of other crop plants. Peas are, however, quite common on Roman sites, for example Goathill Villa and Wookey (Branigan 1976: 133). Peas have not been recovered from early Saxon sites, possibly as a result of poor conditions of preservation and the general lack of all classes of domestic refuse.

Legumes such as *Lens* sp. (lentil) although recovered from several Roman and even late Saxon sites, must be considered either as an imported crop or a species accidentally imported as a crop weed. However, casual cultivation of this species may have taken place in Wessex during the Roman, late Saxon and later periods.

It is always difficult to establish how important legumes were in the overall economy of any region, simply because the nature of processing and utilisation of them leaves less archaeological trace than cereals (Dennell 1972: 157), in particular the brittle rachis hulled forms (Hillman 1978: 169). In the later medieval period where the archaeological information can be compared with the documentary evidence, legumes appear to gain importance with increased quantities being recovered from the archaeological sites and increased quantities being grown on ecclesiastical and other estates, as witnessed by documentary evidence (Titow, pers. comm.). On the basis of these two sources of information it can be proposed that the increase in beans on sites from the 3rd century BC onwards in Wessex also possibly indicates a real increase in bean cultivation. The late Iron Age evidence for beans especially in Hampshire, can be taken as further evidence of crop diversification at this period. It is possible that some settlements such as Owslebury (Murphy 1977: 227) may have specialised in growing bean crops or at least show a preference for their utilisation. The occurrence of Vicia and Lathyrus species on sites and the accumulations found for example just outside the Wessex region at Fishbourne (Grieg 1976: 376) may indicate deliberate collection or even cultivation for human or more likely animal consumption.

Improvements in the size of bean and pea seeds can only be established from the late medieval and post-medieval periods on the basis of the archaeological evidence currently available.

Other Vegetable Crops

It is a widespread belief that many of the vegetable crops such as turnip, parsnip, cabbage, radish and mustard, were imported into Britain during the Roman period (Applebaum 1972: 116). Until such crops are recorded from Iron Age sites in Wessex it is inevitable that they should be considered as Roman introductions. However, future work on sites where preservation conditions are more suited to the survival of vegetable remains may well change this picture in the near future. It is hard to consider that the degree of trading contacts, so clearly demonstrated from ceramic evidence, did not also involve trade in perishable foodstuffs, and the importation of new species.

Unfortunately it is difficult to be as precise about the vegetable crops represented on archaeological sites, as indicated by Applebaum (1972: 116). It is important to consider that seeds of many vegetable crops are unlikely to occur on archaeological sites, since it may be that the leaves or the roots of the crops were consumed. It is only where caches of seeds occur that any indication of utilisation and cultivation as a specific crop can be inferred. However, the problem is compounded since the seeds of some vegetable species cannot be easily separated from their wild progenitors or even closely allied non-cultivated varieties. Thus many vegetable crops and plants grown in Wessex in the past are bound to be under-represented and our impression of past vegetable crop cultivation is bound to be distorted.

Speculation about plants that may have been grown as vegetable crops could be endless. It is best to confine discussion to those species that are still being cultivated or for which there is ethno-historical data from the area being studied. It is therefore possible that many plants represented by their seeds on archaeological sites in Wessex (but by modern criteria treated as weeds) may have formed an important part of the overall crop agriculture.

It is only from Roman and later Saxon sites in Wessex where garderobes and latrine pits have preserved human faecal material that reliable analysis of the non-seed portions of vegetable crops, possibly crops only consumed in dire necessity (cf. Akroyd and Doughty 1964: 119) can be undertaken.

Brassica seeds (species of cabbage, mustard, rape) have been recovered from Iron Age sites often in very large quantities. Evidence from pits (Green forthcoming, a) and from the inside of a pot at Old Down Farm (Murphy 1977: Plate 14) indicates that deliberate collection for some purpose was undertaken The evidence suggests a crop stored for future cultivation or culinary use.

Chenopodium album (Fat Hen) and related species do occur frequently on Iron Age and later sites and may as indicated by Reynolds (1979: 67-69) have constituted a crop plant. Beet (Beta sp.) is first recorded in Wessex from a late Saxon context in Winchester (Green unpublished). Once again waterlogged preservation conditions account for the preservation of this species which may represent leaf beet, beetroot or chard.

Seeds of either wild or cultivated carrot (Daucus carota) have been commonly recovered from late Saxon and medieval sites in Wessex in particular from excavations in Winchester. Other vegetable species recorded from late Saxon sites in Winchester also includes evidence for celery (Green unpublished).

Herbs

Specific plants known to have been grown for their use as herbs have been recorded from Roman and later sites. Coriander (Coriandrum sativum) has been recorded from Roman deposits at Bucknowle Farm, Corfe Castle and from Roman deposits in Winchester. Fennel (Foeniculum vulgare) has only been recorded from late Saxon sites in Winchester. Whereas opium poppy (Papaver somniferum) has been recorded from a wide range of sites from the late Iron Age onwards. This species may represent a common weed or a crop either used as a herb or grown for the oil that its seeds contain.

Fibre and Oil Crops

Finds of flax seeds from late Iron Age contexts have rarely been made in Wessex (Murphy 1977: 226). The sparse evidence though not specifically suggesting cultivation in this region does at least indicate that the crop was available though possibly imported from northern Britain or the continent. If the seed had been used for oil production it should have been recovered from a larger number of sites. However, if cultivated for its fibre, the process of harvesting and retting (Boase 1918: 140-147) reduces its chances of being recovered. It is only from waterlogged sites of the Saxon period (Winchester, Southampton, Gloucester) that this species regularly occurs. It is possible that the biases of preservation essentially confined to carbonised material from Iron Age sites with increasing quantities of waterlogged evidence from Roman and Saxon sites is confusing the observed picture.

Cannabis or hemp (Cannabis sativa) does occur on continental sites from the Roman period onwards (Willerding 1970: 287-385). It also occurs in the British Isles for the first time at this period as the 1979 excavations at Bucknowle Farm, Corfe Castle, Dorset, showed (Green unpublished). This evidence, like all the Saxon information on this plant from Wessex (Winchester, Southampton, Gloucester), comes from waterlogged deposits. Hemp becomes far more common from the late Saxon period, for example from Gloucester (Green 1979a: 186). The pollen evidence as noted by Godwin (1967: 42-49, 1975: 415) indicates that the crop was grown from at least the middle Saxon period onwards in the British Isles and is invariably associated with flax in the pollen spectra. The available documentary evidence suggests that both flax and hemp may well have been imported into the region from the continent in the medieval period. Hemp was certainly imported into Southampton in the 13th century (Platt 1973: 13). It is therefore quite possible that both flax and hemp were imported into southern Britain from at least Roman times onwards.

Fruits and Exotic Crops

For most practical purposes it is impossible to distinguish between those exotic crops grown in Wessex and those that may have been imported from the continent or elsewhere. The range of exotic crops that were available on the continent during the Iron Age (Knorzer 1968: 113-124, Knorzer 1976: 61, Knorzer 1966: 113-124 and Willerding 1970: 287-385) is clear. The lack of such evidence of sites in Wessex is probably biased, since rarely have waterlogged sites for this period been examined and waterlogging or anaerobic soil conditions are responsible for the preservation of most of the continental as well as the later fruit and exotic crop evidence from Wessex. This point has been demonstrated on many waterlogged Roman sites outside Wessex (Willcox 1977: 269-282). Fruit crops are rarely if ever preserved by carbonisation (Green 1979b: 283).

However, it must be considered that delicate fruit crops like mulberries and other soft fruit recovered from many Roman sites in Dorset and Devon (Dewar 1955: 58-60) must have been grown in Wessex.

No direct evidence for the cultivation of grapes exists in Wessex from Roman or even later sites. The Roman evidence from Gloucester (Hurst 1978: 162) has been redated to the Saxon period. Clearly vines were cultivated in Saxon and later times and there is no reason to suppose that they had not been introduced into Wessex during the Roman period. The evidence for grapes becomes abundant from middle and later Saxon sites such as Winchester, Hamwih and Gloucester, though invariably associated with monastic sites and sites inhabited by individuals of high social status.

During the Saxon and later periods in Wessex a wide range of fruit crops seem to have been cultivated, although possible specifically cultivated in hedgerows rather than orchards. These include sloe, bullace, plums, cherries and strawberries. Also, blackberries, and raspberries may have been specifically cultivated rather than just being collected in the wild. Abundant evidence for all these species has been recovered from sites such as Winchester (Green forthcoming c).

The lack of evidence for hazel-nut (*Corylus avellana*) from Iron Age and Roman sites has been commented on (Jones 1980). This species is very commonly recovered from mesolithic, neolithic and Bronze Age sites in Wessex. Broken fragments of this species also occur in large quantities on Saxon and medieval sites. Although evidence from waterlogged contexts of the Saxon period invariably also contains whole nuts which were clearly discarded since they were obviously diseased. Since carbonised and waterlogged fragments are equally common prior to the Iron Age and Roman periods the lack of Iron Age evidence must indicate a fundamental change either in the use of this resource or a change in land management and the importance of this species as a coppicing crop.

Crops of a luxury nature that could travel long distances have been recorded from many Roman waterlogged sites in Wessex and include fig, walnut and pine-nut. The recovery of complete waterlogged pine cones of *Pinus pinea* from Winchester (Murphy: pers. comm.) and carbonised pine nuts (Green unpublished) may indicate that this species was grown in Britain rather than imported. It is not until the later medieval period that many of the exotic imported crops recorded on Roman sites once again occur in the archaeological record, and from their occurrence in documentary sources appear to become important constituents in the diet (albeit of wealthy individuals). Hemp (Humulus lupulus) has only been recorded from very late Saxon and medieval sites in Wessex for example in Winchester. However, the sparse evidence from Wessex and the information from the Graveney boat (Wilson and Conolly 1978: 133-150) probably indicates that this crop was imported rather than being widely grown in the area prior to the late medieval period.

Many crops of an exotic nature that have been recorded from sites in Wessex on the basis of their specific climatic requirements, can never have been grown in the region at any period under discussion. It is instructive that many crops, even imported ones such as rice, only occur late in the British plant record as exemplified by the Wessex evidence from Southampton (Green unpublished). Certainly trade was taking place with the continent in wine and other commodities during the Roman and post-Roman periods and it is also reasonable to assume that perishable food stuffs may also have been traded (Fulford 1977: 63). The evidence does suggest a rather conservative approach to crop imports and the importation of new species that could have been cultivated in Wessex.

Weeds

The last group of plants although not of necessity crops themselves consist of plants that must have constituted the commonest weeds in the cereal crops. It is quite clear that certain weed seeds occur with monotonous regularity on most Iron Age and later sites in the Wessex region. One of the most commonly occuring species is Bromus secalinus/mollis (Brome grass). Recently Hubbard (1975: 201) and Arthur (1977: 273) have suggested that it was possibly a tolerated or useful plant. Hubbard bases his assertions on the lack of correlation between this species and any cereal crop. However, as Murphy has shown (1977: 227) this is quite clearly not the case for sites such as Portway and information from other sites such as Old Down Farm (Green forthcoming, a) and Ashville (cf. Jones 1978: 106) bear this out. Brome grass was a persistent weed of hulled wheat in the Iron Age, and it frequently occurs on sites of the Roman and Saxon periods. The evidence from late Saxon sites indicates that it occurs more frequently as a weed of barley crops at this period (Green 1979d: 232-235). The abundance of Bromus on Iron Age sites is due to its being winnowed from crops and hence its occurrence amongst the chaff elements. The same is true for its occurrence on Roman sites. However, the Saxon evidence for Bromus has invariably been recovered from bulk cereal deposits in particular barley crops that seem not to have been thoroughly cleaned of such impurities. It is evident that the disposal of crop waste products and methods of cleaning changed enormously between the early Roman and later Saxon periods in Wessex. This probably results from the different types of wheat being cultivated, and that the evidence from Iron Age sites comes from settlements of an essentially rural character, whereas the late Saxon evidence comes from developing urban centres.

Another plant that is often quoted as a common weed of Iron-Age crops is Agrostemma githago (Corn Cockle). Examination of the evidence from the Hampshire Iron Age sites does suggest that this particular plant was not the menace as a weed that it was to become during the Saxon and medieval periods. On presence analysis this species has occurred on about 15% of the sites examined and never in high concentrations (i.e. over 10 seeds per 10,000 ml of soil examined). This plant, like Bupleureum rotundifolium, Thorow Wax, (Reynolds 1978: 147) and Polygonum convolvulus (Black Bindweed) does not occur as frequently on Iron Age sites as Reynolds would have us believe (1979: 69). Certainly they seem to be no more common on Iron Age sites than they are on later Saxon and medieval ones where carbonised preservation conditions prevail. Some common weed species have on the other hand been rather overlooked and clearly considerable work needs to be undertaken to understand the role of weeds such as Lithospermum arvense (Corn Gromwell) a common weed of the Iron Age and Roman periods (Jones 1978: 101), but infrequently found on Saxon and later sites except as a site specific weed (Green forthcoming, b).

Discussion

Weed seeds from carbonised deposits, their relationship to specific crops and changes in their distribution through time still need proper investigation. There are, however, one or two noticeable trends. Godwin (1975: 477-478) has already implied an increase in weed and in particular crop weed species from the 3rd century BC in pollen spectra. The same pattern has certainly emerged from the seed evidence at sites such as Old Down Farm (Green forthcoming, a) and the evidence recorded by Jones (1978: 107) from the excavations at Ashville, Oxon. An increase in the range of species involved can be detected as well as a noticeable increase in the actual quantities of weed seeds recovered. It is interesting to note that the evidence from Old Down Farm also shows an increase in the quantity of chaff fragments and cereals through time, possibly indicating increased crop processing activities, or greater attention to the cleaning of crops. However, on the nearby site of Portway the evidence suggests that the opposite is true, on this site all the elements, crop weeds, chaff and cereals decrease in occurrence through time.

Such variation between different sites in the Iron Age possibly reflects site specialisation and the change on some sites from a mixed economy where all crop processing activities took place to one where only certain aspects of production and consumption were undertaken possibly as a result of increased pastoral activity. Site specialisation within a region in terms of agricultural production or possibly specialised agricultural processing may in fact herald or result from other economic and social changes culminating in later periods with the clear distinctions between rural and urban centres. Clearly when data permit, the differences between sites must be more clearly examined to see what patterns of variation may emerge from settlements in the same geographical region, sites that possibly had a complimentary function within the overall economy.

During the later periods under discussion, other differences can be observed. In the Saxon and later medieval periods the range of crops recovered from sites can vary enormously. Noticeable differences do exist between urban and rural sites. For example small rural and semi-urban centres do seem to produce greater evidence of oats and legumes than the large urban centres such as Winchester (Green forthcoming, c). Such a dichotomy must represent the different requirements of these settlements and must indicate the influence and preference placed on production within the immediate cereal trading and servicing area.

Having briefly dealt with inter-site variability it is also important to consider intra-site variability. Clearly the work of Dennell (1972: 150) could be applied more widely to see what variation actually exists between the deposits in different types of feature or for that matter the same type of feature. The only information available at the present time (Murphy 1977: 214-215) from Iron Age sites indicates that barley is commonly found in post-holes and that wheat is confined to pits and ditches. This is supported by Wainwright (1968: 114) on the basis of the evidence from Tollard Royal. Whether or not Wainwright's model of barley being stored in above ground structures and wheat in storage pits is a valid one cannot easily be proved or disproved. Clearly, as indicated by Reynolds (1979: 58), the mechanisms involved between harvesting preservation and final deposition are complex and thus simple models may be totally inadequate for indicating the processes involved. Wainwright's assertions can be criticised on the grounds that the evidence from Iron Age storage pits does not specifically represent stored crops, but invariably represents waste products disposed of within the feature when the pit was no longer required for some other purpose.

Unfortunately until further evidence from well dated postholes is available, and on many sites this is rare due to poor stratigraphy, little can be said about variation in the range of plant remains recovered from features on Iron Age sites. For later periods in particularly Roman and later Saxon sites where specific classes of feature such as cess pits have been recorded, the degree of intra-site variation is often extremely noticeable (Green 1979: 41-44).

However, before such detailed transformation studies can be put into practice there are several very major problems that will have to be overcome. It should be pointed out that carbonised seed densities within deposits rarely reflect large mammal bone or pottery densities, and that carbonised seeds are frequently found in features where bone and pottery is lacking. There are some very noticeable differences in the character of the sites being examined during the time span discussed here. The geographical setting of many of the sites is different and of course preservation conditions vary enormously.

Sites of the earliest Iron Age such as phases from Old Down Farm are represented by a series of pits and postholes and ditches. The same is true of most Iron Age sites in the region, for example Portway, Cowdery's Down, Gussage and even larger settlements such as at Winchester and Danebury. However, away from the chalk, for example on the gravels in Oxfordshire and the limestone in the Gloucester and Avon region pits become less common. A prime example is the lack of pits at Crickley Hill. On some of these sites the only available features that can be sampled are post holes, ring gulleys and ditches. Thus to make direct comparisons between these different types of site, those with pits and those without, is almost impossible. Pits invariably contain most of the carbonised plant component recovered and the absence of such features is a severe loss to intra-site and inter-site interpretation.

Roman sites vary considerably in character, there is scant evidence from a range of villa sites typified by Branigan (1976: 134) and Fowler (1976: 191-192). Villa sites in Hampshire have to date received very little attention largely as a result of the past and even present bias of many Romanists towards structures rather than environmental aspects (cf. Frere 1978: ix-x). Only one villa site has been systematically examined for plant remains at Barton Court Farm, Abingdon, Oxon. (Jones, pers. comm.). Added to this is the general lack of information from Roman urban settlements. However, some urban settlements of a specific character have been examined for example the important evidence being recovered from the cremation cemetery outside the north gate at Winchester (Green, unpublished). However, such evidence at the moment is not easily compared with the information from the Roman phases at Portway, Owslebury, Old Down Farm, and sites examined by Monk and Murphy (1977: 68-72; Fasham and Monk 1978: 363-371) at Micheldever Wood and Winnel Down to the north and north-west of Winchester.

Sites of the early Saxon period on the Hampshire chalkland have been examined in some detail. Sites such as Chalton and Cowdery's Down have been extensively sampled. However, when only postholes and timber slots remain, the nature of the surviving evidence is bound to be distorted and difficult to interpret (cf. Addyman 1976: 315). Even when grubenhäuser have been examined (Old Down Farm and Chalton) the evidence recovered has been exceedingly sparse. Chalton is the only middle Saxon site in this region that has so far produced pits with cereal accumulations (Green unpublished). Cereal accumulations have however, been recovered from the settlement site examined by Murphy at Stratton Park (Murphy 1977: 76-77) a site of early Saxon date (Fasham: pers. comm.). However, the evidence from these rural sites cannot easily be compared with the information from the earliest contexts at Winchester or at Hamwih, sites which have provided the largest sample of middle and later Saxon plant remains from the Wessex region (in the case of Winchester from pits and wells associated with monastic and religious occupation).

Conclusion

It has not been the aim of this paper to discuss the effects of growing different crops on the societies being studied. Such aspects discussed by Champion (1980: 352) and Bradley (1978: 32-34) clearly lie outside the scope of this paper.

The archaeological evidence for crops from Wessex during the Iron Age, Roman and Saxon periods is now far more complex than previously indicated, especially in the light of those species that were previously thought to have been late Iron Age introductions, but that were clearly present from the earliest Iron Age. Likewise many crops considered as introductions during the Saxon period clearly were present from Roman times, although no evidence for continuity of cultivation has yet been established.

Major changes in crop agriculture, the balance between the different species being cultivated, can be observed during the 3rd and 2nd centuries BC. Similar changes in the cereal crops grown were also taking place towards the end of the Roman period. Such changes in crop cultivation may reflect wider patterns of economic changes as indicated by Reece (1980: 90).

It has been shown that within the Wessex area defined in this paper that differences existed not only in the range of crops grown through time but the range grown throughout the region at a single point in time. It is particularly difficult to distinguish agricultural activities particularly associated with rural or urban areas until the mid to late Saxon period. Crops were quite clearly grown and processed in the Roman and the later medieval towns. Differences between the crops represented on the different urban sites can be seen both in the Roman and the late Saxon periods. Differences between the crops represented on rural sites occupying the same geographical niche often within a mile or so of one another has also been established for the Iron Age and later periods.

It is to be hoped that this paper has indicated the range and volume of the current work in Wessex and the nature of the problems in identifying and interpreting past crop cultivation from the archaeological record.

Acknowledgements

I would like to thank the following people for their help and advice during the preparation of this paper and for their allowing me to use unpublished data in this synthesis: Gordon Hillman, Martin Jones, Mick Monk, Peter Murphy, and Jane Renfrew. I would also like to thank Tim Champion and Jennie Coy for constructive comments. None of these people should in any way be held responsible for the views and information contained in this paper.

Bibliography

Addyman, P. V. (1976) Archaeology and Anglo-Saxon Society. In: Problems in Economic and Social Archaeology (eds G. de G. Sieveking, Longworth, I. H. and Wilson, K. E.) pp. 310-322.

Applebaum, S. (1972) Crops and plants. In: The agrarian history of England and Wales 43-1042 A.D. (ed Finberg, H. P. R.) Cambridge: University Press pp. 108-121.

Arthur, J. R. B. (1977) The plant remains. In: Excavations at Bishopstone Sussex (ed Bell, M.) Sussex Archaeological Collections 115 273-275.

Akroyd, W. R. and Doughty, J. (1964) Legumes in human nutrition. The Food and Agricultural Organisation of the United Nations Nutritional Studies 19.

Boase, W. N. (1918) Flax and flax fibre cultivation. The Scottish Journal of Agriculture 1 140-147.

Bradley, R. (1975) Salt and settlement in the Hampshire Sussex borderland. In: Salt: Study of an Ancient Industry (ed de Brisay, K. W.) pp. 20-25.

Bradley, R. (1978) The Prehistoric Settlement of Britain. London: Routledge.

Branigan, K. (1976) Villa settlement in the west country. In: The Roman West Country (eds Branigan, K. and Fowler, P. J.) Newton Abbot: David and Charles pp. 120-121.

Begler, E. B. and Keatinge, R. W. (1979) Theoretical goals and methological realities: problems in the reconstruction of prehistoric economies. World Archaeology 11 208-226.

Brailsford, J. (1949) Excavations at Little Woodbury, Parts IV & V. Proceedings of the Prehistoric Society 15 156-168.

Champion, T. C. (1979) The Iron Age: Southern Britain and Ireland. In In: Introduction to British Prehistory (eds Megaw, J. V. S. and Simpson, D. D. A.) pp. 344-432.

Cunliffe, B. W. (1978) Iron Age Communities (2nd Edition) London: Routledge and Kegan Paul.

Dennell, R. W. (1972) The interpretation of plant remains: Bulgaria. In: Papers in Economic Prehistory (ed Higgs, E.) pp. 149-159.

Dennell, R. W. (1976) Prehistoric crop cultivation in southern England. The Antiquaries Journal 56 11-23.

Dennell, R. W. (1978) Early Farming in South Bulgaria from the VI to the III Millennia B.C. B.A.R. 45.

Dennell, R. W. (1977) In: Gatcombe: The Excavation and Study of a Romano-British Villa Estate, 1967-1976. (ed Branigan, K.) B.A.R. 44.

Dewer, H. S. (1965) Note on a new range of buildings and a Roman well at Low Ham, Somerset. Dorset Notes and Queries 27 58-60.

Evans, A. and Jones, M. (1980) The plant remains. In: Gussage All Saints, an Iron-Age Settlement in Dorset (ed Wainwright, G. J.) Department of the Environment (London) Archaeological Report 10 pp. 172-175.

Fasham, P. J. and Monk, M. A. (1978) Sampling for plant remains from Iron Age pits: some results and implications. In: Sampling in Contemporary British Archaeology (eds Cherry, J. F., Shennan, S. and Gamble, C.) B.A.R. 50 pp. 363-371.

Field, N. H., Mathews, C. C. and Smith, I. F. (1964). New Neolithic sites in Dorset and Bedfordshire with a note on the distribution of Neolithic storage pits in Britain. Proceedings of the Prehistoric Society, 30 352-381.

Fowler, P. J. (1976) Farms and field in the Roman west country. In: The Roman West Country (eds Branigan, K. and Fowler, P. J.) Newton Abbot: David and Charles pp. 162-182.

Frere, S. (1978) Editorial. Britannia 9, IX-X.

Fulford, M. (1977) Pottery and Britain's trade in the late Roman period. In: Pottery and Early Commerce (ed Peacock, D. P. S.) pp. 35-76.

Godwin, H. (1956) History of the British Flora (1st Edition). Cambridge: University Press.

Godwin, H. (1967) The ancient cultivation of hemp. Antiquity 41 42-49.

Godwin, H. (1975) History of the British Flora (2nd Edition). Cambridge: University Press.

Green, F. J. (1979a) The plant remains. In: Excavations at 1 Westgate Street, Gloucester (eds Heighwey, C. M., Garod, A. P. and Vince, A. G.) Medieval Archaeology 23 186-190.

Green, F. J. (1979b) Medieval Plant Remains: Methods and Results of Archaeobotanic Analysis from Excavations in Southern England with Especial Reference to Winchester and Urban Settlements of the 10th-15th Centuries. Unpublished M.Phil. Thesis, Southampton University.

Green, F. J. (1979c) Phosphatic Mineralisation of seeds from archaeological sites. Journal of Archaeological Science 6 279-284.

Green, F. J. (1979d) Collection and interpretation of botanical information from medieval urban excavations in southern England. In: Festschrift Maria Hopf (ed Korber Grohne U.) Archaeo-Physika 8 39-55.

Green, F. J. (Forthcoming, a) The plant remains from Old Down Farm.

Green, F. J. (Forthcoming, b) The plant remains from Newbury.

Green, F. J. (Forthcoming, c) In: Winchester Studies X (ed Renfrew, J. M.).

Green, F. J. (Forthcoming, d) Problems of interpreting differentially preserved plant remains from excavations on medieval urban sites (Proceedings of the Urban Environment conference) York: C.B.A.

Green, F. J. (In Preparation) The plant remains from excavations in the northern Suburbs of Winchester.

Green, F. J. (In Preparation) The plant remains from Abbotstone Down.

Grieg, J. (1971) 'Seeds'. In: Excavations at Fishbourne 1961-1969 (ed Cunliffe, B.) Volume 11, The Finds. Reports of the Research Committee of the Society of Antiquaries of London 27 376.

Halstead, P., Hodder, I. and Jones, G. (1978) Behavioural archaeology and refuse patterns: A case study. Norwegian Archaeological Review II 118-131.

Harding, D. W. (1974) The Iron Age in Lowland Britain London: Routledge and Kegan Paul.

Helbaek, H. (1940) Studies on prehistoric and Anglo-Saxon cultivated plants in England. Proceedings of the Prehistoric Society 6 176-178.

Helbaek, H. (1952) Early crops in southern England. Proceedings of the Prehistoric Society 18 194-233.

Hillman, G. (1978) Remains of crops and other plants from Carmarthen (Church Street). Cambrian Archaeological Association Monographs and Collections 1 107-111.

Hillman, G. (1978) On the origins of domestic rye Secale cereale: The finds from aceramic Can Hasan III in Turkey. Anatolian Studies 28 157-174.

Hillman, G. (Forthcoming) Grain processing at 3rd century Wilderspool. In: Excavations at Wilderspool (eds Hinchliffe, J. and Williams J. H.) London: HMSO.

Hillman, G. (Forthcoming) Spelt kilning at Catsgore, Somerset. In: Excavations at Catsgore, Somerset 1970-73 (ed Leech, R. H.) Bristol: CRAAGS

Hubbard, R. N. L. B. (1975) Assessing the botanical component of human palaeoeconomies. Bulletin of the Institute of Archaeology (London) 12 197-205.

Hurst, H. (1978) Saxon and early medieval viticulture at Gloucester. The Antiquaries Journal 43 162.

Jesson, K. and Helbaek, H. (1944) Cereals in Great Britain and Ireland in Prehistoric and early historic times. Det Kongelige Danske Videnskaernes Selskab Biologiske Skrifter III 2.

Jones, B. (1975) The north western interface. In: Recent Work in Rural Archaeology (ed Fowler, P. J.) Bradford-on-Avon: Moonraker Press pp. 93-105.

Jones, M. (1978) The plant remains. In: The Excavation of an Iron-Age Settlement, Bronze Age Ring Ditch and Roman Features, at Ashville Trading Estate, Abingdon, Oxfordshire 1974-70 (ed Parrington M.) Oxford Archaeological Unit Report 1, C.B.A. Research Report 28 pp. 93-110.

Jones, M. (1980) Carbonised Cereals from Grooved-Ware Contexts. Proceedings of the Prehistoric Society 46.

Knorzer, K. H. (1966) Uber funde Römisher importfrüchte in Novaesium, (Neusss Th.). Bonner Jahrbucher 166 433-443.

Knorzer, K. H. (1968) 6000 Jährige Geschichte der Getreideaahrung im Rheinland. Decheniana 119 113-124.

Knorzer, K. H. (1976) Datierung Durch Grossrestanalyse. Folio-Quaternaria 47 60-63.

Liversidge, J. (1968) Britain in the Roman Empire. London: Routledge and Kegan Paul.

Monk, M.A. (1978) The Plant Economy and Agriculture of the Anglo-Saxons in Southern Britain: With particular reference to the Mart Settlements at Southampton and Winchester. (Unpublished) M.Phil. Thesis, Southampton University.

Murphy, P. L. (1977) Early Agriculture and Environment on the Hampshire Chalklands, Circa 800 B.C. - 400 A.D. (Unpublished) M.Phil. Thesis, Southampton University.

Platt, C. (1973) Medieval Southampton: The Port and Trading Community A.D. 1000-1600. London: Routledge and Kegan Paul.

Reece, R. (1980) Town and Country: The End of Roman Britain. World Archaeology 12 77-92.

Reid, C. (1917) Plants, wild and cultivated. In: The Glastonbury Lake Village (eds Bulleid, A. and Gray, H. St. G.) Glastonbury Antiquarian Society, 2 625-630.

Reynolds, P. J. (1978) Archaeology by experiment: a research tool for tomorrow. In: New approaches to our Past (eds Darvill T., Parker Pearson M., Smith R. W. and Thomas R. M.) pp. 139-155.

Reynolds, P. J. (1979) Iron-Age Farm: The Butser Experiment. London: British Museum.

Rivet, A. L. (1964) Town and Country in Roman Britain. London.

Stanes, R. (1969) In: The South West and the Land (eds Havinden, M., King C. M. and Stanes R.) Exeter Papers in Economic History 2 43-55.

Vancouver, C. (1810) General View of the Agriculture of the County of Hampshire. London.

Wacher, J. (1978) Roman Britain. London: Dent.

Wainwright, G. J. (1968) The Excavation of a Durotrigian Farmstead near Tollard Royal on Cranbourne Chase, Southern England. Proceedings of the Prehistoric Society, 34 102-147.

Willcox, G. H. (1977) Exotic plants from Roman waterlogged sites in London. Journal of Archaeological Science 4 269-282.

Willerding, U. (1970) Vor- und Frühgeschichtliche Kulturpflanzenfunde in Mitteleuropa. Sonderdruck Auf Neue Ausgrabungen und Furschungen in Niedersachsen 5 287-375.

Wilson, D. G. and Conolly, A. P. (1978) Plant remains including the evidence of hops. In: The Graveney Boat: a 10th Century Find from Kent. (ed Fenwick, V.) B.A.R. 53 pp. 133-150.

8. IRON AGE, ROMANO-BRITISH AND ANGLO-SAXON ANIMAL HUSBANDRY - A REVIEW OF THE FAUNAL EVIDENCE

Mark Maltby

To review the evidence for animal husbandry over a period that covers the majority of two millennia should be a daunting task. After all, the collection and recording of animal bones began, albeit spasmodically, in the late 19th century and a few detailed analyses of certain aspects of faunal studies were published (e.g. Pitt-Rivers 1892). It is therefore sad to relate that the past century has seen so little advance in our knowledge of the pastoral economy in Britain between the Iron Age and the Saxon periods. The lack of progress has placed severe limitations upon any attempts to synthesize the information from any of the periods. Richard Bradley was accurate with this assessment of the prehistoric data:

> "Despite a widespread reluctance to take any notice of bones, these remain the most informative evidence of pastoral activity. Few sites have been dug with any attempt to secure adequate samples, and for fewer still are there any figures for species composition, meat weight, age or sex. Little attention has been paid to the social, ritual or purely physical factors which can impede understanding of this material, and the majority of the acceptable data relates to a limited period and a limited region" (Bradley 1978: 35).

The best information in fact comes from Iron Age sites in southern England. Yet it is interesting to note that a recent review of this period for southern Britain included only ten sites which possessed detailed published bone reports (Champion 1979: 427). Even those included several that by modern standards are limited in scope. Among them are reports from Stanwick, Yorkshire (Fraser 1954) and the Iron Age levels of Coygan Camp in south Wales (Westley 1967), which consist of brief summaries of the species present but few other details. Similar limitations are found in the short faunal reports from Sutton Walls, Herefordshire (Cornwall 1953) and the hill fort at Rainsborough, Northamptonshire (Banks 1967). Much more information was published about the animal bones from the hill fort at Grimthorpe (Jarman et al. 1968), including a list of the bone elements identified to the major species, a summary of the metrical analysis of sheep and cattle bones and a discussion of the ageing data. Unfortunately only 735 fragments from all deposits were assigned to species and this and all the other reports mentioned above were based on small samples of less than 1,000 fragments identifiable to species thus limiting the amount of detailed information that could be acquired from them. The other sites listed by Champion produced larger samples but in some cases the reports are limited. The animal bones from the late Iron Age village at Glastonbury (Bulleid & Gray 1917) preserved extremely well in the deposits and in areas were found in great quantities. Not all the bone, however, was used in the analysis and the sample was selected mainly for complete and measurable bones. The report was concerned principally with this metrical analysis.

Over 1,300 fragments were identified to the major domestic species from the bones recovered from the excavations at Hawk's Hill, Surrey (Carter et al. 1965). The report included a summary of the results of ageing of the teeth of cattle and sheep. On the other hand, no list of bone elements was published nor any details of metrical analysis. The report from Eldon's Seat, Encombe, Dorset (Cunliffe & Phillipson 1968) did include a more detailed record of the types of bone fragments represented. It also summarised the ageing evidence from the mandibles. In addition there is a brief discussion of possible butchery practices but this analysis lacks any consideration of the role of preservation in the formation of the faunal sample and the possibility of lateral variation of the material. The lack of intra-site analysis can also be raised as a criticism of the report from the hill fort of Croft Ambrey, Herefordshire (Whitehouse & Whitehouse 1974). Nor is there any record of the types of bone identified in the substantial Iron Age sample, although three methods of quantification were employed to obtain relative frequencies of the major species. The report did include, however, a detailed analysis of tooth ageing data, particularly of sheep/goat, a discussion of butchery methods and a list of each measurement taken.

The most comprehensive faunal analysis listed by Champion was that of the material from the excavations at the Ashville Trading Estate, Abingdon, Oxfordshire (Wilson 1978). The substantial sample was examined in much greater detail than in many of the earlier reports and the analysis included discussions of preservation of the material; the intra-site analysis of bones recovered from pits and ditches; an examination of the frequencies of different parts of the carcass; a detailed discussion of ageing evidence; a summary of the metrical analysis; an excellent discussion of the ageing data and a detailed record of any observed pathology and complete or partial skeletons.

To the above list we may add the report from late Iron Age levels at Hardingstone, Northamptonshire (Gilmore 1969) and the one for the bones from the early Iron Age promontory fort at Budbury, Wiltshire (Westley 1970), which included 2,500 fragments identified to species, although the report contained little other information. More recent reports have appeared notably for the hill fort at Winklebury (Jones, R. 1977). This report included a breakdown of fragments into elements of the body and a detailed discussion of the butchery evidence. The largest published Iron Age sample to date belongs to Gussage All Saints, Dorset (Harcourt 1979), in which 15,500 fragments were identified. The report included a summary of metrical analysis of the major species, a comparison of two methods of ageing data, an assessment of the relative meat weights represented for each species and a discussion of pathological and butchery evidence. Unfortunately, once again there was no detailed intra-site analysis nor any quantification of the types of bone represented in any of the periods.

In addition to these reports there are several smaller samples of limited value and several more important assemblages as yet unpublished that will be incorporated into this review. Even with these, however, the data base for the Iron Age is very limited.

At first sight the picture from the Romano-British period appears more optimistic. In his comparative survey King (1978) was able to list over 90 sites which had some quantification of Roman material. With the addition of several reports since then, we have over 100 sites where evidence of animal bones is available. However, although we possess a good quantity of sites, the quality of the samples themselves leaves much to be desired. Most of the reports are little more than cursory appendices with a minimum of detail and are often divorced from the discussion of other aspects of the site. Of the reports listed by King only a dozen are based on more than 2,000 fragments identified to species and once again several of these are recorded and discussed inadequately. Three of them (Rotherley, Woodyates and Woodcuts, all on Cranbourne Chase, Dorset) were excavated in the late 19th century and their faunal reports were concerned mainly with metrical analysis (Pitt-Rivers 1892). The same can be said of the Corstopitum (Corbridge) report (Meek & Gray 1911). The published reports of the large samples from Old Winteringham and the villa at Winterton, Lincolnshire (Higgs & Greenwood 1976) contain few details other than the number of fragments identified. The early reports from the villa at Shakenoak Farm, Oxfordshire (Brodribb et al. 1968; 1971; 1972) suffer from the same limitation, although the later reports by Cram (1973; 1978) are much more comprehensive and the analyses of intra-site variability, ageing data, measurements, butchery and pathology are accompanied by detailed tabulations of the data.

The Romano-British levels from Coygan Camp contained over 4,700 fragments identified to species and the brief report lists these fragments according to locality but only gives few other details (Westley 1967). The report on over 2,000 identifiable fragments from the early Roman fortress at Longthorpe, near Peterborough included a full list of the measurements taken, the number of the different skeletal elements represented and a brief discussion of butchery and ageing evidence (Marples 1974). The 1970-75 excavations at Vindolanda produced a large sample of bone and the analysis of it contained a detailed breakdown of the skeletal elements, a summary of ageing data, metrical analysis of some of the more commonly recovered bones and some discussion of the pastoral economy of the settlement (Hodgson 1977). The report from Fishbourne (Grant 1971) is broadly similar in range and detail. The report from Portchester Castle (Grant 1975) examined the largest Romano-British faunal collection published to date (nearly 29,000 fragments were identified). Apart from only a brief summary of the metrical data, the rest of the analysis covers aspects of species representation, butchery, ageing and pathology in impressive detail.

Since King's publication several other fairly substantial Romano-British faunal samples have been analysed from excavations at Exeter (Maltby 1979), the fort at Watercrook, Cumbria (Fifield 1979), Dorchester, Oxfordshire (Grant 1978), Nazeingbury, Essex (Huggins 1978) and Frocester Court, Gloucestershire (Noddle 1979). In addition, several important collections are being studied but even when these are published, the body of data is still depressingly small considering the number of excavations that have taken place on such sites. Our combined knowledge derived from archaeological data and classical sources is still meagre, as is demonstrated in several comparatively recent general reviews of the period (e.g. Wacher 1980; Rivet 1964). Applebaum

(1972) in his more detailed study of Romano-British agriculture relied heavily on a few small faunal samples for the basis of his evidence of the pastoral economy.

Information about the animal bones from the Anglo-Saxon period has been enhanced by the recent publication of the first of the reports from Hamwih (Southampton), which with nearly 50,000 fragments identified to species confortably exceeds the previously published data of this date (Bourdillon & Coy 1980). The analysis covers comprehensively aspects of quantitative analysis, ageing, metrical data, butchery and pathology. The Saxon levels at Portchester Castle produced another large sample of nearly 23,000 fragments subdivided into three phases and published in similar detail to their Roman counterparts (Grant 1976). The substantial animal bone assemblage from North Elmham Park, Norfolk, has recently been published in full (Noddle 1980). Apart from these reports the archaeological evidence for animal husbandry is once again poor. Clutton-Brock (1976) has compared aspects of faunal assemblages from five sites in southern England (North Elmham, Thetford and Sedgeford in Norfolk, Sandtun, Kent and Mawgan Porth, Cornwall) paying particular attention to species representation and metrical analysis of the major domesticates. Noddle (1975) has also produced a comparative survey of Saxon and medieval material from eight sites in southern England, including North Elmham, Hereford and New Wintles of Saxon date. Bone reports with some quantification have been produced for Dinas Powys (Haglund-Calley and Cornwall 1963; Alcock 1975), Maxey (Seddon et al. 1964), Shakenoak Farm (Brodribb et al. 1968; 1972), Glastonbury Tor (Harcourt 1970), Cadbury Congresbury (Noddle 1970), Whitehall (Chaplin 1971), Walton, Aylesbury (Noddle 1976), Cheddar (Higgs et al. 1979), St. Peter's Street, Northampton (Harman 1979), Yeavering (Hope-Taylor 1979), Durham (Rackham 1979) and Ramsbury (Coy 1980 in press). Considering the diversity of settlement type, the large timespan and the wide geographical range of these sites, the number of samples is again limited. In addition some of these samples are small or not easily comparable. Some documentation of the Saxon period concerning pastoral husbandry has survived but the evidence again is piecemeal (e.g. Finberg 1972).

It is clear therefore that the data base for this study is a poor one. What questions can such material answer? That sort of hypotheses can be tested? It should already be apparent that there are severe limitations imposed by our insufficiently detailed knowledge of temporal, regional, ecological and settlement variability. No area of England or Wales can claim to have an adequately sampled sequence of animal bones that take these important variables into account. In addition, it will be shown that various archaeological and methodological problems exist in the interpretation of the material that have rarely been considered in attempted reconstructions of pastoral economies, whether of a single settlement, region or period. Inevitably, therefore, the questions we can attempt to answer at the moment are of a limited nature. There is now at least sufficient evidence to observe the presence or absence of various domestic species in these periods. To some extent it is possible to monitor the fluctuations in species representation, observe trends in the exploitation of the principal domestic animals and examine changes in stock size. In some cases it is possible to reassess assumptions that have been made about these topics in the light of more recent evidence. In some instances we can at least begin to place the observed changes (and evidence of stability) into a temporal and regional framework and begin perhaps to construct more sophisticated models to explain developments in the pastoral economy by relating the archaeozoological evidence to other aspects of archaeological and historical evidence. The rest of this article will therefore attempt to review our present knowledge of the major aspects of the faunal record and in conclusion will try and suggest ways this information can be used in the development and testing of models of a more general nature.

The Domestic Stock

The apparently simple question of what domestic species were kept in the Iron Age, Romano-British and Anglo-Saxon periods is not as straightforward as it seems. There can be no doubt about the importance of cattle and sheep throughout and to a lesser extent pig, horse and dog. The problem lies with the possible presence of other domestic species, in particular the goat, the donkey, the mule and poultry.

The Goat

In most general reviews of the periods involved where animal husbandry receives a mention, goats are regarded as one of the principal domestic species kept (e.g. Bradley 1978: 35; Wacher 1980: 109; Finberg 1974: 77). Has this been demonstrated archaeologically? The problem for the archaeozoologist has always been the difficulty of distinguishing the bones of goat from sheep from fragmentary material. It has been shown, however, that several bones can be assigned to species provided the diagnostic features have survived (Boessneck et al. 1964; Boessneck 1969). Metrical analysis particularly on the metapodia has been used successfully to separate the species (e.g. Payne 1969). Of course not all fragments can be distinguished and the recording of fragments designated as "ovicaprine" or "sheep/goat" will remain a feature of faunal reports. Nevertheless, it should now be possible to obtain some idea of whether goats were present in any numbers at least from the larger samples. Certainly from excavations of Iron Age sites in England, the present evidence would suggest that goats were kept at most in very small numbers. Although goat was represented in all three phases at Gussage All Saints, Dorset, only 25 specimens were identified in a sample of over 15,000 fragments and many of these were horn cores that had been sawn from the skull (Harcourt 1979: 153). At Eldon's Seat, Dorset no caprine features were noted on 12 horn cores and 96 metapodia (Cunliffe & Phillipson 1968: 227). Further east in Hampshire, several samples covering all periods of the Iron Age, notably from Winklebury (Jones, R. 1977: 60), Winnall Down (Maltby n.d. 1), Balksbury 1973 (Maltby n.d. 2) and the banjo enclosure in Micheldever Wood (Coy n.d.) all produced negligible amounts of goat bones in substantial samples. No goat bones were identified from Hawk's Hill, Surrey compared to 18 positively identified as sheep (Carter et al. 1965: 41). From the Ashville excavations the presence of goat was indicated by a single humerus fragment and none of the horn cores recovered were assigned to goat (Wilson 1978: 111). Only sheep were identified at Grimthorpe (Jarman et al. 1968: 182). The situation at Croft Ambrey is less clear. Horn cores and a complete skeleton of goat were identified together with several other fragmentary bones. However, the authors assumed that the majority of the ovicaprine sample belonged to sheep, using the samples from Eldon's Seat and Grimthorpe as parallels; (Whitehouse & Whitehouse 1974: 215). Finally, only about half a dozen goat horn cores were selected for measurement from the large ovicaprine sample from Glastonbury (Bulleid & Gray 1917: 659). None of the largest Iron Age samples investigated to date have therefore produced substantial numbers of goat bones and several have no evidence of the species at all. It is possible that more goats were kept but were not utilised for meat and therefore are not found in deposits containing bones mainly derived from butchery and cooking waste but this is perhaps unlikely. It certainly seems that goats were exploited for meat only rarely and in some cases may not have been kept at all.

In the Roman period the pattern is similar, although goats are generally slightly better represented. Again, restricting the review to the larger samples, metrical analysis of sheep and goat metacarpi distinguished two specimens of goat from 19 of sheep at Vindolanda (Hodgson 1977: 11). All eight distal metapodia on which metrical analysis was undertaken at Watercrook, Cumbria belonged to sheep (Fifield 1979: 309). Only one incomplete skull of goat was identified at Corstopitum (Meek & Gray 1911: 119). Apart from these military sites in northern England goat bones have been found only rarely elsewhere. The villa at Shakenoak Farm, Oxfordshire produced some evidence of goat. Indeed it was stated on the basis of the number of horn cores identified that sheep and goats were kept in the proportion of six goats to two sheep (Cram 1978: 123). This would be unusual and the metrical proportions of the few complete metapodia (Cram 1978: 152-153) are all more typical of sheep than goat. Reliance on horn core evidence alone in distinguishing the two species can bias the results towards goat since all goats possess horns whereas some sheep do not. In addition the larger goat horns may have been more highly valued for working and the distribution of their horn cores may be influenced by this factor. The Roman deposits at Exeter produced six goat horn cores and ten of sheep. Of the measurable distal metapodia on the other hand, only one belonged to goat and 12 to sheep (Maltby 1979: 41). In the excavations of the early Roman fort at Longthorpe no goat bones were identified in a ovicaprine sample of nearly 600 bones (Marples 1974: 123). A minimum of six goats were identified from two late Roman wells at Tripontium, Warwickshire but this assemblage was unusual and included several complete skeletons of various species (Noddle 1973).

Classical references to the importance of goats in Britain appear to be restricted to Diocletian's price-fixing edict, which referred to the birrus Britannicus—a cloak reputed to be made from goat's hair (Rivet 1964: 123). The archaeological evidence suggests that goats were still of little importance in Roman Britain. They certainly provided very little of the meat diet.

The Anglo-Saxon period does at least provide some documentary evidence for the keeping of goats (Finberg 1972: 497, 515) but references are rare (Finberg 1974: 77). Archaeological evidence would suggest that they continued to play only a small role as meat producers, although it is conceivable that they were kept mainly for their milk and skins and not butchered for meat. Goat bones were found consistently in small numbers at Hamwih (Bourdillon & Coy 1980: 111) but there is no evidence that large numbers of goats were exploited at any of the settlements excavated to date.

A wider selection of samples may reveal regional variation in the goat's distribution or importance but as the evidence now stands it would be misleading to regard the goat as a principal stock animal in any of the periods.

Mule and Donkey

Most equid bones found on archaeological sites have been identified as horse. There is perhaps evidence, however, that both mules and donkeys were present from at least Roman times. A mandible of a mule has been identified from 2nd century AD levels from the excavations at the Billingsgate Buildings in the City of London. Although this specimen may have belonged to an imported animal, its presence is interesting since there is good documentary evidence for the mule's employment as a draught animal and a beast of burden in the Roman Empire (Armitage & Chapman 1979). Donkey has also been recorded occasionally on Romano-British sites, for example at Tripontium (Noddle 1973) and possibly at the villa at Frocester Court, Gloucestershire (Noddle 1979: 52). Because of the similarities of the skeletons of ponies, donkeys and mules, it is possible that mules and donkeys have been misidentified in some samples. Future studies of equid material should have regard for their possible occurrence.

Poultry

The problem of differentiating the bones of the various species of poultry and their respective wild versions makes this discussion rather difficult. More research is needed on the metrical analysis and morphological distinctions of their skeletons before a clear picture can emerge. Bird bones are fragile and do not survive well but the rare occurrence of bird bones including possible domesticates on Iron Age sites (Table 1) is probably significant, since all the sites listed contained some deposits that preserved bone well. Only Winklebury contained substantial numbers of domestic fowl bones but these include the bones of two skeletons (Jones, R. 1977: 64). On all the other Iron Age sites on which bird bones have been quantified, those of domestic species were rare or absent and there is no certainty even that they belonged to the domesticated varieties. Julius Caesar mentioned that the Britons kept chickens and geese but had a taboo on eating their flesh (Rivet 1964: 125). Certainly there is as yet very little archaeological evidence that poultry was eaten by the Iron Age inhabitants of southern England. They may have provided eggs and feathers but contributed very little, if at all to the meat diet.

In all the major Romano-British samples examined to date only domestic fowl were present consistently in any numbers (Table 1). Once again it is not clear how many of the few bones of geese or duck that have been found belonged to domesticated birds. This situation had changed by the Anglo-Saxon period. In addition to domestic fowl, domestic geese have been found in quantity in most of the larger samples. Domestic duck/mallard has, however, been rarely found with the possible exception of Portchester Castle (Eastham 1976). Domestic geese were also mentioned in some documentary records (Clutton-Brock 1975: 388). The keeping or at least the eating of geese appears on present evidence to have begun on a large scale only in this period. A greater regional and chronological selection of samples should be able to clarify the areas and date of this development.

TABLE 1: RECORDED OCCURRENCES OF BONES OF POULTRY

Site	Source	Fowl	Goose	Duck
a) Iron Age				
Ashville	Bramwell (1978)	R	A	R
Balksbury 1973	Maltby (n.d. 2)	A	R	R
Budbury	Westley (1970)	R	R?	A
Glastonbury	Bulleid & Gray (1917)	A	?	A
Gussage All Saints	Harcourt (1979)	R?	R?	R?
Winnall Down (R17)	Maltby (n.d. 1)	A	R	R
Winklebury	Jones, R. (1977)	P	A	A
b) Romano-British				
Baylham House	Maltby (n.d. 3)	R	R	R
Coygan Camp	Westley (1967)	R	R	?
Exeter	Maltby (1979)	P	R	R
Fishbourne	Eastham (1971)	P	R	R
Frocester Court	Bramwell (1979)	P	R	R
Longthorpe	Marples (1974)	P	R	R
Portchester Castle	Eastham (1975)	P	R	R
Shakenoak Farm	Marples (1972, 1973)	P	R	R
Watercrook	Fifield (1979)	R	A	A
c) Anglo-Saxon				
Durham	Rackham (1979)	P	P	A
Hamwih	Bourdillon & Coy (1980)	P	P	R
Hereford	Noddle (1975)	R	R	?
North Elmham	Noddle (1975)	P	P	?
Portchester Castle	Eastham (1976)	P	P	P
Ramsbury	Coy (1980 in press)	P	R	R

Fowl = domestic fowl; Goose = domestic goose/grey lag goose (*Anser anser*); Duck = domestic duck/mallard (*Anas platyrhynchos*); A = absent; R = rare (less than 1% of identified fragments of all species); P = present (more than 1% of identified fragments).

Quantitative Assessments of Species Representation

The interpretation of the relative number of animal bones of different species has frequently been a principal component of faunal studies. Its main objectives have been the estimation of how much each species contributed to the meat diet and the assessment of the relative numbers of the different stock kept. There have been several attempts to generalise about the changes in the composition of the domestic herds in a region or within a period and these are worthy of consideration here.

Cunliffe (1978: 183-185) has stated that in broad terms there appears to have been a gradual increase in the numbers of sheep relative to cattle during the first millennium BC in southern England. He compared the bone samples from the Cranbourne Chase sites of South Lodge Camp, Martin Down Camp and the Angle Ditch, in which the percentage of cattle varied between 48-67% (Pitt-Rivers 1898) and the late Iron Age village at Glastonbury (Bulleid & Gray 1917) where sheep bones were reported to outnumber cattle by a ratio of 17 to 1. Cunliffe also noted that the percentage of sheep increased in the later phase at Eldon's Seat (Cunliffe & Phillipson 1968). He linked the increase of sheep with the spread of downland arable in the Iron Age and the need to manure this land to maintain large scale grain production. In addition Cunliffe suggested that the large variation in the relative number of pig bones on Iron Age sites may be a reflection of ecological variation. The more suitable habitats for pigs would have been near woodlands rather than the open downland. Hence pig bones in assemblages on settlements on downlands were low.

Bradley (1978: 37-38) also pointed out the variability of pig remains on Iron Age sites citing Glastonbury and Croft Ambrey as extreme examples of low and high representation of the species in relation to sheep and thus questioned Clark's (1947) hypothesis that sheep gradually replaced pigs throughout prehistoric Britain as the woodlands were cleared. Like Cunliffe, Bradley suggests that regional and environmental factors were important considerations in the pig's importance. He also considered that there was a general increase in the proportion of sheep from the later Bronze Age to the later Iron Age.

The most comprehensive survey of Romano-British faunal material has been published by King (1978). His main conclusions from data from over a hundred sites can be summarised as follows. There was a distinct trend away from the keeping of sheep in the Roman period, probably due to the presence of more settlements in areas more suitable for cattle and pigs. Assemblages with more than 30% sheep bones were limited mainly to the lowland area of England and to dry, light soils. Secondly, the more "Romanised" settlements such as villas, roadside settlements, towns and forts tended to have fewer sheep than the native sites which maintained the Iron Age pattern. Pig bones were more common on "Romanised" settlements than on native sites, again partially indicating the presence of more settlements near woodland but also the probable influence of taxation and other cultural factors, such as the Roman's high regard for pork. Many military deposits contained high proportions of cattle bones and, finally, Anglo-Saxon faunal material was not markedly different from Roman except for the high proportion of pig bones on a few sites. There was a trend which saw the increase in the proportion of sheep again towards the high levels attained in the Middle Ages.

In her comparison of faunal samples from five Anglo-Saxon sites, Clutton-Brock (1976) contrasted the apparent importance of pigs as seen in the documentary sources and their poorer representation on archaeological sites. She suggested that their carcasses may have been commonly boned and salted for bacon and their bones were thus not found among other kitchen waste. She also concluded that cattle and sheep became gradually more important during the period.

Problems of Quantification (a) Methodology

These are examples of broad comparisons of species representation. More limited assessments have often been made of data from different phases of one site or between the material from a few contemporary sites. How reliable are these comparisons? It is important to emphasise that the methods used in the estimations of the relative abundance of each species from archaeological samples need not necessarily produce the same results. At Hamwih for example, cattle provided 52.5% of the domestic stock fragments, 72.1% of the weight of the identified bones and 31.4% of the minimum number of individuals estimated from the mandibles (Bourdillon & Coy 1980: 84-85). Corresponding fluctuations in the percentage of other species were found. Similar variations in results have been noted from samples from Portchester Castle (Grant 1975: 379-382) and Exeter (Maltby 1979: 6). None of the methods necessarily reflects accurately the proportion of different species kept. Indeed, as Bourdillon & Coy (1980: 86) have pointed out, any ratios derived from archaeozoological data reflect only the dead animals. Any reconstruction of the living herds must also take into account the age structures and life expectancies of the different species. Nevertheless, although absolute reconstructions are probably impossible employing these methods of quantification, comparisons of relative proportions of animals can theoretically be attempted provided the methods of analysis used are compatible. This has not always been the case. The use of the Glastonbury data to emphasise the importance of sheep in the late Iron Age is particularly unfortunate in this respect. A close consideration of the Glastonbury report reveals that the bones used in the estimation of species representation were selected from a much larger total sample (Bulleid & Gray 1917: 642). Particular interest was paid to the metrical analysis of complete skulls and longbones. Of the 181 bones of cattle recorded, a high proportion can be accounted for in the measurement tables and it is clear from the discussion of cattle in the text that complete limb bones were preferentially selected. Yet it was previously stated that most of the cattle bones were fragmentary (Bulleid & Gray 1917: 652-654, 641). It is also clear that a large number of the 3, 013 sheep bones were complete, although many belonged to immature animals and were not subsequently used in the metrical analysis. Unlike cattle, "one noticeable feature amongst the limb-bones is the almost entire absence of breakages for the purposes of extracting the marrow" (Bulleid & Gray 1917: 655). The bias in the selection of complete bones would therefore have favoured sheep and it is misleading to compare the relative percentages of species represented in the Glastonbury sample with those from elsewhere, where more fragmentary material was counted. Similarly, Payne (1972b: 79-80) has noted that the methods of counting bones vary between individual faunal analysts and these methods are not always explained, thus making direct comparisons between the samples difficult.

Payne (1972a: 1975) has also demonstrated that poor recovery techniques can bias faunal samples towards the collections of large mammal bones. Although the results from sieving experiments from Greece were much more dramatic than for example, Hamwih (Bordillon & Coy 1980: 82-83), the possibility that poor of hurried rescue excavations may have biased some samples in favour of cattle and horse in relation to sheep, pig and smaller species must still be taken into consideration in the comparison of results.

Problems of Quantification (b) Intra-site variability

Another problem in assessing the importance of individual species is that of variation of samples within a settlement. By whatever means quantification is done most excavations produce a tiny fraction of the bones originally deposited and recoverable from a settlement. This can result in misleading interpretations of species representation. The case of the multiperiod settlement on Winall Down (R17) near Winchester, Hampshire is a good example of where such factors can be shown to be important. The site was excavated and rigorously sampled by the M3 Archaeological Rescue Committee under the direction of Peter Fasham. Phase 3 of the settlement was dated from the 7th century to early 3rd century BC and consisted of a subrectangular enclosure ditch with a single entrance, approximately 25 pits and a few postholes and scoops scattered mainly within the enclosure and a large area of quarry scoops and pits to the north of the enclosure. The ditch was excavated at regular intervals by 31 sections and in total 78.5 metres representing 29.4% of the circuit were removed. 21 of the pits produced animal bones and the quarry produced a large number of fragments scattered in its area.

TABLE 2: Bone Fragments of Major Domesticates from Phase 3 Deposits of Winnall Down

Species	Ditch	Pits	Quarry	Others	Total
Cattle	297 (61.4%)	142 (25.0%)	245 (50.9%)	15	699 (44.4%)
Horse	51 (10.5%)	50 (8.8%)	60 (12.5%)	4	165 (10.5%)
Sheep/Goat	105 (21.7%)	314 (55.4%)	148 (30.8%)	22	589 (37.4%)
Pig	31 (6.4%)	61 (10.8%)	28 (5.8%)	3	123 (7.8%)

Table 2 lists the simple fragment counts including shaft fragments and loose teeth of the major domesticates (excluding bones belonging to partial skeletons) for all deposit types. Overall, cattle fragments outnumbered sheep but there was a large range of variation in their percentages in the different deposits. Cattle were much better represented in the quarry and ditch than in the pits. The contrast between the feature types was shown reasonably consistently in the individual pits and ditch sections, although two pits produced abnormal samples (Maltby n.d. 1). The reasons for the variations were a combination of differential preservation and disposal practices. Sheep/goat and pig bones survived less well in the ditch, which contained a higher percentage of eroded bones than the pits. The sample of sheep/goat from the ditch was dominated by loose teeth and shaft fragments of tibia and radius, whereas the more fragile parts of the skeleton were much under-represented. Cattle bones survived better in the ditch than sheep/goat bones because most of them belonged to mature animals and were thus more resilient to erosion. In addition there was evidence that the ditch was used in places as a depository for cattle bones that had been stripped of meat and then dumped. The bones from the quarry were of a different nature again being poorly preserved and containing a high proportion of unidentifiable fragments, loose teeth and other dense fragments of all species, whereas the more porous and less hardy elements were under-represented compared to other feature types (Maltby n.d. 1).

This example shows how much intra-site variability can affect the overall representation of species. The final fragment percentages (Table 2) were dependent on the amount excavated from the different deposits. Had all the ditch been excavated, for example, and the densities and types of bones from the excavated samples been typical of the rest of the ditch, the estimated overall percentage figures would have changed to 53.1% cattle, 10.5% horse, 29.4% sheep/goat and 7.1% pig, representing a 8.7% increase in cattle and a 8% reduction in sheep/goat. Similar variations would be found using other methods of quantification. The problems are compounded if these results are compared to the Phase 4 deposits (Middle Iron Age), from which the vast majority of the bone was recovered from pits. Excluding skeletons, nearly 4,600 bones were recovered from 83 pits. Of the major species, 29% of the fragments belonged to cattle, 9% to horse, 53% to sheep/goat and 10% to pig. As in the previous phase, there was a fair degree of variation in the faunal assemblages in the pits. There was a tendency for outlying pits to contain higher proportions of cattle and horse bones, although any intra-site comparisons in this phase are hindered by the problem of not knowing which pits were open contemporaneously. Was there any change in the relative numbers of animals kept? Overall, the proportion of sheep/goat fragments increased in the latter phase but if only the contents of the pits are compared there was very little difference in species representation. Clearly more detailed examination of the bone types represented is needed before the question can be resolved (Maltby n.d. 1). The example does illustrate, however, the problem of using percentages of bone fragments in comparisons of species representation on one site, where there is clear evidence that different contexts are reflecting either different patterns of waste disposal or presenting different conditions of post-depositional preservation.

There are several other instances where it can be shown that differential deposition of bones has biased samples in favour of a particular species. The late 1st century AD deposits from the infill of the legionary ditch at Exeter produced a concentration of cattle bones consisting predominantly of mandible, skull and metapodia fragments deposited as waste from primary butchery of cattle carcasses. This assemblage contrasted markedly from other samples recovered elsewhere in Roman Exeter and was heavily biased towards cattle (Maltby 1979: 11). A similar concentration of cattle jaws, skull and metapodia was found in an early 2nd century AD pit at Aldgate in London. As at Exeter, the horn cores and most of the meat bones had been removed elsewhere (Watson 1973). A preponderance of cattle horn cores and metapodia and, to a lesser extent, other cattle bones was discovered in late 4th century levels at Angel Court, London (Clutton-Brock & Armitage 1977).

Cattle horn cores dominated a 4th century AD sample from Kingston Hill Farm, Oxfordshire (Wilson 1976). A pit from a 2nd century AD military deposit at Little Chester, Derbyshire contained a high proportion of smashed bones of young cattle (Askew 1961). This assemblage has been interpreted as the remains of bones broken up in the production of bone grease and has parallels with military sites on the continent (King 1978: 225). Concentrations of pig metapodials and phalanges thrown away in quantity as waste have biased samples from Roman Exeter (Maltby 1979: 11-13) and Nazeingbury, Essex (Huggins 1978). These examples all provide an interesting insight into butchery and marketing practices and there are other cases where redistribution of carcasses can be demonstrated. They are all clear cases where the bones recovered do not represent a cross-section of the animals kept. Of course, they can be regarded as atypical samples and omitted from calculations of species representation but we are then faced with the problem of deciding what constitutes a typical sample. The number of bones of each species recorded is dependent upon the method of quantification used, excavation techniques, preservation conditions, discard practices, butchery practices, redistribution of the carcasses and the ages of the animals exploited. All these processes impede the translation of the faunal data into realistic statistics of species present. To understand bone samples better, we have to take all these factors into consideration by studying the types of bone represented and by taking note of the effects of intra-site variability.

Problems of Quantification (c) Inter-site variability

The possible incompatibility of the contents of faunal samples has to be considered particularly when the assemblages from different sites are compared. The problem of sample variability is illustrated in Tables 3-4. These list the principal elements recorded for cattle and sheep/goat respectively in a selection of samples. In order to standardise the figures, the number of fragments of each element is expressed as a percentage of the most commonly occurring element. It is clear that there is a wide range of variation in the composition of the samples of both species and few are very similar. The reasons for the variability can be attributed to one or a combination of the processes outlined above. Whatever the causes, is it correct to compare the total number of fragments obtained from such heterogeneous assemblages and assume that a comparison of the overall fragment figures will reflect the variation in the relative numbers of each species eaten or kept? It is possible, that in some cases such variability may not have biased the figures but this has never been demonstrated satisfactorily. It may be possible to monitor large scale change in species representation by such methods but any subtler variations are difficult to interpret, since, as we have seen, fragment percentages can be affected greatly by other factors. Grayson (1980) and Dennell (1980) are perhaps correct in their suggestion that we should abandon attempts to be too precise in this type of quantification and be satisfied with obtaining some order of ranking of species instead. Certainly the usual methods of analysis are unlikely to produce much more accurate results. There is a need to develop a better understanding of the nature of faunal samples. Ethnoarchaeological studies have shown the complexities involved in the formation of bone assemblages (Maltby 1980). Our interpretations of such material to date have been oversimplistic and our methods inadequate. In this

TABLE 3: Variation of Major Cattle Fragments from a Selection of Iron Age, Romano-British and Anglo-Saxon Samples

Sample/Source	MRP	Jaw	Scp	Hum	R/U	Fem	Tib	Mc	Mt	Tar	Phl
Exeter GS 55-75 AD	29	100	86	66	48	72	69	41	48	31	41
GS 100-200 (Maltby	34	100	50	77	85	50	94	32	41	21	27
CC 200-300 1979)	23	100	57	13	35	35	22	39	13	35	91
GS 300-400	58	100	86	66	72	88	85	31	64	50	81
TS 300-400	84	100	56	31	32	33	52	45	41	41	58
Winnall Down Ph.4 (Maltby	110	100	47	51	72	41	56	21	30	35	26
Phase 6 n.d.1)	91	100	52	52	55	45	46	30	55	68	39
Winklebury (Jones, R. 1977)	69	100	68	29	57	41	52	25	39	29	65
Baylham House 120-200 AD Pits	115	100	37	57	55	43	69	33	48	36	52
Layers (Maltby n.d. 3)	54	100	13	59	52	48	82	17	59	54	78
Longthorpe (Marples 1974)	170	33	21	14	16	14	17	26	12	22	100
Shakenoak Farm Site K Gp.J	37	57	54	22	60	11	30	46	46	78	100
Site K Gp.O (Cram	63	40	21	14	44	8	16	54	64	73	100
Site K Gp.Q 1978)	51	45	20	18	49	4	28	37	55	100	96
Site K Gp.S	60	40	18	7	32	5	33	30	28	78	100
Brancaster (Wall pers. comm.)	226	73	52	17	35	18	21	44	47	29	100
Vindolanda (Hodgson 1977)	445	49	77	20	31	12	20	49	64	57	100
Hamwih (Bourdillon & Coy 1980)	1294	82	75	61	77	59	69	36	38	64	100
Minimum percentage		33	13	7	16	4	16	17	12	21	27
Maximum percentage		100	86	77	85	88	94	54	64	100	100
Mean		79	49	37	50	35	48	35	44	50	75
Standard deviation		27	24	23	18	24	26	10	16	23	28

MRP = number of fragments of most represented bone. Other figures are percentage of most represented bone. Jaw = mandible; Scp = scapula; Hum = humerus; R/U = radius/ulna; Fem = femur; Tib = tibia; Mc = Metacarpus; Mt = metatarsus; Tar = tarsals/carpals; Phl = phalanges.

TABLE 4: Variation of Major Sheep/Goat Fragments from a Selection of Iron Age, Romano-British and Anglo-Saxon Samples

Sample/Source	MRP	Jaw	Scp	Hum	R/U	Fem	Tib	Mc	Mt	Tar	Phl
Exeter GS 55-75 AD	29	42	38	42	66	62	100	24	69	10	14
GS 100-200 (Maltby	68	51	12	22	62	32	100	29	49	6	3
CC 200-300 1979)	17	6	12	35	65	59	76	12	24	100	59
GS 300-400	70	16	27	39	73	57	100	26	39	4	4
TS 300-400	19	79	32	37	58	21	100	32	21	0	0
Winnall Down Ph.4 (Maltby	138	100	23	35	85	50	83	51	65	26	33
Phase 6 n.d.1)	114	100	6	20	46	13	75	34	50	11	18
Winklebury (Jones, R. 1977)	175	100	24	33	83	55	98	39	57	35	84
Baylham House 120-200 AD Pits	133	56	17	35	38	41	100	35	46	1	5
Layers (Maltby n.d.3)	88	73	18	34	48	25	93	51	100	15	6
Longthorpe (Marples 1974)	44	64	55	93	91	61	100	86	61	48	75
Shakenoak Farm Site K Gp.J	29	100	0	0	24	4	35	28	41	0	7
Site K Gp.O (Cram	26	85	0	4	31	0	42	58	100	4	4
Site K Gp.Q 1978)	14	100	0	36	29	0	50	57	79	21	21
Site K Gp.S	21	48	0	14	29	0	19	100	76	0	0
Brancaster (Wall pers. comm.)	88	100	33	41	65	14	60	47	80	16	49
Vindolanda (Hodgson 1977)	96	100	70	34	54	23	45	47	33	7	5
Hamwih (Bourdillon & Coy 1980)	890	64	66	59	100	57	83	45	45	30	24
Minimum percentage		6	0	0	24	0	19	12	21	0	0
Maximum percentage		100	70	93	100	62	100	100	100	100	84
Mean		71	24	34	58	33	76	45	58	19	23
Standard deviation		30	22	20	23	23	27	22	23	24	27

MRP = number of fragments of most represented bone. Other figures are percentages of most represented bone. Jaw = mandible; Scp = scapula; Hum = humerus; R/U = radius/ulna; Fem = femur; Tib = tibia; Mc = metacarpus; Mt = metatarsus; Tar = tarsals/carpals; Phl = phalanges.

respect much of the debate about the relative merits of fragment analysis or minimum numbers analysis is irrelevant because neither method on its own is suitable to deal with this type of data. To obtain an accurate assessment of species representation, we have to use only those bone elements that were butchered, distributed, disposed of and preserved in similar ways. Alternatively some kind of weighting of the figures has to be devised. In either case, the correct procedure is first to establish how the samples have been formed before any attempt is made to interpret species composition.

How does this criticism affect the general statements that have been made about species representation cited above ? Most of those comparisons rely on figures produced by different faunal analysis, who often did not take such considerations into account. Since in some instances details of bone elements are not given nor in most cases are there any discussions of preservation or intra-site variability, it is impossible to be certain whether the samples used in the generalisations are directly comparable. Accordingly the theory that sheep became more important during the Iron Age remains essentially untested. The variation in the number of pig bones in Iron Age assemblages may be an accurate indicator of the amount of woodland available for pannage. It would, however, be comforting to establish that no other factors, particularly differential preservation, were involved in this variability. Several of King's (1978) conclusions about Romano-British assemblages may well stand the test of time and more detailed comparisons of the material. Certainly the abundance of cattle and pig bones on some types of site looks at least superficially to be interesting and worthy of further research.

It is, however, apparent that the results of species quantification have been unsatisfactory. General comparisons have employed figures of species representation that have been produced by inadequate sampling and methodology. This is the major area of faunal studies where old assumptions need to be cast aside and replaced by a new approach. Not until then shall we be able to improve on the vague and sometimes misleading statements that have been made about stock numbers.

The Exploitation of Domestic Stock

It is believed that animal bone studies can establish the principal reasons why the various domestic species were kept and how intensively they were exploited. Cattle, for example, can be bred mainly as working animals, or as dairy producers or simply as providers of meat or a combination of all three. The kill-off patterns and the relative numbers of males, females and castrates kept will vary according to the particular regime of husbandry practised. By studying the ageing and sexing evidence of faunal samples, it is possible to investigate these topics. Once again there are problems in transforming the archaeozoological data into general statements about herd structures and exploitation patterns. Nevertheless some general patterns have begun to emerge from the ageing data of the major domestic species.

Sheep

Sheep produce meat, skins, manure, milk and wool. The relative importance of each plays some part in the manner in which they are husbanded. Both Cunliffe (1978: 183) and Bradley (1978: 36-37), using the published evidence available to them, concluded that sheep were exploited mainly for meat in some areas during the Iron Age and principally for wool in others. Rivet (1964: 123-124) and Applebaum (1972: 214-215) emphasised the importance of wool production on sheep rearing in the Romano-British period and Clutton-Brock (1976: 382) has suggested that large numbers of sheep were kept to provide enough wool for profitable trading in the Anglo-Saxon period.

Payne (1973) has shown that the kill-off patterns of sheep populations raised principally for wool, meat or milk should be quite different from each other. A substantial number of ewes and wethers would be allowed to reach maturity if wool production was predominant, to enable several annual growths of the fleece to be collected. On the other hand, only the animals selected for breeding would be required to reach maturity in a system that intensively exploited sheep for meat and the emphasis would be on the fattening and culling of young animals. Theoretically it should be possible to relate the ages of the bones found on archaeological sites to the regime of exploitation, although as Payne has pointed out, flocks are not usually kept for a single product, particularly in subsistence economies.

Several methods of ageing analysis have been used on British faunal assemblages in recent years. Those which employ epiphyseal fusion data will not be considered here, since differential preservation plays a dominant role in the survivial of the fusion points and makes this method of analysing age structures most unreliable. This leaves the various methods of analysing age through the evidence of tooth eruption and wear. Three methods have been used quite frequently. Ewbank et al. (1964) divided the tooth eruption sequence of complete mandibles into 26 stages. Payne (1973) employed the evidence of tooth wear as well as eruption in his recording method and for overall analysis divided the dental development into eight stages. Grant (1975) also used both the evidence of eruption and wear of the mandibular cheek teeth. By giving numerical values to each of five stages of eruption and 15 wear stages of the molars and by adding together the scores for each molar, an overall numerical value (n.v.) can be obtained for the mandible. A high numerical value indicates a lot of wear on the teeth and implies that the mandible belonged to a mature animal. None of these methods provides an absolute estimate of age. We have only an imprecise knowledge of the rates of dental development of sheep in these periods. Modern estimates are derived from improved breeds whose development is much faster than the sheep studied here, although by how much is unclear. Little is known of variation in the tooth eruption of sheep during the periods themselves. Most studies have used estimates derived from the data of Silver (1969), which may not be totally applicable. Nevertheless the ageing methods do provide an opportunity to observe relative changes in the number of mandibles at the different stages of development. There are some problems in comparing the results of the three methods and Hamilton (1978) has found that the methods produced results that differed by up to 10% in the estimations of cumulative mortality rates when applied to the same sample. Although sheep and goat mandibles are very difficult to tell apart, the dominance of sheep on all these sites is so great (see above p. 159) it seems reasonable to consider that the vast majority of the mandibles belonged to sheep.

Ageing Data from the Iron Age

Several samples from Iron Age settlements in southern England have now been examined. The analysis of two middle Iron Age samples from Hampshire, Winnall Down and Balksbury (material from the 1973 excavations directed by Geoffrey Wainwright) have produced very similar results (Fig. 8: 1). The diagram shows the number of mandibles scoring each of Grant's (1975) numerical values and includes estimated values from incomplete molar rows. It is important to emphasise that these values do not represent equal lengths of time. Changes in the toothwear stages of older mandibles are generally slower, although that is somewhat of an oversimplification. As a guide, the first molar is in wear by c. n.v. 8, the second molar at c. n.v. 18 and the third molar at c. n.v. 30. In both samples, therefore, there was a concentration of very young mandibles, very few that were assigned values of 15-30 and then a broad concentration of older mandibles with fully erupted tooth rows. Several points need to be made about these samples. On both sites they were recovered almost exclusively from pits and contexts that preserved the mandibles extremely well. It is certain that had the samples derived from deposits that were less favourable to bone preservation, the survival rate of the youngest and most fragile jaws would have been seriously impaired. Secondly, it is uncertain whether these samples, although quite substantial, represent an accurate cross-section of the sheep kept by the inhabitants of the settlements. It is possible that there was redistribution of stock or carcasses between settlements. It is also conceivable that the mandibles in these pits were biased towards these age groups because of the particular disposal strategies employed by the inhabitants. A wider range of samples from contemporary neighbouring settlements is required to test these possibilities.

Assuming for the time being, however, that the samples are representative, what inferences can be drawn about sheep husbandry? The concentrations of young mandibles include those belonging to neonatal mortalities and those of lambs with their first molar not fully erupted or only in an early stage of wear. Absolute ageing is problematic but, even allowing for very slow eruption rates, it seems likely that the majority of these mandibles belonged to animals that died under a year old. Given the poor quality of the stock, a high rate of neonatal deaths is to be expected. Payne's models (1973: 282-284) allow for up to 30% of the lambs born each year to die of natural causes. High rates of young mortalities were prevalent in England in the Middle Ages (Miller & Hatcher 1978: 217). The older lambs represented at Winnall Down and Balksbury, although they were butchered for meat, were certainly not kept alive long enough to reach an optimum age and weight for culling for meat. In fact there were very few sheep of that age, as the low number of mandibles with numerical values of 15-30 indicates (Fig. 8: 1). Superficially, the observable age pattern fits more closely to Payne's model of milk exploitation, in which in addition to natural mortalities a high percentage of the flock are slaughtered in their first year leaving a few rams but mainly ewes for breeding purposes and their milk. Alternatively, it is possible to view the ageing pattern as evidence for a very low level of efficiency in sheep husbandry, in which only the stock selected for breeding were allowed to mature. This may indicate that there was a shortage of winter fodder for sheep or at least no incentive nor necessity to overwinter a significant proportion of the stock. In either case,

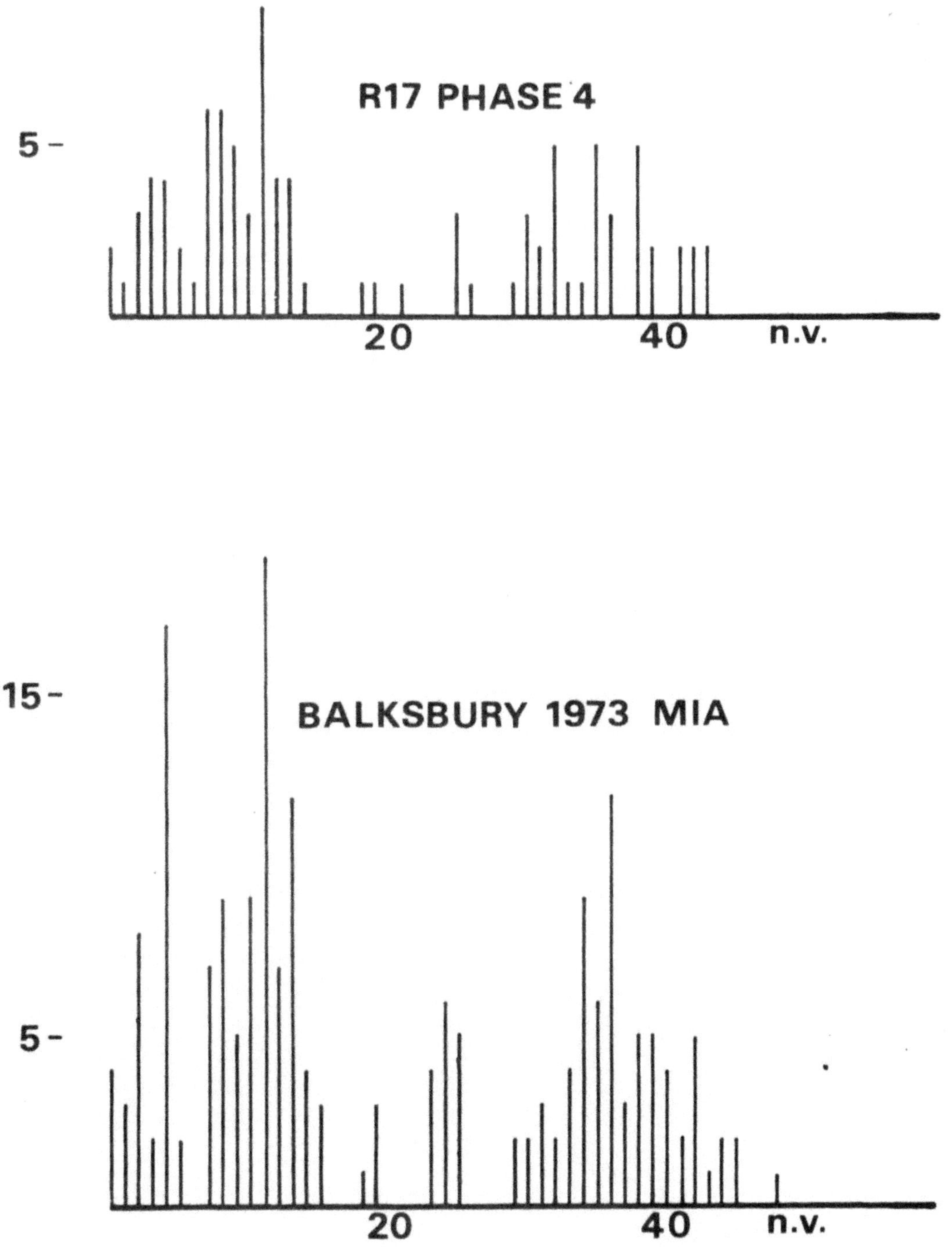

Fig. 8.1 Ageing data from sheep/goat mandibles from middle Iron Age samples from Winnall Down and Balksbury 1973. n.v. = numerical value. Method of analysis follows Grant (1975). Estimated values are included.

although wool would have been provided by the older animals, the apparently high rates of immature mortalities suggest that wool production was not of primary importance in the exploitation of sheep at these settlements.

The smaller samples from the earlier Iron Age deposits from Winnall Down and Balksbury produced similar results to the one described above and these have parallels with other samples from southern England. The large samples from Phases 1-2 at Gussage All Saints contained a high proportion of sheep mandibles with, at most, only the first molar in wear, a low percentage of jaws with only the first and second molars in wear and a larger group with fully erupted tooth rows (Harcourt 1979: 152). The sample from Croft Ambrey contained a larger number of adult sheep but again a low number of mandibles at the stage when only the first and second molars are in wear (Whitehouse & Whitehouse 1974: 218-219). Similar results were obtained from the samples from Ashville and the late Iron Age deposits at Barton Court Farm, Oxfordshire (Hamilton 1978: 129).

Of the other large Iron Age samples, those from Eldon's Seat and Hawk's Hill cannot be compared directly with these since the data were grouped in a different way. The analysis of the mandibles from Barley, Hertfordshire, produced rather different results (Ewbank et al. 1964). Although mandibles of first year animals and adult stock still formed the largest groups, c. 20% of the sheep represented by complete mandibles were killed between the early wear stages of the second and third molars. Two late Iron Age samples from southern England also contained a higher percentage of mandibles af this age. At Gussage All Saints, 21% of the mandibles were at this stage of development (Harcourt 1979: 152) and the excavations of the banjo enclosure in Micheldever Wood, Hampshire produced roughly equal numbers of these and those of the youngest age group (Coy n.d.). Whether these samples provide evidence for a change in sheep husbandry that resulted in the culling of relatively more second and third year animals remains to be tested on other sites in the area.

Obviously the number of Iron Age samples is inadequate to provide information about the possibility of local, regional or temporal variation. The samples that have been examined so far do, however, have certain common traits. In all of them the number of first year mortalities represented is high compared to samples from later periods. Such jaws are small and fragile and are likely to be under-represented. Apart perhaps from the samples from Barley, Gussage All Saints and Micheldever Wood, all the Iron Age sites have produced few mandibles of sheep killed at an age when they would have provided a lot of meat for the amount of fodder they required. Efficient meat production was not, it seems, a characteristic of Iron Age sheep husbandry. How important wool production was depends of the interpretation of the relative number of young and old animals represented in the samples. Certainly all the adult animals could have provided wool and the occurrence of spindle whorls and loomweights at many sites testifies to textile manufacture. However, if first year mortalities were as high as suggested at sites such as Balksbury (Fig. 8: 1), the emphasis would have been more on the maintenance of a viable breeding stock rather than large scale wool production. Other sites have produced a greater proportion of adult animals. If it could be demonstrated that this was not merely a reflection of poorer preservation of the young mandibles, the case for the importance of wool would be strengthened. If not, an alternative hypothesis would regard sheep husbandry to be of a low standard, geared towards subsistence activities only, providing meat, milk, manure and wool but not at a commercial level.

Ageing Data from the Romano-British Period

Romano-British deposits have revealed fairly consistently a change in emphasis in sheep exploitation. The mandibles from Portchester Castle (Grant 1975) provide a good example (Fig. 8:2). The number of mandibles with numerical values of 20-30 is high. A lot of the animals represented therefore were killed between the early wear stages of the second molar and the full eruption of the third molar. This corresponds roughly to Stages D and E in the method of Payne (1973: 293). If ageing estimates are correct, the sample shows an emphasis on the kill-off of second and third year animals for their meat. Much fewer first year animals are represented apart from some neonatal mortalities which were recovered mainly from wells (Grant 1975: 397-398). Samples from all types of Roman settlement have produced similar concentrations of mandibles of this age. These include urban centres such as Exeter (Maltby 1979: 42), military sites, for example Vindolanda (Hodgson 1977: 16), villas, for example Shakenoak Farm (Cram 1978: 128-135) and other rural sites, for example, Balksbury (Maltby n.d. 2). Generally fewer concentrations of mandibles belonging to first year mortalities have been found, although some early Roman sites have produced them in some numbers, for example, Winnall Down (Maltby n.d. 1), Baylham House, Suffolk (Maltby n.d. 3) and the military deposits at Margidunum, Nottinghamshire (Harman 1969: 101). The change in emphasis to the culling of second and third year animals may therefore have been a gradual one, although there is a need to study a much wider range of samples before the pattern can be better understood. Redistribution of stock and animal products undoubtedly took place in this period and it is conceivable that certain age classes will be found more commonly at some settlements than others. Nevertheless the present incomplete evidence suggests that meat production became more important in sheep management and there was a greater level of efficiency in sheep farming. A few samples have produced a relatively high number of mature animals, notably the villa at Barton Court Farm (Hamilton 1978: 129), the later levels at Fishbourne (Grant 1971: 384) and, to a lesser extent, Shakenoak Farm (Cram 1978: 128-135) and the Romano-British levels at Balksbury (Maltby n.d. 2). Once again it will be interesting to observe whether this pattern is typical of rural settlements and whether urban and military centres tended to attract a higher proportion of younger animals raised principally for their meat.

It is uncertain how important wool production became in the Romano-British period. Apart from the examples listed above, there is no evidence that adult animals were kept in numbers significantly above the level required to maintain the breeding stock, although investigations of other villa and rural assemblages may alter the picture. Change of emphasis in sheep farming may have occurred within the period but there is insufficient material to test this.

Ageing Data from the Anglo-Saxon Period

Evidence from Anglo-Saxon sites is even more limited but it can be shown that wool production became more important in some areas during the latter part of the period. The sample from the middle-late Saxon deposits at

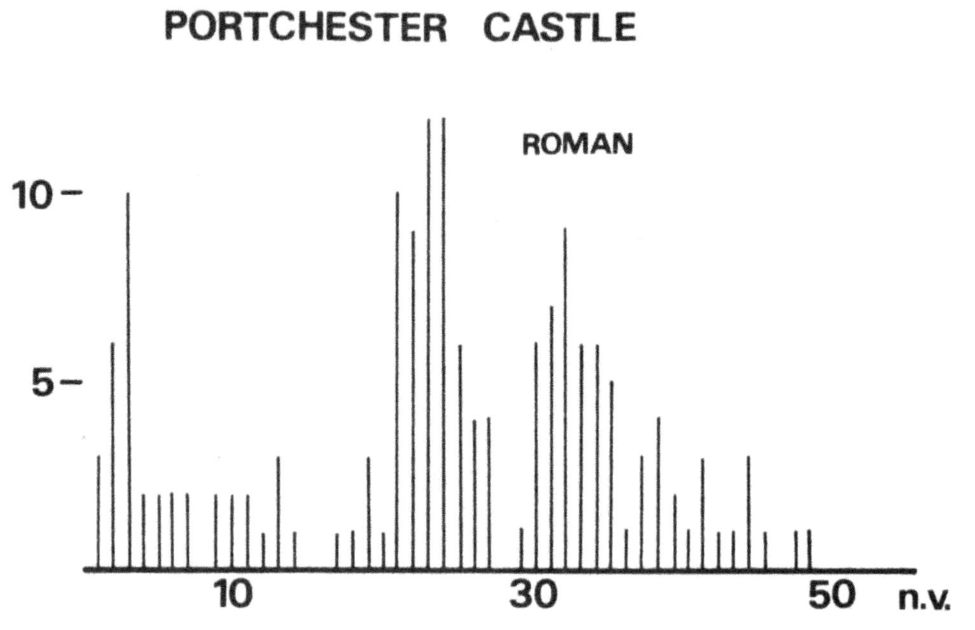

Fig. 8.2 Ageing data from sheep/goat mandibles in Romano-British deposits at Portchester Castle (Groups 1-5). Data adapted from Grant (1975). n.v. numerical value. Method of analysis follows Grant (1975). Estimated values are included.

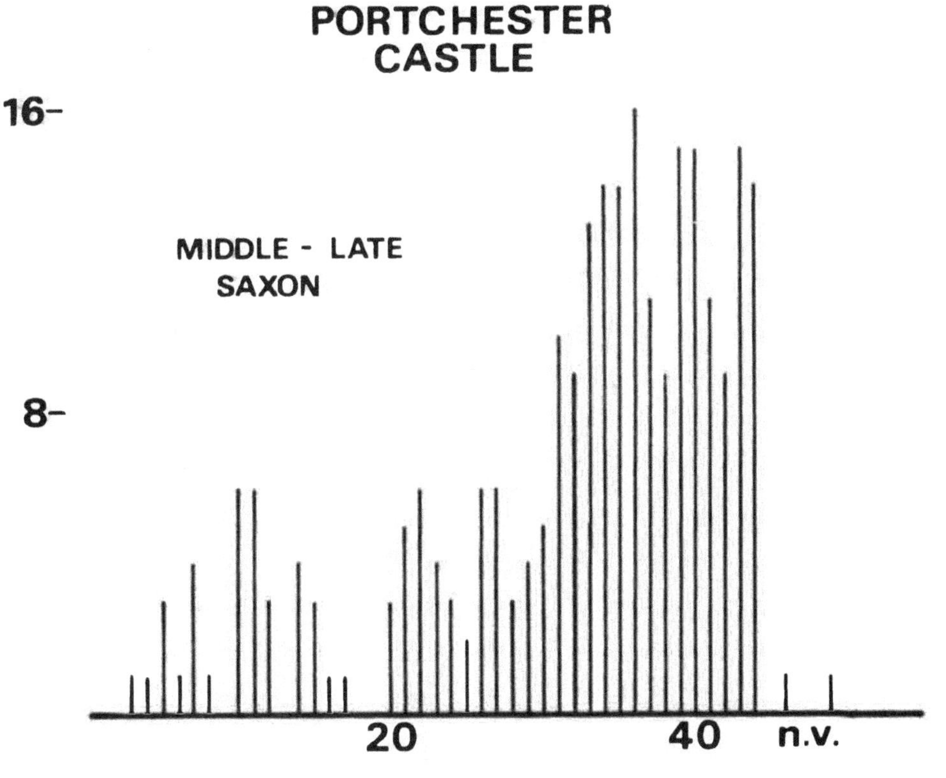

Fig. 8.3 Ageing data from sheep/goat mandibles in middle-late saxon deposits at Portchester Castle. Data adapted from Grant (1976). n. v. = numerical value. Method of analysis follows Grant (1975). Estimated values are included.

Portchester Castle (Fig. 8:3) included a concentration of mandibles with numerical values of over 30, indicating that the majority belonged to mature animals. The smaller sample from the late Saxon period at Portchester was similar (Grant 1976: 278). The contrast between this sample and the one from the Roman occupation of the fort (Fig. 8:3) is marked. If these jaws accurately reflect the kill-off pattern, meat production was now only a secondary consideration to wool production. Of course it is again possible that the sheep represented at Portchester Castle do not contain a cross-section of the sheep kept in the area. It is interesting to note, however, that the urban deposits at Hamwih also produced a large proportion of mature animals (Bourdillon & Coy 1980: 87). Most of the mandibles from the late Saxon ironworking site at Ramsbury, Wiltshire also belonged to mature animals (Coy 1980 in press). Several other samples, particularly from North Elmham, Norfolk (Noddle 1975: 257), the urban deposits from the St. Peter's Street excavations, Northampton (Harman 1979: 331), Walton, Aylesbury (Noddle 1976: 277) and Durham (Rackham 1979: 53) have produced high percentages of adult sheep. Only the sample from Sedgeford, Norfolk has not produced this pattern (Clutton-Brock 1976: 382). Although the number of samples is small, they have been derived from a wide range of settlement types in several parts of the country. It is tempting to equate this evidence with a large scale increase in the importance of wool production to enable extensive wool and cloth trading to take place. The origins of this development may lie in the early Saxon period but there is as yet little information about samples of that date, although a vast sample from West Stow, Suffolk is currently being studied.

It is therefore possible to see long term changes in sheep exploitation. Once the possibility of regional, ecological and cultural variability has been investigated the pattern will be better understood. At present there seems to have been an underlying trend from a low level of subsistence husbandry in the Iron Age, through improvements and emphasis on better meat production during the Romano-British period, to the development of wool production as the most important component of sheep husbandry by the late Saxon period.

Cattle

Ageing Data from the Iron Age

Analyses of the herd structures of cattle in Iron Age faunal samples have been limited by the fact that most of the largest and best studied assemblages have produced comparatively little ageing and sexing data. The picture of cattle exploitation is therefore very restricted. To take the evidence of tooth eruption and wear only, a few samples have produced a relatively large number of immature mandibles. In particular the sample from Phase 1 (Early Iron Age) at Gussage All Saints contained a very high percentage (36%) of jaws with at most only the deciduous premolars and first molar in wear (Harcourt 1979: 151). Many of these must have belonged to calves under a year old and some were neonatal mortalities. Apart from these, most of the mandibles belonged to adult animals. Only 24 mandibles could be examined from the Ashville excavations (Hamilton 1978: 132), using the method of Grant (1975). Most of these had numerical values of less than 20 and accordingly did not have the second molar fully erupted. Sexing of the metapodia and distal radius suggested that most of the adult animals were cows (Wilson 1978: 135). Although mandibles of young calves were represented in both the early and middle Iron Age samples from Balksbury, the majority had n.v. of over 30 (Fig. 8:4) and most of these had fully developed tooth rows, belonging to animals at least five years old and probably substantially older in many cases. A similar pattern was discerned from the Iron Age samples from Winnall Down (Maltby n.d. 1) and Eldon's Seat, Dorset (Cunliffe & Phillipson 1968: 229). Another pattern of ageing appears in two late Iron Age samples. At Barton Court Farm, Oxfordshire, there was a concentration of mandibles with the first two molars in wear but the third molar unerupted (Hamilton 1978: 132), belonging to animals under three years of age on modern estimates (Silver 1969: 296). A small sample from 1st century AD deposits from the excavations at Baylham House, Suffolk (Maltby n.d. 3) also contained a number of mandibles at a similar stage of development, indicating the culling of young cattle for meat. Other Iron Age faunal assemblages are difficult to compare because the methods of ageing are not fully explained. The variability of these samples could be the result of a combination of factors and it is premature to suggest significant changes in cattle husbandry during the Iron Age. Other factors such as regional variability, redistribution of stock, differences in disposal strategies and possible sampling, preservation and recovery biases could be involved. These factors cannot be examined on the existing data. A high kill-off of young calves and a predominance of adult cows in some assemblages may imply that dairying was an important element of cattle husbandry in some communities but much more sexing evidence is required before we can place any confidence in such statements.

Ageing Data from the Romano-British Period

Evidence from the Romano-British period is better documented, although the same limitations apply with regard to the possible explanations for the observed patterns. At least two recurring patterns can be observed from the studies of mandibular tooth eruption. The first contains mandibles of cattle of all ages, although mature individuals usually predominate. Examples can be found on villa and other rural sites. At Barton Court Farm a sample of 34 mandibles of Roman date included 16 which did not have the third molar erupted, including five of young calves (Hamilton 1978: 132). The mandibles from Sites C and K at the villa at Shakenoak Farm, Oxfordshire, dated mainly to the 3rd and 4th centuries, contained 12 mandibles with at most the third molar in an early stage of wear and only six mandibles with fully erupted tooth rows (Cram 1973: 1978). The even smaller sample from Fishbourne also contained a relatively large number of immature cattle (Grant 1971: 385). The 2nd century deposits at the roadside settlement at Baylham House, Suffolk included eight cattle mandibles that had not reached the stage when the third molar was fully erupted. 13 other jaws had fully erupted tooth rows. The late Roman deposits at Balksbury produced ten mandibles with fully erupted tooth rows but seven that had not reached dental maturity (Fig. 8:4). Small samples of cattle mandibles from the roadside settlements of Margidunum, Nottinghamshire (Harman 1969: 101) and at Scole, Norfolk (Jones, G. 1977: 210) also contained a relatively high proportion of immature specimens. Finally, the 4th century deposits from several sites in Exeter showed the presence of a significant number of immature cattle (Maltby 1979: 30, 155). Most of the young cattle represented in these samples were of a reasonable

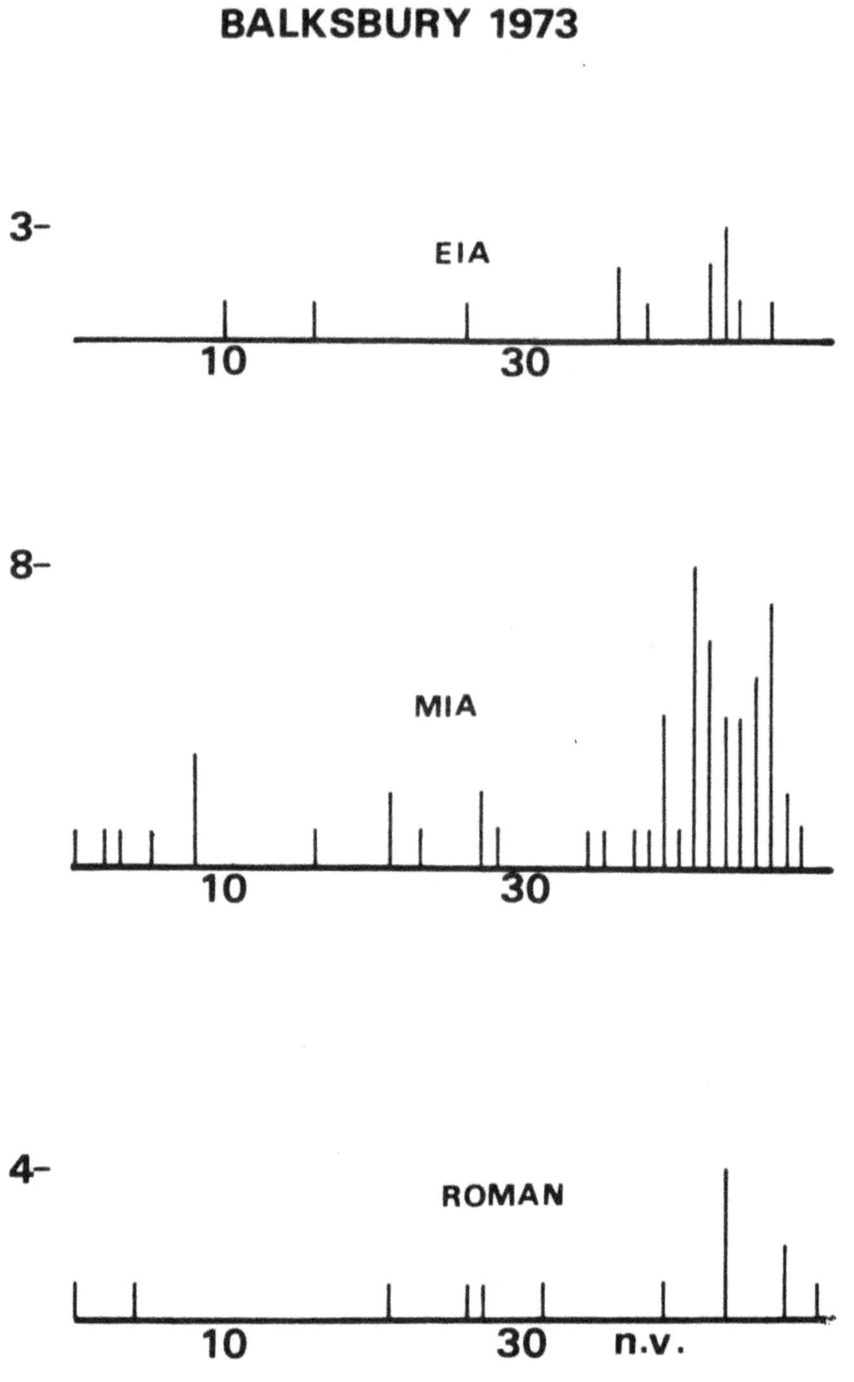

Fig. 8.4 Ageing data from cattle mandibles in early Iron Age, middle Iron Age and Romano-British contexts from Balksbury 1973. n.v. = numerical value. Method of analysis follows Grant (1975). Estimated values are included.

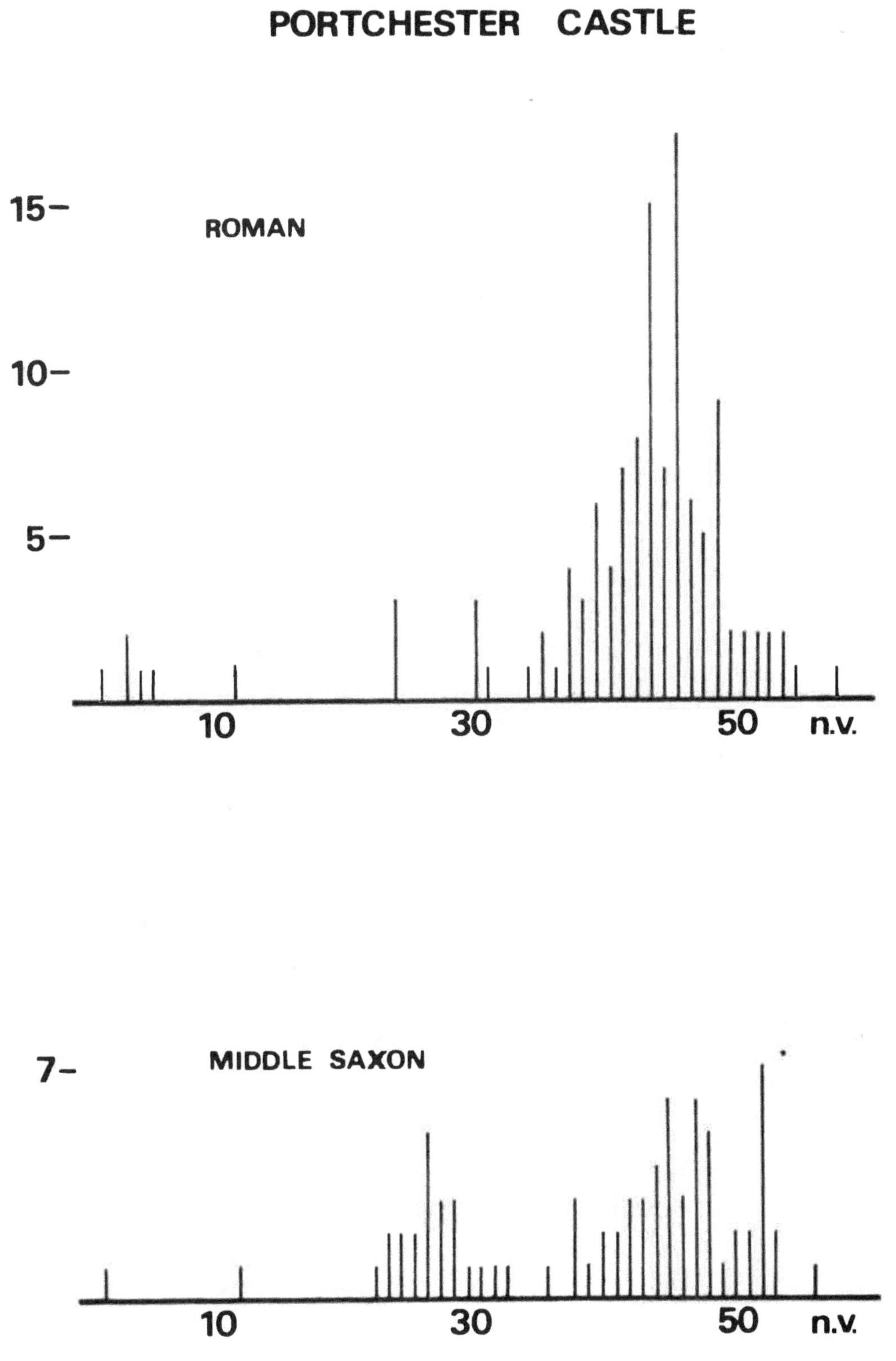

Fig. 8.5 Ageing data from cattle mandibles in Romano-British and middle-late Saxon deposits at Portchester Castle. Data adapted from Grant (1975, 1976). n.v. = numerical value. Method of analysis follows Grant (1975). Estimated values are included.

size for culling for meat, although not always fully grown. In most cases these were animals not required for breeding, working or dairying that were fattened for slaughter for their meat and hides.

The second pattern of ageing is typified by the collection of mandibles from the Roman levels at Portchester Castle (Grant 1975). The great majority of mandibles recovered in the large sample belonged to adult animals and included very few jaws with numerical values of under 35 (Fig. 8:5). Most of the cattle were therefore over five years of age and some had heavy toothwear which indicates that quite old animals were slaughtered. The predominance of mature cattle recurs in several other samples derived from military and urban sites. From the north of England, a large percentage (78.9%) of the 147 ageable mandibles from the excavations at Vindolanda possessed the third molar in wear (Hodgson 1977: 12). 132 ageable mandibles from deposits dated to the 1st to 3rd centuries at Exeter included only 10 that definitely did not have the third molar in wear (Maltby 1979: 155-156). Analysis of the mandibles of 4th century date from the excavations at Angel Court, Walbrook in London produced similar results (Clutton-Brock & Armitage 1977: 92). Of the 155 mandibles of late Roman date examined from the excavations at 1, Bleachfield Street, Alcester, Warwickshire, under the direction of Paul Booth, only 10 did not have the third molar in wear (Maltby n.d. 4). Attempts to distinguish the sexes of the adult cattle by metrical analysis of the metapodia have been made on several of the samples and in each case it appears that cows were more commonly represented (Grant 1975: 401; Maltby 1979: 33-34; Clutton-Brock & Armitage 1977: 92), although the reliability of the various techniques of metrical analysis has yet to be established.

The evidence from all the Romano-British samples investigated to date would suggest that most cattle were allowed to reach maturity and it is unlikely that they were all raised simply for their meat, even allowing for slow rates of growth. Working and dairy cattle were probably important elements in the economy. Explanation of the variation in the number of immature cattle represented remains tentative. It is interesting, however, that the heaviest concentrations of adult cattle have so far appeared only on military and urban settlements. Organisation of cattle marketing and the need to provision these centres with meat may have resulted in the supply of particular types and age groups of cattle. Most of the other settlements may have been more self-sufficient and therefore their deposits included a higher percentage of immature cattle not required for breeding or working but also not in demand for redistribution to other centres. In Exeter, the increase in the number of immature cattle in the 4th century deposits coincided with evidence for a change in the settlement pattern and the presence of stock enclosures associated with houses within the walls, perhaps indicating an increase in the farming element of the population (Maltby 1979: 90). The evidence may also imply that there was a collapse of the former supply network of cattle brought to the town for slaughter. Most of the other settlements which contained higher proportions of immature animals were villas, rural settlements or settlements that did not need to be supplied with a large number of cattle brought in for slaughter from elsewhere. The possible dichotomy between the cattle represented on rural, urban and military sites remains a topic for further investigation.

Ageing Data from the Anglo-Saxon Period

Of the few Anglo-Saxon samples examined, the age distribution of the mandibles from the middle-late Saxon levels at Portchester Castle (Grant 1976: 276) appears typical. Most cattle were mature, some having heavy wear on the teeth and including a high proportion of working, dairy and breeding stock. There was, however, a smaller but substantial group of mandibles in which the third molar was not erupted. These probably belonged to immature animals killed for their meat possibly between three and four years of age (Fig. 8:5). A similar distribution occurred in the large sample of mandibles from Hamwih and the sexing evidence suggested that the majority of the immature specimens were males not required for breeding or working (Bourdillon & Coy 1980: 105-108). Another example of this age distribution has been discovered in the sample from the St. Peters Street excavations in Northampton (Harman 1979: 331).

Cattle provided the most meat in all periods and it is clear that they were also much valued as beasts of burden and possibly as dairy animals. Improvements of ageing and sexing methods and the availability of a wider range of samples should enable a more detailed and accurate assessment of the relative importance of these bovine products to be made. Cattle were undoubtedly the most valuable of the domestic stock, although possibly not always the most numerous. Their importance may have resulted in the herds being regarded as objects of visible wealth and social prestige. Cattle trading no doubt took place in all periods and there is evidence from the Romano-British period at least of large-scale redistribution. The mechanisms and control of such trade remain to be considered in more detail.

Pigs

The number of alternative exploitation strategies for domestic pigs is limited by the fact that almost all their value lies in their meat and lard. Their high reproduction rates enable a substantial kill off of young animals to take place and consequently relatively few animals are required to reach maturity. Thus the presence of large numbers of immature bones is to be expected and indeed is a feature of all the larger quantified assemblages under review. What is less certain is how intensive this exploitation was, how quickly the stock were fattened for slaughter, whether there were seasonal peaks in the culling of pigs and whether fattening processes relied solely on pannage or whether some sty husbandry was practised.

Evidence from the Iron Age is flimsy but there seems to have been no great intensity of pig exploitation on the settlement at Ashville, where 37% of 30 mandibles had the third molar in wear (Wilson 1978: 135) nor at Gussage All Saints, where between 33-47% of the mandibles had also reached that stage of development (Harcourt 1979: 153). The Roman liking for suckling pig is well known (White 1970: 318-320) and on some Romano-British sites it has been argued that there was a relatively high percentage of first year mortalities (Maltby 1979: 57; Grant 1971: 383). The better representation of pigs on most "Romanised" sites (King 1978: 216) may also be significant here. Yet, even at Exeter and Fishbourne the proportion of mandibles with the third molar in wear was quite high and such jaws must have belonged to pigs at least two years old and possibly older, if reported 18th century tooth eruption rates are a more appropriate comparison than modern estimates (Silver 1969: 298-299). At Portchester Castle, the pig mandibles from both the Roman (Grant 1975: 398) and Saxon assemblages (Grant 1976: 279-290) included many with the permanent premolars just coming into wear and the third molar erupting and also a fair proportion with the third molar in an early stage of wear. Possibly most of these were second and third year cullings and there was certainly no very intensive kill off of younger pigs, although the smaller jaws may be under-represented. Similarly 26.6% of the pig mandibles from Hamwih had all the cheek teeth in wear (Bourdillon & Coy 1980: 112). Refinement of the ageing methods may lead to further information about the peak periods of slaughter but, apart from a few of the Romano-British assemblages, it appears that most pigs were not killed until at least their second year and the intensity of pig husbandry was never particularly high by modern standards.

Horse

The presence of horse bones is attested from archaeological sites throughout the periods under consideration. Indeed, there is now abundant evidence that horsemeat made an important contribution to the diet in at least some parts of England during the Iron Age. Horse bones butchered for meat have been recorded, for example, at Ashville (Wilson 1978: 119, 122, 125), Tollard Royal, Dorset (Bird 1968: 147), Winnall Down (Maltby n.d. 1) and Balksbury (Maltby n.d. 2). On some of these sites horse bones are very well represented and the importance of horses as working and riding animals has been pointed out by several authors (e.g. Harding 1974: 87; Bradley 1978: 38). Certainly most horse bones recovered on Iron Age sites belonged to mature animals and examinations of toothwear have shown that many of them were over 10 years of age. Meat production was not the primary purpose of their exploitation. The lack of immature specimens at Gussage All Saints led Harcourt (1979: 158) to suggest that no breeding of horses was practised but that they were rounded up periodically when certain animals were selected for training. As horses are not suitable for working until three years of age, this type of round-up would have saved the expense of rearing and feeding the foals and allowed the processes of natural selection to take place. This hypothesis remains to be tested against a wider range of samples and several other explanations could account for the absence of very young horse bones from Iron Age deposits. For example, Champion (1978: 384) has pointed out that the evidence for the production of horse gear at Gussage All Saints implies that a specialist was resident there, It is not beyond the bounds of possibility that these were also specialist horse-breeders on some settlements. If so, only adult horses may have been redistributed to other settlements. Nevertheless the importance of horses primarily as transport animals and secondarily as producers of meat appears consistent.

The poor representation of horse of Romano-British and Anglo-Saxon settlements may simply be the result of the decline in the horse's importance as a producer of meat. Certainly few horse bones have been found in Romano-British deposits in assemblages derived mainly from butchery waste and records of butchery on horse bones are rare. This does not necessarily imply that horses had become less important. Occasional discoveries of butchered horse bones have been made on Anglo-Saxon sites, for example, at Sedgeford, Norfolk (Clutton-Brock 1976: 383) and Hamwih (Bourdillon & Coy 1980: 105) but again, where found, adult horses predominated the samples and their value as transport and pack animals continued to be the predominant feature of their exploitation.

Metrical Analysis

Measurement of bones have been made in attempts to make assessments about the relative sizes of the stock, to distinguish between the sexes and to monitor the possible importation of new stock. Unfortunately, because of the haphazard nature of many of these studies and the variation of measurements taken by different analysts, such work has not produced the results that it has the potential to obtain. The advent of archival and computer recording and some attempts at standardisation of the measurements taken (e.g. von den Driesch 1976) will improve the situation. For the time being, however, any analysis of past work is limited to comparisons of particular measurements. Single measurements are themselves unreliable because a combination of genetic, nutritional and sex factors can produce variations in size. Only by the study of several measurements from one bone and by comparisons of measurements from different bones will we be able to obtain a better understanding of the stature and variety of the domestic stock. That, however, is for the future. This review will concentrate on an assessment of the evidence to date.

Cattle

Earlier studies of prehistoric and early historic cattle bones from Britain have been made. Jewell (1962) emphasised the diminutive form of Iron Age cattle and showed that similar small cattle were present during the Roman period. He also noted that bones of larger cattle appeared in the latter period and concluded that a larger breed of cattle was imported. Hodgson (1968) found that the same trends were apparent in the samples he compared. These surveys can now be supplemented by data obtained from more recent excavations. Metrical analysis of the maximum length of the astragalus from Iron Age, Romano-British and Anglo-Saxon samples (Table 5) confirm the appearance of larger cattle in the Romano-British period. Apart from the small sample from Corstopitum and the two samples from Exeter, all the Romano-British assemblages produced a higher mean measurement than the Iron Age samples. Although the Romano-British and Anglo-Saxon samples contained some astragali as small as those from Iron Age sites, they also included others that were significantly larger than any found in the earlier assemblages. The same trend can be observed on other bones. Although most of the Romano-British samples listed in Table 5 are dated to the latter part of that period, there is some evidence that larger cattle were present in some areas during the early years of occupation. The astragali from Winnall Down, Hampshire, included large specimens only in Phase 6 deposits dated to the 1st and 2nd centuries AD (Fasham pers. comm.). None of the astragali from the Iron Age features (Phases 3 and 4) attained the size of the largest Romano-British specimens (Fig. 8:6). More late Iron Age material is required for comparison but the present evidence suggests that the appearance of larger cattle did coincide with the Roman invasion. The possibility of the importation of cattle is therefore strong, although much more detailed analyses than these are needed to confirm this. Improvements in the size of the native stock may also have occurred during the Romano-British period but there may have been regional variations. Comparisons of the cattle bones from Exeter and most other Romano-British assemblages illustrate the small size of cattle brought for slaughter there. Comparisons of the maximum proximal width of metatarsi from Exeter and Alcester demonstrate this (Fig. 8:7). Regional variation in the size of cattle

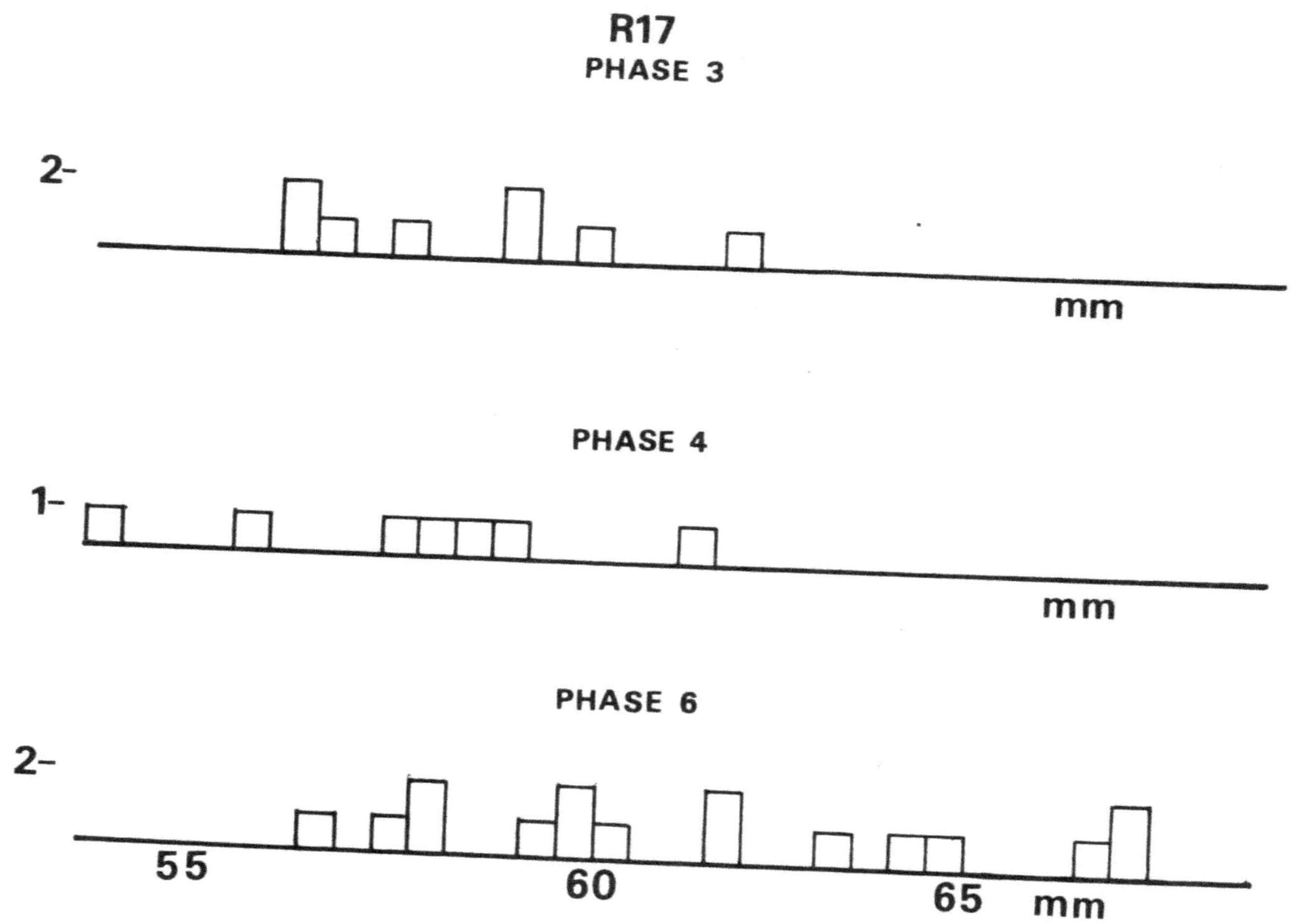

Fig. 8.6 Maximum lengths of cattle astragali from early Iron Age, middle Iron Age and early Romano-British deposits at Winnall Down, Hampshire.

TABLE 5: METRICAL ANALYSIS OF THE MAXIMUM LENGTH OF CATTLE ASTRAGALI

Site	Date	N	Range	Mean	Source
Exeter	55-300	14	50.7-59.6	55.2	Maltby (1979)
Catcote	Iron Age	14	51-63	57.0	Hodgson (1968)
Gussage All Saints	Iron Age	54	54-62	57.0	Harcourt (1979)
Winnall Down	MIA	7	53.1-61.0	57.3	Maltby (n.d. 1)
Croft Ambrey	Iron Age	20	55-63	57.7	Whitehouse (1974)
Balksbury 1973	MIA	12	55.0-63.1	57.9	Maltby (n.d. 2)
Appleford	Iron Age	8	55-60	58.0	Wilson (1978)
Winnall Down	EIA	8	55.8-61.6	58.0	Maltby (n.d. 1)
Corstopitum	Roman	9	53-63	58.0	Hodgson (1968)
Exeter	300-400	18	54.3-62.0	58.3	Maltby (1979)
Ashville	Iron Age	18	53-64	58.5	Wilson (1978)
Barley	Iron Age	13	54.1-62.1	58.5	Jarman et al. (1968)
Grimthorpe	EIA	8	56.3-61.5	59.5	Jarman et al. (1968)
Hamwih	Mid Saxon	167	49.2-71.5	60.9	Bourdillon & Coy (1980)
Baylham House	100-200	10	56.0-65.8	61.3	Maltby (n.d. 3)
Alcester	Late Roman	30	53.9-67.6	61.4	Maltby (n.d. 4)
Shakenoak Farm	Late Roman	44	53-72	61.6	Cram (1978)
Winnall Down	Early Roman	16	56.1-68.4	61.6	Maltby (n.d. 1)
Ramsbury	Mid Saxon	6	51.5-66.5	61.9	Coy (1980 in press)

All measurements in millimetres. EIA = Early Iron Age; MIA - Middle Iron Age.

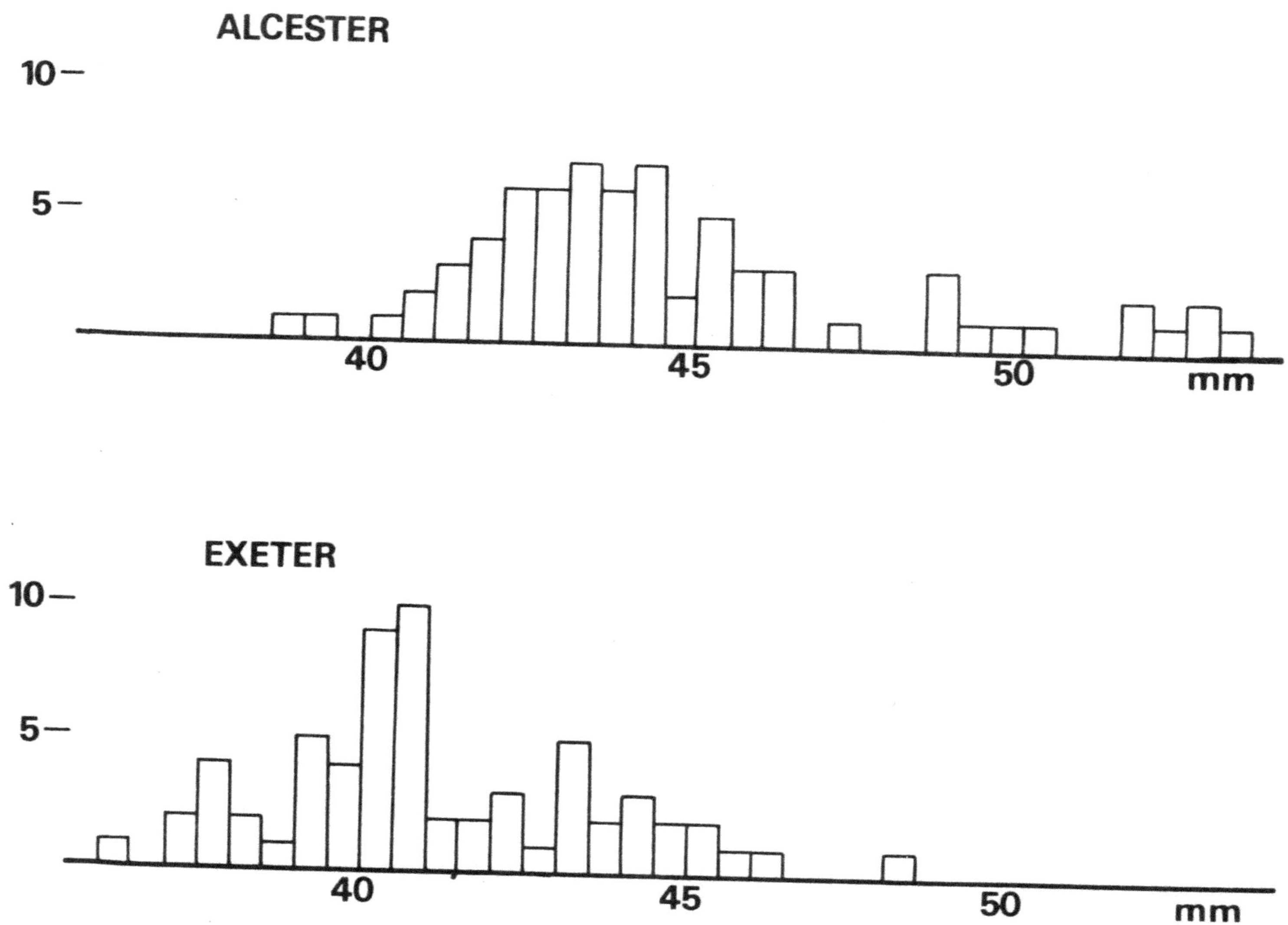

Fig. 8.7 Maximum proximal widths of cattle metatarsi from Romano-British samples from Alcester and Exeter.

in Roman Britain can therefore be demonstrated. Very small samples of measured cattle bones from Dinorben (Gardner and Savory 1964) and Coygan Camp (Westley 1967: 193) in Wales also produced no evidence for the presence of large cattle. It is possible that there were environmental constraints on the size and type of cattle kept in some parts of Britain but the explanation may be more complex than this. It has been shown that large cattle appeared in Roman occupied territory in central and eastern Europe but were not found in contemporary settlements outside the area of Roman occupation (Bökönyi 1974: 128-133; Boessneck and von den Driesch 1978: 31-33). There were also variations in the number of large cattle found at different types of settlement within the Roman provinces (Bökönyi 1974: 130). The degree of Roman influence on cattle farming in Britain may therefore account for the variability in cattle size observed in faunal samples. South-west England may have been outside the area where larger cattle were introduced or bred.

Sheep

The most common measurement taken consistently on ovicaprine assemblages has been the maximum distal width of the tibia. The significance of this measurement as an indicator of size has yet to be adequately demonstrated and the size of the bone is to an extent determined by sexual dimorphism (Noddle 1980: 396). Nevertheless there has been some consistency in the results obtained to date. Table 6 shows the range and means of distal tibiae measurements from 25 samples. Apart from the small sample from Grimthorpe, the means from the Iron Age samples fall at the lowest end of the scale, with very few specimens measuring more than 25 mm. Specimens of this size and larger have been found more commonly in Romano-British and Anglo-Saxon samples, although the smallest specimens were as small as the earlier ones. The means from the Romano-British samples fall between 22.8-25.5 mm. Six of the seven samples with the highest means in the list originated from Anglo-Saxon sites in East Anglia and southern England (25-2-26.3 mm). Other measurements of sheep bones support these observations, indicating an increase in the average size of sheep in the Romano-British and Anglo-Saxon periods. Possible introductions of stock could have taken place in both periods but improvements in the existing stock by better husbandry could also account for the variability. There is some evidence for an increase in the size of sheep within the Romano-British period on some sites. The average size of distal tibiae increased, for example, at Alcester and Frocester Court, although these samples are far from adequate (Table 6).

There is certainly evidence for regional variation in sheep size in the Romano-British and Anglo-Saxon periods. The bones from Exeter showed only marginal increases in mean size and were consistently smaller than most of the specimens from late Roman Alcester, for example (Fig. 8:8). Sheep in the south-west may have continued to be smaller after the Romano-British period, if the small Dark Age sample from Mawgan Porth, Cornwall is typical (Table 6).

The initial increase in size of some of the flocks in the Romano-British period may have parallels with the developments in cattle husbandry. The introduction or development of larger stock in some areas may be an indication of the rather nebulous concept of "Romanisation". If so, there may be a

TABLE 6: METRICAL ANALYSIS OF THE MAXIMUM DISTAL WIDTH OF SHEEP TIBIAE

Site	Date	N	Range	Mean	Source
Winklebury	Iron Age	20	16.9-23.9	21.2	Jones, R. (1977 archive)
Winnall Down	MIA	12	21.1-23.7	22.3	Maltby (n.d.1)
Barley	Iron Age	11	19.9-26.0	22.3	Jarman et al. (1968)
Balksbury 1973	MIA	19	20.4-24.0	22.4	Maltby (n.d.2)
Mawgan Porth	C9-11th.	7	21.0-25.0	22.6	Clutton-Brock (1976)
Winnall Down	EIA	7	21.3-24.0	22.7	Maltby (n.d.1)
Croft Ambrey	Iron Age	10	21.0-24.0	22.7	Whitehouse (1974)
Balksbury 1973	Roman	7	21.0-27.3	22.8	Maltby (n.d.2)
Frocester Court	100-300	12	20.0-25.0	22.8	Noddle (1979)
Exeter	55-100	21	21.3-29.2	23.1	Maltby (1979)
Exeter	100-300	30	21.4-25.9	23.3	Maltby (1979)
Alcester	100-200	9	21.1-26.0	23.6	Maltby (n.d.4)
Exeter	300-400	15	22.3-27.0	23.9	Maltby (1979)
Winnall Down	Early Roman	8	21.9-25.6	23.9	Maltby (n.d.1)
Grimthorpe	EIA	5	22.0-25.5	24.0	Jarman et al. (1968)
Frocester Court	Late Roman	13	23.0-27.0	24.0	Noddle (1979)
Shakenoak Farm	Late Roman	26	22.0-28.0	24.5	Cram (1978)
Baylham House	100-200	22	21.6-28.8	24.5	Maltby (n.d.3)
Sedgeford	Mid Saxon	29	22.0-28.0	25.2	Clutton-Brock (1976)
Alcester	Late Roman	59	21.2-29.4	25.5	Maltby (n.d.4)
Ramsbury	Mid Saxon	12	22.7-28.0	25.9	Coy (1980 in press)
Hamwih	Mid Saxon	267	21.8-30.0	25.9	Bourdillon & Coy (1980)
North Elmham	Mid Saxon	191	23.0-29.0	26.1	Noddle (1980)
North Elmham	Late Saxon	71	22.0-29.5	26.2	Noddle (1980)
Thetford	Late Saxon	13	21.0-29.0	26.3	Clutton-Brock (1976)

All measurements in millimetres. EIA = Early Iron Age; MIA = Middle Iron Age.

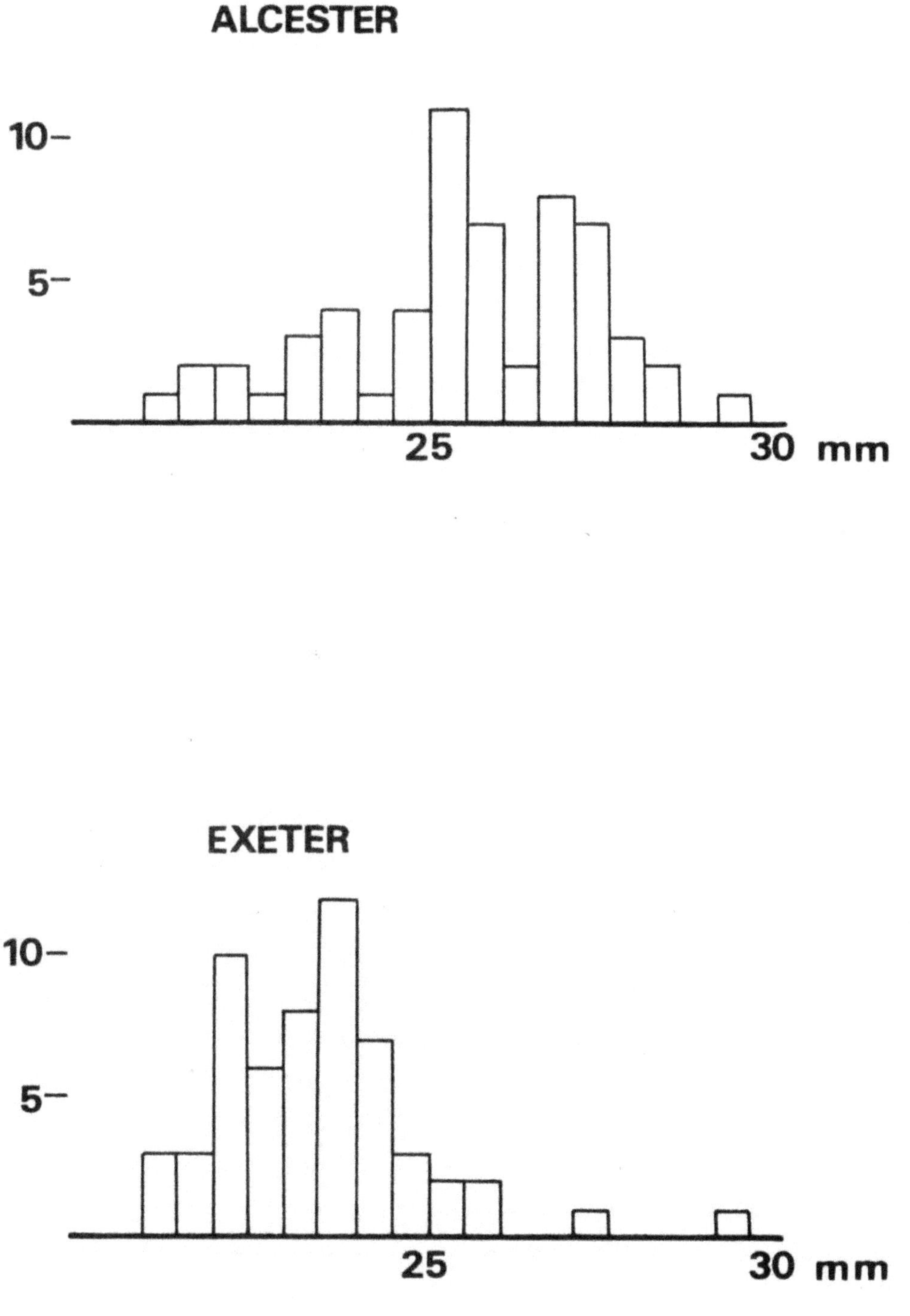

Fig. 8.8 Maximum distal widths of sheep/goat tibiae from Romano-British deposits at Alcester and Exeter.

parallel with the situation in the Roman provinces of central and eastern Europe, where similar changes in sheep size took place (Bökönyi 1974: 178).

Horse

Iron Age samples have consistently produced evidence for the keeping of small ponies ranging from 10-14 hands high. Examples have come from Ashville (Wilson 1978: 135), Gussage All Saints (Harcourt 1979: 153) and Winklebury (Jones, R. 1977: 64). Metrical analyses of horse bones of Romano-British and Anglo-Saxon date have been limited by their rare occurrence on most sites. At present it seems that the smallest ponies represented in the Iron Age samples do not appear on later sites and some bones of larger horses have been recovered, although few appear to have been over 14 hands (e.g. Wilson 1978: 117-118; Bourdillon and Coy 1980: 104-105; Clutton-Brock 1976: 383).

Dogs

Harcourt's (1974) review of the evidence for dogs on British archaeological sites included an extensive survey of metrical analysis. More recent studies have tended to support his observations. Measurements of Iron Age specimens revealed relatively little range in variation in the shape of the skulls and estimates of shoulder heights using conversion factors from the lengths of limb bones ranged between 35-58 cm, mostly in the upper range, on material derived from 28 sites (Harcourt 1974: 163). The range has been increased slightly by the discovery of a partial skeleton of a dog in a late Iron Age deposit at Ashville that had an estimated shoulder height of 60 cm (Wilson 1978: 125). Although articulated skeletons of dogs appear more frequently on Iron Age sites than those of other species, there is no doubt that dog meat was consumed at some settlements. Butchery marks on dog bones made during the disarticulation of the skeleton and the stripping of meat have been observed, for example, at Ashville (Wilson 1978: 122), Winnall Down (Maltby n.d. 1) and Balksbury (Maltby n.d. 2). The latter two sites also produced a relatively large number of bones of very young puppies, a phenomenon that also occurred at Gussage All Saints (Harcourt 1979: 154). It is possible that the numbers of dogs were controlled by killing some of the newborn puppies, although natural neonatal mortality could have accounted for some of these deaths.

The Roman period saw a significant increase in the variation of the types of dog kept. Estimates of shoulder heights ranged from 24-72 cm and skull shapes showed much greater variation (Harcourt 1974: 164-166). The smallest dogs were regarded as lap dogs and the largest as hunting dogs. The variation in skull shape appears to have decreased in the Anglo-Saxon period, if the specimens from the 11 sites examined by Harcourt are typical. The range in shoulder heights was similar (23-71 cm) but the majority of specimens belonged to large individuals.

Conclusions

Methodological Questions

It is obvious that any conclusions derived from faunal studies are handicapped by the lack of suitable data. Whatever aspect has been considered, the same problems have been found. There is a limited number of adequate samples and there are substantial gaps in our knowledge in some periods and regions. Incompatibility of archaeozoological techniques has made inter-site comparisons difficult. There has been a general lack of understanding of consideration of the problems of the formation of faunal samples and studies of intra-site variability have been few and far between. Finally, and perhaps most disturbing, there is the inadequacy of many of the past and current archaeozoological methodologies to deal efficiently with the data.

The number of samples will increase significantly during the next decade. This will improve the situation only if they are studied in a more rigorous way. Attempts to come to terms with differential preservation of animal bones and other aspects of intra- and inter-site variation must be made. Taphonomic studies have begun to investigate some of these problems (e.g. Binford and Bertram 1977) and the potential complexities of the derivation of animal bones on archaeological sites have been demonstrated in several ethnoarchaeological studies (e.g. Binford 1978). In this country, attempts at studying faunal samples in such detail are in their infancy. They involve much more rigorous recording and analysis of animal bones than has commonly been the case. These may be time-consuming and laborious tasks, yet they are necessary. Any bone recovered from a site has been the subject of several modifications of natural or cultural origin. Understanding of the composition of the herds and their exploitation cannot be achieved from animal bones without such studies. They involve the careful analysis of each bone fragment recovered to obtain information about preservation, butchery and gnawing. It is also hoped that more research will be undertaken to obtain more reliable estimates of ageing, sexing and the types of animals kept. More detailed metrical analyses have a part to play in the latter studies but other approaches, such as those of chemical analysis and observations of morphological criteria on bones (Noddle 1978) may also prove of value. Brothwell (this volume) has already shown the potential information that can be derived from the study of pathological data. It is important to draw upon as many lines of investigation as possible, in order to make full use of the inherently imperfect data studied by faunal analysts.

From Appendix to Integration

The purpose of this article has been to show the state of progress of faunal studies and to point out the limitations. Faunal analysis is a specialism within archaeology. Its independence, reflected by the inclusion of separate animal bone appendices in many site reports, is understandable and to a certain extent justified. It is the lack of integration of faunal evidence with the rest of the archaeological record that has been the major drawback of much previous work. Faunal analysis is only one source of information that can be drawn upon to study pastoral husbandry. Bradley (1978: 45-53), for example, has discussed the evidence available from the artefacts, field evidence and facilities that may have been associated with pastoral activities in the prehistoric period. Nor should pastoral farming be looked at in isolation from arable farming but few attempts have been made to integrate faunal and floral data from British sites. A notable exception has been the work at Gatcombe (Branigan 1977), where there has been an attempt to assess the potential agricultural output from the villa estate and a consideration of the operation of the complex internal and external economies of the estate. To be fair, several current projects are making concerted efforts along similar lines.

For it is indeed the integration of faunal studies that is urgently required. It is hoped that even the limited faunal analyses to date have shown—or at least have the potential to show—changes and developments in stock-keeping practices within regions and through time. Explanations of these changes obviously cannot be made without reference to other aspects of the archaeological and historical records. In the same way, developments in pastoral farming have to be studied within the framework of broader models or hypotheses. The pastoral economy had a bearing upon and was affected by many environmental, economic and behavioural factors. The relationship between these factors can be complex. As Clarke (1978: 116) asserted:-

> "The dynamic equilibrium between the economic subsystem and the other subsystems is stabilized by continually adjusting the mutual values and states of these networks. The economy must be kept as nearly as possible in equilibrium with the information and constraints from taboos, social organisation and division of labour by age and sex, individual personalities, and the available capacity or range of variety of material culture."

Whether or not one studies such processes in terms of culture systems, the underlying principle remains the same. The attitude that there are two types of archaeology—"environmental" and "cultural"—is unnecessarily divisive, since the two elements have always been incontrovertibly linked.

It is already possible to show how the results of faunal analysis can provide useful information about many wider issues. In a recent volume the impact of the Roman invasion on native British communities was discussed and various lines of investigation were pursued by several authors, including pottery studies, industry and trading patterns, coinage, urban development and rural settlement patterns (Burnham and Johnson 1979). Similar studies on animal bones could shed light on the impact the invasion had upon the pastoral economy. The need to supply the troops of the invading army, the possible introduction of new stock and the introduction of new farming techniques may all have had significant repercussions on husbandry practices. There is already some evidence for changes in the exploitation patterns of sheep and cattle in the Romano-British period. In addition, the appearance of some larger stock and evidence for the supply of specific age groups of cattle to some military and urban settlements may all be seen to be direct or indirect consequencies of the Roman invasion, although the details of the mechanisms involved in those developments need to be ascertained. How great Roman influence was in different parts of the province has also to be determined. It is interesting to note, for example, that the apparent absence of larger sheep and cattle in the south-west peninsula occurs in an area where there are virtually no villas (Fox 1973: 171-174; Wacher 1980: 123) and thus supports the view that "Romanisation" had little impact in that area (Rivet 1964: 116). In areas where villas and native settlements are both found, it should be possible to develop the work of King (1978) to test whether the apparent dichotomy in wealth and power of the inhabitants of Roman Britain in the late 1st and 2nd centuries (e.g. Todd 1978: 204-205) is also reflected in the faunal remains.

The relationship between other developments in settlement patterns and animal husbandry has not as yet been investigated in detail. For example, the changes observed in the settlement pattern in some parts of England in the late Iron Age (Miles, this volume) may coincide with changes in species exploitation but again too few samples of that date have been analysed. Similarly, it should be possible to use faunal analyses to test the models for change in late Roman Britain. Reece (1980) has suggested that in the 4th century town life in the true sense of the word had virtually disappeared and that there was a change from the town-based economy prevalent in the 2nd century to a villa and village-based economy in the 4th century. Such fundamental changes in settlement and economic patterns must also have had some impact on the agricultural basis of the economy. As Reece points out (1980: 78), towns contained a large element of population not directly involved with agricultural production but engaged in administration or trading activities. There was therefore a greater need for the organisation of food supplies. In support of this, analysis of faunal material of 1st and 2nd century date in Exeter produced evidence for large-scale butchering of cattle carcasses and concentrations of cattle and sheep killed at specific ages for their meat. In contrast, the 4th century deposits from Exeter have so far revealed no comparable large-scale butchery evidence, a greater range of cattle killed at different ages and evidence for stock enclosures within the city walls. All these factors may indicate the collapse of the former system of supply (Maltby 1979). Examination of faunal material from other urban sites ought to be able to test how typical the samples from Exeter are.

The above are just a few examples taken from the Romano-British period to show how faunal analysis can be incorporated into testing more general hypotheses. There are many more examples, some of which derive from observations made on aspects of faunal studies discussed earlier in this review. Space precludes detailed examination of these topics, and in any case the present limited available data make several alternative explanations possible. It is hoped, however, that the current position of faunal studies has been shown. Some of the basic groundwork towards the understanding of the relative representation, introduction and exploitation of different domestic species has been done. It remains, on the one hand, to broaden our knowledge of the many poorly represented regions, periods, settlement types and the less common domestic species. More important, however, is the incorporation of faunal studies into the wider framework of prehistoric and early historic studies and to develop methodologies better equipped to cope with these broader issues. To do this, we need to develop a planned regional approach to faunal studies. It will then be possible to compare the data from different types of site against a common background of environment and agricultural potential. It will also be possible to tackle the problems of sample variability constructively, so that all samples, including quite small ones, have significance. In some areas such regional projects have begun and only by the integration of such multi-period research with other aspects of the archaeological record will questions relating to the development of animal husbandry be answered.

Acknowledgements

I would like to thank Paul Booth, Peter Fasham, Neil Loughlin and Geoffrey Wainwright for allowing me to incorporate data from their excavations and Shelagh Wall and Jennie Coy for the supply of unpublished faunal data. Tim Champion, Jennie Coy and Clive Gamble provided valuable and constructive comments to earlier drafts of this paper. The figures were drawn by Martin Oake.

Bibliography

Alcock, L. (1975). Dry bones and living documents. In: The Effect of Man on the Landscape: the Highland Zone (eds Evans J. G., Limbrey S. and Cleere H.) Council for British Archaeology Research Report 11; pp. 117-122.

Applebaum, S. (1972). Roman Britain. In: The Agrarian History of England and Wales. Volume I. (ed Finberg H. P. R.) Cambridge: University Press pp. 1-277.

Armitage, P. L. and Chapman, H. (1979). Roman mules. The London Archaeologist 3 339-346.

Askew, S. (1961). Appendix. In: Webster, G., An excavation on the Roman site at Little Chester, Derby. Derbyshire Archaeological Journal 81 107-108.

Banks, J. W. (1967). Human and animal bones. In: Avery, M., Sutton, J. E. G. and Banks, J. W., Rainsborough, Northants., England: excavations 1961-5. Proceedings of the Prehistoric Society 33 302-305.

Binford, L. R. (1978). Nunamiut Ethnoarchaeology. New York: Academic Press.

Binford, L. R. and Bertram, J. B. (1977). Bone frequencies and attritional processes. In: For Theory Building in Archaeology (ed Binford, L. R.) New York: Academic Press pp. 77-153.

Bird, P. F. (1968). Animal bones from Tollard Royal. In Wainwright, G. J., Excavation of a Durotrigian farmstead near Tollard Royal in Cranbourne Chase, Southern England. Proceedings of the Prehistoric Society 34 146-147.

Boessneck, J. (1969). Osteological differences between sheep (Ovis aries Linné) and goat (Capra hircus Linné). In: Science in Archaeology (eds Brothwell, D. and Higgs, E. S.) London: Thames and Hudson pp. 331-358.

Boessneck, J. and Driesch, A. von den (1978). The significance of measuring animal bones from archaeological sites. In: Approaches to Faunal Analysis in the Middle East (eds Meadow, R. H. and Zeder, M.A.) Harvard University: Peabody Museum Bulletins 2 pp. 25-39.

Boessneck, J., Müller, H. H. and Teichert, M. (1964). Osteologische Unterscheidungsmerkmale zwischen Schaf (Ovis aries Linné) und Ziege (Capra hircus Linné). Kühn-Archiv 78 1-129.

Bökönyi, S. (1974). History of Domestic Mammals in Central and Eastern Europe. Budapest: Akadémiai Kiadó.

Bourdillon, J. and Coy, J. P. (1980). The animal bones. In: Excavations at Melbourne Street, Southampton, 1971-76. (ed Holdworth, P.) CBA Research Report 33 pp. 79-121.

Bradley, R. (1978). The Prehistoric Settlement of Britain. London: Routledge and Kegan Paul.

Branigan, K. (1977). Gatcombe: the excavation and study of a Romano-British villa estate, 1967-1976. BAR 44.

Bramwell, D. (1978). The bird bones. In: The Excavation of an Iron Age Settlément, Bronze Age Ring-ditches and Roman Features at Ashville Trading Estate, Abingdon (Oxfordshire) 1974-76 (ed. Parrington, M.) Oxfordshire Archaeological Unit Report 1. CBA Research Report 28 p. 133

Bramwell, D. (1979). Bird remains. In Gracie, H. S. and Price, E. G., Frocester Court Roman villa: second report. Transactions of the Bristol and Gloucestershire Archaeological Society 97 61-62.

Brodribb, A. C. C., Hands, A. R. and Walker, D. R. (1968). Excavations at Shakenoak Farm, near Wilcote, Oxfordshire. Part 1: Sites A and D. Oxford: privately printed.

Brodribb, A. C. C., Hands, A. R. and Walker, D. R. (1971). Excavations at Shakenoak Farm, near Wilcote, Oxfordshire. Part II: Sites B and H. Oxford: privately printed.

Brodribb, A. C. C., Hands, A. R. and Walker, D. R. (1972). Excavations at Shakenoak Farm, near Wilcote, Oxfordshire. Part III: Site F. Oxford: privately printed.

Bulleid, A. and Gray, H. St. G. (1917). The Glastonbury Lake Village. Glastonbury Antiquarian Society, volume 2.

Burnham, B. C. and Johnson, H. B. (1979). Invasion and Response. The Case of Roman Britain. BAR 73.

Carter, P. L. and Phillipson, D., with Higgs, E. S. (1965). Faunal report. In Hastings, F. A., Excavations of an Iron Age farmstead at Hawk's Hill, Leatherhead. Surrey Archaeological Collections 62 40-42.

Champion, T. C. (1979). The Iron Age (c. 600 BC-AD 200). A. Southern Britain and Ireland. In: Introduction to British Prehistory (eds Megaw, J. V. S. and Simpson, D. D. A.), Leicester: University Press pp. 344-432.

Chaplin, R. E. (1971). The Study of Animal Bones from Archaeological Sites. New York: Seminar Press.

Clark, J. G. D. (1947). Sheep and swine in the husbandry of prehistoric Europe. Antiquity 21 122-136.

Clarke, D. L. (1978). Analytical Archaeology. London: Methuen; second edition.

Clutton-Brock, J. (1976). The animal resources. In: The Archaeology of Anglo-Saxon England. (ed Wilson, D. M.) London: Methuen pp. 373-392.

Clutton-Brock, J. and Armitage, P. L. (1977). Mammal bones from Trench A. In T. R. Blurton, Excavations at Angel Court, Walbrook, 1974. Transactions of the London and Middlesex Archaeological Society 28 88-97.

Cornwall, I. A. (1953). Animal bones. In Kenyon, K. M., Excavations at Sutton Walls, Hertfordshire, 1948-1951. Archaeological Journal 110 79-83.

Coy, J. P. (1980 in press). The animal bones. In Haslam, J., Excavations of a Mid-Saxon iron smelting site at Ramsbury, Wiltshire. Medieval Archaeology, 24.

Coy, J. P. (n.d.). Animal bones from the Micheldever Wood Banjo Enclosure, Hampshire. R27 M3 Motorway Rescue Excavations.

Cram, C. L. (1973). The animal bones. In Brodribb, A. C. C., Hands, A. R. and Walker, D. R., Excavations at Shakenoak Farm, near Wilcote, Oxfordshire. Part IV: Site C. Oxford: privately printed 145-164.

Cram, C. L. (1978). Animal bones. In Brodribb, A. C. C., Hands, A. R. and Walker, D. R., Excavations at Shakenoak Farm, near Wilcote, Oxfordshire. Part V: Sites K and E. Oxford: British Archaeological Reports pp. 117-178.

Cunliffe, B. (1978). Iron Age Communities in Britain. London: Routledge and Kegan Paul; second edition.

Cunliffe, B. and Phillipson, D. W. (1968). Excavations at Eldon's Seat, Encombe, Dorset. Proceedings of the Prehistoric Society 34 191-237.

Dennell, R. W. (1979). Prehistoric diet and nutrition: some food for thought. World Archaeology 11 121-135.

Driesch, A. von den (1976). A Guide to the Measurement of Animal Bones from Archaeological Sites. Harvard University: Peabody Museum Bulletins 1.

Eastham, A. (1971). The bird bones. In: Excavations at Fishbourne 1961-1969. (ed Cunliffe, B.) Reports of the Research Committee of the Society of Antiquaries of London 27; volume II pp. 388-393.

Eastham, A. (1975). The bird bones. In: Excavations at Portchester Castle, vol. I: Roman. (ed Cunliffe, B.) Reports of the Research Committee of the Society of Antiquaries of London 32; pp. 409-415.

Eastham, A. (1976). Bird bones. In: Excavations at Portchester Castle, vol. II: Saxon. (ed Cunliffe, B.) Reports of the Research Committee of the Society of Antiquaries of London 33; pp. 287-296.

Ewbank, J. M., Phillipson, D. W., Whitehouse, R. D. and Higgs, E. S. (1964). Sheep in the Iron Age: a method of study. Proceedings of the Prehistoric Society 30 423-426.

Fifield, P. W. (1979). Animal bones. In Potter, T. W., Romans in North-West England. Cumberland and Westmorland Antiquarian and Archaeological Society Research Series 1; 299-311.

Finberg, H. P. R. (1972). Anglo-Saxon England to 1042. In: The Agrarian History of England and Wales. Volume I. (ed Finberg, H. P. R.) Cambridge: Cambridge University Press; pp. 385-525.

Finberg, H. P. R. (1974). The Formation of England 550-1042. London: Hart-Davis, MacGibbon.

Fox, A. (1973). South-West England 3500 BC-AD 600. Newton Abbot: David and Charles.

Fraser, F. C. (1954). Animal bones. In Wheeler, R. E. M., The Stanwick Fortifications. Reports of the Research Committee of the Society of Antiquaries of London 17; 57-58.

Gardner, W. and Savory, H. N. (1964). Dinorben. Cardiff: National Museum of Wales.

Gilmore, F. (1969). The animal and human skeletal remains. In Woods, P. J., Excavations at Hardingstone, Northants., 1967-8. Northampton: Northamptonshire County Council; 43-55.

Grant, A. (1971). The animal bones. In: Excavations at Fishbourne 1961-1969. (ed Cunliffe, B.) Reports of the Research Committee of the Society of Antiquaries of London 27; volume II, pp. 377-388.

Grant, A. (1975). The animal bones. In: Excavations at Portchester Castle, vol. I. Roman. (ed Cunliffe, B.) Reports of the Research Committee of the Society of Antiquaries of London 32; pp. 378-408, 437-450.

Grant, A. (1976). The animal bones. In: Excavations at Portchester Castle, vol. II. Saxon. (ed Cunliffe, B.) Reports of the Research Committee of the Society of Antiquaries of London 33; pp. 262-287.

Grant, A. (1978). Animal bones. In Bradley, R., Rescue excavation in Dorchester-on-Thames. Oxoniensa 43 32-36.

Grayson, D. K. (1980). On the quantification of vertebrate archaeofaunas. In: Advances in Archaeological Method and Theory. Volume 2. (ed Schiffer, M.) New York: Academic Press; pp. 199-237.

Haglund-Calley, L. and Cornwall, I. W. (1963). Report on the Dinas Powys animal bones. In: Dinas Powys: an Iron Age, Dark Age and Early Medieval Settlement in Glamorgan. (ed Alcock. L.) Cardiff: University of Wales Press; pp. 192-194.

Hamilton, J. (1978). A comparison of the age structure at mortality of some Iron Age and Romano-British sheep and cattle populations. In: The Excavation of an Iron Age Settlement, Bronze Age Ring-ditches and Roman Features at Ashville Trading Estate, Abingdon, (Oxfordshire) 1974-76. (ed Parrington, M.) Oxfordshire Archaeological Unit Report 1 CBA Research Report 28; pp. 126-133.

Harcourt, R. A. (1970). Animal bones. In Rahtz, P., Excavations on Glastonbury Tor, Somerset 1964-6. Archaeological Journal 127 56-63.

Harcourt, R. A. (1974). The dog in prehistoric and early historic Britain. Journal of Archaeological Science 1 151-176.

Harcourt, R. A. (1979). The animal bones. In Wainwright, G. J., Gussage All Saints: an Iron Age Settlement in Dorset. London: Department of Environment Archaeological Report 10; 150-160.

Harding, D. W. (1974). The Iron Age in Lowland Britain. London: Routledge and Kegan Paul.

Harman, M. (1969). The animal bones. In Todd, M., The Roman settlement at Margidunum: excavations 1966-8. Transactions of the Thoroton Society of Nottinghamshire 73 96-103.

Harman, M. (1979). The mammalian bones. In Williams, J. H., St. Peter's Street, Northampton, Excavations 1973-1976. Northampton Development Corporation Archaeological Monograph 2; 328-332.

Higgs, E. S. and Greenwood, W. (1976). Fauna. In Stead, I. M., Excavations at Winterton Roman Villa. London: Department of the Environment Archaeological Reports 9; 301-303.

Higgs, E. and Greenwood, W. with Garrand, A. (1979). Faunal Report. In: Rahtz, P. The Saxon and Medieval Palaces at Cheddar BAR 65 pp. 354-362.

Hodgson, G. W. I. (1968). A comparative account of the animal remains from Corstopitum and the Iron Age site at Catcote near Hartlepools, Co. Durham. Archaeologia Aeliana 46 127-162.

Hodgson, G. W. I. (1977). Vindolanda II. The Animal Remains 1970-1975. Hexham: Vindolanda Trust.

Hope-Taylor, B. (1977). Yeavering: an Anglo-British Centre of Early Northumbria. London: Department of the Environment Archaeological Reports 7.

Huggins, P. J. (1978). Excavations of Belgic and Romano-British farm with Middle Saxon cemetery and churches at Nazeingbury, Essex, 1975-6. Transactions of the Essex Archaeological Society 10 29-117.

Jarman, M. and Fagg, A., with Higgs, E. S. (1968). Animal remains. In I. M. Stead, An Iron Age hill-fort at Grimthorpe, Yorkshire, England. Proceedings of the Prehistoric Society 34 182-189.

Jewell, P. A. (1962). Changes in size and type of cattle from prehistoric to medieval times in Britain. Zeitschrift für Tierzüchtung und Züchtungsbiologie 77 159-167.

Jones, G. (1977). Zoological evidence. In Rogerson, A., Excavations at Scole, (1973). East Anglian Archaeology Report 5; 209-213.

Jones, R. T. (1977). Animal bones. In Smith, K., The excavation of Winklebury Camp, Basingstoke, Hampshire. Proceedings of the Prehistoric Society 43 58-69.

King, A. (1978). A comparative survey of bone assemblages from Roman Sites in Britain. Institute of Archaeology Bulletin 15 207-232.

Maltby, J. M. (1979). Faunal Studies on Urban Sites: the Animal Bones from Exeter 1971-1975. (Exeter Archaeological Reports 2). Sheffield University: Department of Prehistory and Archaeology.

Maltby, J. M. (1980). Modern refuse and ancient behaviour. Nature 284 215-216.

Maltby, J. M. (n.d.1). The animal bones from Winnall Down, Hampshire. R17, M3 Motorway Rescue Excavations.

Maltby, J. M. (n.d.2). The animal bones from the excavations at Balksbury, Hampshire, 1973.

Maltby, J. M. (n.d.3). The animal bones from the excavations at Baylham House, Suffolk.

Maltby, J. M. (n.d.4). The animal bones from the excavations of 1, Bleachfield Street, Alcester, Warwickshire.

Marples, B. J. (1972). Miscellaneous bones of birds and small animals. In Brodribb, A. C. C., Hands, A. R. and Walker, D. R., Excavations at Shakenoak Farm, near Wilcote, Oxfordshire. Part III: Site F. Oxford: privately printed; 131.

Marples, B. J. (1973). Miscelleneous bones of small mammals and birds. In Brodribb, A. C. C., Hands, A. R. and Walker, D. R., Excavations at Shakenoak Farm, near Wilcote, Oxfordshire. Part IV: Site C. Oxford: privately printed; 164-165.

Marples, B. J. (1974). Report on the animal bones. In Frere, S. S. and St. Joseph, J. K. S., The Roman fort at Longthorpe. Britannia 5 122-128.

Meek, A. and Gray, R. A. (1911). Animal remains. In Forster, R. H. and Knowles, W. H., Corstopitum, report on the excavations of 1910. Archaeologia Aeliana 7 78-125.

Miller, E. and Hatcher, J. (1978). Medieval England: Rural Society and Economic Change 1086-1348. London: Longman.

Noddle, B. A. (1970). The animal bones. In Fowler, P. J., Gardner, K. S. and Rahtz, P. A., Cadbury Congresbury, Somerset, 1968. Bristol: 87-40.

Noddle, B. A. (1973). Animal bones from two wells at Tripontium. In Cameron, H. and Lucas, J., Tripontium: second interim report. Transactions and Proceedings of the Birmingham and Warwickshire Archaeological Society 85 136-142.

Noddle, B. A. (1975). A comparison of the animal bones from eight medieval sites in southern Britain. In: Archaeozoological Studies. (ed Clason, A. T.) Amsterdam: North Holland Publishing Co.; pp. 248-260.

Noddle, B. A. (1976). Report on the animal bones from Walton, Aylesbury excavations 1973-4. Records of Buckinghamshire 20 269-287.

Noddle, B. A. (1978). Some minor skeletal differences in sheep. In: Research Problems in Zooarchaeology (ed Brothwell, D. R., Thomas, K. D. and Clutton-Brock, J.) Institute of Archaeology Occasional Publication 3; pp. 133-141.

Noddle, B. A. (1979). The animal bones. In Gracie, H. S. and Price, E. G., Frocester Court Roman villa: second report. Transactions of the Bristol and Gloucestershire Archaeological Society 97 51-60.

Noddle, B. A. (1980). The animal bones. In Wade-Martins, P., North Elmham Park. East Anglian Archaeology Report 9; volume 2 375-412.

Payne, S. (1969). A metrical distinction between sheep and goat metacarpals. In: The Domestication and Exploitation of Plants and Animals (eds Ucko, P. J. and Dimbleby, G. W.) London: Duckworth pp. 295-305.

Payne, S. (1972a). Partial recovery and sample bias: the results of some sieving experiments. In: Papers in Economic Prehistory. (ed Higgs, R. S.) Cambridge: University Press; pp. 49-64.

Payne, S. (1972b). On the interpretation of bone samples from archaeological sites. In: Papers in Economic Prehistory. (ed Higgs, E. S.) Cambridge: University Press; pp. 65-81.

Payne, S. (1973). Kill off patterns of sheep and goats: the mandibles from Asvan Kale. Anatolian Studies 23 281-303.

Payne, S. (1975). Partial recovery and sample bias. In: Archaeozoological Studies. (ed Clason, A. T.) Amsterdam: North Holland Publishing Co., pp. 7-17.

Pitt-Rivers, A. H. (1892). Excavations in Bokerley Dyke and Wansdyke. Excavations in Cranbourne Chase, volume III. London: privately printed.

Pitt-Rivers, A. H. (1898). Excavation of the South Lodge Camp. Excavations in Cranbourne Chase, volume IV. London: privately printed.

Rackham, J. (1979). Animal resources. In Carver, M. O. H., Three Saxo-Norman tenements in Durham City. Medieval Archaeology 23 47-54.

Reece, R. (1980). Town and country: the end of Roman Britain. London: Hutchinson; second edition.

Rivet, A. L. F. (1964). Town and Country in Roman Britain London: Hutchinson; 2nd Edn.

Seddon, D., Calvocoressi, D. and Cooper, C. (1964). In Addyman, P. V., Dark Age Settlement at Maxey, Northants. Medieval Archaeology 8 69-71.

Silver, I. A. (1969). The ageing of domestic animals. In: Science in Archaeology (eds Brothwell, D. and Higgs, E. S.) London: Thames and Hudson; 283-302.

Todd, M. (1978). Villas and Romano-British society. In: Studies in the Romano-British Villa. (ed Todd, M.) Leicester: University Press; pp. 197-208.

Wacher, J. (1980). Roman Britain. London: Dent.

Watson, J. P. N. (1973). The bones from Pit 15. In Chapman, H. and Johnson, T., Excavations at Aldgate and Bush Lane House in the City of London, (1972). Transactions of the London and Middlesex Archaeological Society 24 51-53.

Westley, B. (1967). In Wainwright, G. W., Coygan Camp: a prehistoric Romano-British and Dark Age Settlement in Carmarthenshire. Cardiff: Cambrian Archaeological Association; 190-194.

Westley, B. (1970). The animal bones from Budbury. In Wainwright, G. W., An Iron Age Promontory Fort at Budbury, Bradford-on-Avon, Wiltshire. Wiltshire Archaeological Magazine 65 152-153.

White, K. D. (1970). Roman Farming. London: Thames and Hudson.

Whitehouse, R. and Whitehouse, D. (1974). The fauna. In Stanford, S. C., Croft Ambrey. Hereford: privately printed; 215-221, 238-242.

Wilson, R. (1976). The animal bones. In Parrington, M., Roman finds and animal bones from Kingston Hill Farm, Kingston Bagpuize, Oxon. Oxoniensa 41 67-69.

Wilson, R. (1978). Methods and results of bone analysis/General conclusions and discussion of the bone sample. In: The Excavation of an Iron Age Settlement, Bronze Age Ring-ditches and Roman Features at Ashville Trading Estate, Abingdon (Oxfordshire) 1974-76. (ed Parrington, M.) Oxfordshire Archaeological Unit Report 1, CBA Research Report 28 pp. 110-126, 133-139.

9. THE SIGNIFICANCE OF DEER REMAINS AT OCCUPATION SITES OF THE IRON AGE TO THE ANGLO-SAXON PERIOD

Annie Grant

The purpose of this discussion paper is to examine the relationship between the agricultural communities of the Iron Age, Roman and Saxon periods and the animals, particularly the deer, that were living wild in Britain. Unfortunately, there are as yet very few published faunal analyses from sites dated to the Saxon period and consequently the discussion of this period can only be very limited.

A survey of the evidence, available from faunal analyses of material recovered from occupation sites dated to the period under consideration, shows that at the vast majority of sites, deer remains are very rare or entirely absent. Typical are sites such as the Iron Age hillfort of Danebury, in Hampshire (Grant in prep.) where less than 0.5% of the bones were deer bones, the Roman fort at Stanwick in Yorkshire (Fraser 1954) where approximately 2% of the bones were deer bones and the early Saxon occupation at Dinas Powys in Wales (Haglund-Cally and Cornwall 1963) where 1% of the bones were deer bones.

It must be stressed that there are several problems in establishing the numbers of deer remains present in archaeological contexts. First there is the problem of the identification of deer bones. Red deer and cattle bones, and roe deer and sheep bones may be confused and it is not usually possible to ascertain the accuracy of the identifications, especially where sites were excavated several years ago. Secondly, in many of the reports of animal bone remains, especially in those published before the 1960s, details of the numbers of bones found are omitted and the relative importance of the speçies identified is not clear. Thirdly, when deer remains are mentioned, a distinction is not always made between deer bone and deer antler, the significance of which is shown below. These problems are further compounded by problems arising from small-scale excavations, excavations planned to look for defences rather than occupation areas and by the fact that by no means have all areas of Britain been given equal archaeological attention. Thus this paper does not attempt a comprehensive summary of the evidence, although a large number of bone reports from sites all over Britain have been examined. The discussion generally uses as examples sites that were excavated relatively recently and where the faunal analysis was appropriately detailed.

Although a variety of wild animals could have been hunted in Britain, including wild boar, wolves, foxes, badgers and the smaller wild mammals, references in the literature to finds of bones from these animals are extremely rare. The general assumption is (e.g. Megaw and Simpson 1979: 352) that hunting did not play a significant part in providing meat for agricultural communities and that most meat consumed was obtained from domestic animals.

This paper does not attempt to quarrel with this as a general assumption but tries to examine the implications of that assumption and to examine the situation in more detail.

Deer are likely to have been of interest to a community for several possible reasons. They could be a source of food, a source of raw material (particularly antlers), the prey in sporting activities and a pest inflicting damage on crops. Any possible magical or religious significance of deer is not considered here.

The absence of deer bones at a site may be due to the fact that the community did not choose to hunt for deer (for whatever reason) or that they could not hunt and catch deer because there were no deer living in the local environment. First, then, we must establish the likely distribution of deer in Britain in the period under discussion.

Fallow deer, although present in Britain before the last glaciation, seem to have become extinct in Britain and much of Europe by the post-glacial period. They were reintroduced, but the exact time of their reintroduction is disputed. Although the Romans are popularly credited with their reintroduction, in fact there is very little evidence for their presence in significant numbers before the Norman Conquest. This problem has been discussed in detail by the Chapmans (1975). For the purpose of this paper the assumption has been made that fallow deer were either absent or rare in Britain in the Iron Age, Roman and Saxon periods.

Red and roe deer are indigenous species and, being creatures of the woodlands and forests, were undoubtedly widely distributed in Britain following the last glaciation. Their present more restricted distribution is largely a result of the progressive reduction of their numbers, associated with the destruction of their natural habitat. The clearance of the forest cover, particularly on the chalk and limestone soils and the Breckland sands, is well attested from the neolithic period onward (Evans 1975) but very extensive areas of forest must have remained, particularly on the heavier soils. There can have been few agricultural communities that did not have access to areas of woodland that could have supported herds of deer. Red and roe deer will adapt to more open environments too, as is shown by their presence on the open moorlands of Northern England and Scotland at the present day, but this is not their natural environment, clearly seen in the fact that on moorland, the deer rarely achieve the size of woodland and forest animals (Ritchie 1920). Further evidence of their widespread distribution is given in the presence of deer remains, even if only in small numbers, on sites scattered over much of Britain in a wide variety of environments. There is certainly no evidence then that the scarcity of deer bones was necessarily a result of the scarcity of the live animals.

If deer were present close to occupation sites, the question is, why were they not hunted as they had been in for example the mesolithic period in Britain? The obvious answer to this question is that when domestic animals were introduced to Britain, the settled agricultural communities or even the pastoralists had neither the need nor the time for hunting to supplement the protein diet of the community. It would appear that the animal protein needs of the population must have been largely satisfied by the domestic animals that the population had themselves raised, tended or traded for and that this was a more efficient way of producing food than hunting. Meat may have had a relatively minor role in the diet if agriculture was able to supply sufficient cereal and vegetable products, The scarcity of deer bones may testify to the efficiency of the farming systems of the agricultural communities and to the advantages of keeping animals that provide not only meat but milk, wool, manure and traction.

If this then is the explanation of the absence of scarcity of deer bones at the majority of sites so far excavated, it is clearly important to examine the reason for the presence of more significant numbers of deer bones at a minority of sites.

At the Iron Age and Roman site of Dun Mor Vaul (Noddle 1974) the percentage of deer bones recovered was over 20% of the total number of bones, Substantial numbers of seal bones were also found and in only one phase of occupation did the percentage of the bones of domestic animals exceed 64% of the total—in the final phase of occupation only 28% of the bones were those of domestic animals. Dun Mor Vaul is situated on Tiree, an island in the Inner Hebrides that lies in a rather exposed position. Although the analysis of the pollen from the site (Pilcher 1974) implies that there was agricultural activity on the island and the analysis of the grain (Renfrew 1974) gives evidence for the cultivation of grain, it is likely that stock rearing and agriculture may not have been as easy or as productive as at the more sheltered and fertile mainland sites. At Skitten (Platt 1948) in Caithness, a county where even today farming is difficult, red deer bones were described as being more common than usual. At these sites, in relatively harsh environments, hunting and perhaps gathering too might have been necessary to supplement the food provided by agricultural activities. The choice of hunted animals would of course depend on the local environment. At some of the Cornish sites and especially at sites of this period in the Scillies, although some deer bones have been found, seal bones are common. At Nour-nour (Turk 1968) the bones of seal are as common as those of sheep. This situation is by no means found at all coastal sites however. At Portchester Castle, a coastal site in Hampshire, there seems to have been very little exploitation of the sea's resources, at least in the Roman and Saxon periods (Grant 1975a, 1975b). Here the position, although coastal, is close to good arable and grazing land which could have provided all the food that was required.

It is therefore suggested that comparatively large numbers of the bones of deer or of other wild, hunted animals, may indicate an environment where agriculture and stock rearing were difficult, because of poor soil or a severe climate. The hunted animals would be necessary to supplement the food produced by agricultural activities.

Other sites where deer bones were more common than usual for the period are found in inland regions in good agricultural areas, and here a rather different explanation is suggested. At the Roman villa sites of Chalk (Eastham 1972) and Latimer (Hamilton 1971) and at the Roman town of Clausentum (Cornwall 1953) there was a fairly marked increase in the percentage of deer bones found in the deposits dated to the latest phases of occupation at the sites. At both Chalk and Latimer, the increase in the percentage of deer bones coincides with a period of abandonment and squatter activity. At Chalk, there was

was an increase in the percentage of deer bones from 5% in the period of the final occupation of the villa to 13% in the 4th century, when the villa had been abandoned, but rubbish was still accumulating, indicating some sporadic or squatter occupation. At Latimer, the percentage of deer bones increased from 4% in the first phase of occupation to 19% during a period of abandonment and squatter occupation from 280 to 290 AD and to 26% in the last villa phases. In the two phases of post villa occupation, dated to the very end of the Roman occupation, the percentage of deer bones was 14% and 21%. At Clausentum, deer bones were absent in the earliest occupation layers, but were present in the occupation levels of the 4th and 5th centuries.

It is suggested that during periods of partial abandonment and casual occupation, the established agricultural systems that would have been the mainstay of the villas' economies may have broken down. Although much of the meat supply could have been provided by domestic animals, perhaps animals still living on the villa estates, or perhaps animals traded for, there was a need to supplement the meat supply by hunting. This phenomenon, on the limited evidence available, seems to occur most frequently in the late Roman period. There is a belief that the 4th century was a particularly prosperous period for the British countryside (Frere 1978: 318). However, by the end of the 4th and the beginning of the 5th centuries, due to the withdrawal of Roman arms, roads and drains fell into disrepair and the vital relationship between town and villa would have become much more difficult, if not impossible to maintain. This would have an effect on farming as much as on any other aspect of life; relationships between villas and markets, movements of stock and foodstuffs and the management of the villa estates may well have been adversely affected. Thus the need to supplement the diet by hunting may have become more pressing than at any previous time. It should be noted that at some late Roman sites there is no evidence for this phenomenon. At Portchester (Grant op. cit.), the very small number of deer bones in the late Roman occupation debris has already been noted.

If an increase in deer bones can indeed be seen as an indication of disruption to an established agricultural system, it is interesting to note that there is no evidence to suggest that there were any major differences in the relationship of wild animals to domestic animals at sites occupied before and after the Roman Conquest. This adds further weight to the evidence from sites such as Odell (Grant forthcoming) that the Roman Conquest and occupation did not disrupt the existing agricultural system, but in fact had surprisingly little impact on the rural economy. Unfortunately, there is very little evidence indeed for the immediate post-Roman period. The percentage of deer bones found in Saxon contexts at Portchester Castle is slightly higher than in the Roman contexts, but at Latimer, there is no evidence of an increase in the percentage of deer remains in the post-villa phases, although the figures are still quite high.

We have so far discussed environmental and economic factors as explanations for 'atypical' numbers of deer remains. Social factors may also be expected to influence the nature of the faunal remains. In this contexts it is interesting to note the differences between the faunal assemblages from prehistoric and early historic sites and those from the medieval period, where, in many more instances, deer bones form a significant part of the animal bone remains. For example, at Portchester Castle (Grant 1978), 14% of the animal bones from the late medieval layers were deer bones and at Lyveden in Northamptonshire (Grant 1971), over 20% of the bones found in 13th to 14th century contexts were deer bones. In the medieval layers excavated in the town of Bedford, deer bones were generally rare, but were most common in the occupation debris associated with the castle (Grant 1979). Certainly environment and economy may still be important factors, but if we compare the bone assemblages from the medieval layers at Portchester with those from the late Roman and Saxon occupation, the differences are so marked as to be unlikely to be explained solely by environmental changes over the period of occupation. Documentary evidence for the medieval period in England shows us that deer hunting was a popular sport and deer were even contained in parks, areas of woodland enclosed by fences (Cantor 1970). The deer of the Royal parks were prohibited to anyone but the king, but many of the nobility and wealthy had their own private parks. High percentages of deer at Portchester and in the castle occupation at Bedford may indicate the pursuance of a pastime that wealth and privilege made possible. Deer bones at a medieval peasant occupation may however evidence the meagre living gained by the inhabitants. Small scale agriculture and stock rearing, with the profits milked by the landlord might tempt the peasants to risk the highest penality to supplement their diet.

Deer hunting for sport did not begin in the Middle Ages and we know from documentary evidence from the Greek and Roman periods that hunting was a popular sport in the Classical world. Hunting scenes are depicted in mosaics and the Celtic literature also suggests that hunting may have been popular in the Celtic world (Ross 1970). However, as in the Middle Ages in Europe, it seems to have been predominantly the sport of the aristocracy. We may not be surprised to find little evidence of hunting at the small Iron Age farmsteads or the small rural Roman villas, but it is perhaps surprising to find so little evidence for hunting at the obviously important Iron Age sites such as Danebury or Maiden Castle (Jackson 1943). It is even more surprising that there were so few deer bones at Roman sites such as Fishbourne (Grant 1971), where there was so much other evidence for the adoption of a Roman-style life of luxury and leisure (Cunliffe 1971). In fact the figures for deer bone remains at Fishbourne show that deer remains are most common in the construction phase, when any livestock rearing and agriculture may not have been fully established and the occupants of the villa site may have been mainly construction workers who could have hunted deer to supplement their diet and wages. Strabo's much quoted reference to the hunting dogs that he says were one of Britain's major exports before the Conquest, is confusing in the face of so little material evidence for hunting in the Iron Age.

Harcourt (1979) has pointed out that deer bones may be under-represented at occupation sites, as only the meat from a hunted animal may be brought back to a settlement, the waste bones being left at the kill site. Detailed analysis of the type of deer bones found at settlement sites may help to clarify this possibility, but one cannot help thinking that it might be easier to transport a dead animal whole than as a collection of joints of meat. On the East Coker Mosaic a deer is shown being carried on a pole between the shoulders of two men. The four legs of the deer are tied together and the animal is hanging from the pole by its legs (see Burke 1978: fig. 57).

It would thus appear that although the presence of deer bones may be a social indicator, especially in the medieval period, we have as yet, no clear evidence of the hunting of deer primarily for sport at sites that were evidently the dwellings of persons of importance in Iron Age and Roman times. We may be seeing important cultural differences between, for example the Romanised Britons and the Romans themselves, but also evidence that although medieval deer hunting was ostensibly a luxury, it may have fulfilled an important economic function that was not necessary or appropriate in Iron Age or Roman Britain.

So far, we have only considered the role of wild animals in the food producing economy. Deer do have another function apart from their use as providers of meat and sport and that is as the source of what may have been an important raw material—antler. If the faunal remains analyses in excavation reports have few references to deer bones, in the pages of the small finds reports are frequent references to finds of antler particularly of red deer. Furthermore, in the instances where sufficient detail is given by the authors and the appropriate evidence is available, it would appear that the majority of the antlers found had been naturally shed rather than removed from dead animals. At Burgh Castle in Norfolk, 62% of the large collection of basal antler fragments had been shed (Grant, forthcoming) and at Lynch Farm, Peterborough, only shed antler was found (Wilson 1975). This is not a phenomenon confined to the later prehistoric period. At Blackpatch flint mine, Curwen (1922) reports that of the red deer antler picks found, only one had a portion of skull adhering and all the others were shed antlers. At Durrington Walls, 287 of the antlers found had been cast and only 29 were from dead or killed animals (Harcourt 1971). The particular significance of this information is in the implied seasonality of the collection of deer antler and perhaps therefore also of the manufacture of antler tools and objects. Red deer (most of the antlers are from red deer rather than roe deer) shed their antlers in March and April and it would seem likely that the shed antlers would have been collected at about that time of year. It is known that the deer will themselves eat their antlers as a source of extra minerals and antlers are also eaten by insects such as ants. They would also be in far better condition for tool manufacture if collected soon after they had been shed than if they had been left exposed to the elements. It may not be too far fetched to suppose that expeditions were made to areas of forests known to be inhabited by deer to collect a supply of antlers.

The particular advantage that antler has over bone, which may account for the effort expended in collecting antler when bone would have been easily available from slaughtered domestic animals, lies in the internal structure of antler. Although bone and antler are essentially made of the same basic substance, antlers are denser and therefore stronger than bone. Antlers have no haemopoietic tissue in the cancellous bone and so a large space inside the centre of the bone is not required (Halstead 1974). Shed antler will often be more suitable for tool manufacture than unshed antler because it is only when the antler has attained its final shape near the end of the growth cycle that it undergoes the internal reconstruction and calcification that are necessary to give it strength. Thus there is the further possibility that if, when deer were hunted, both meat and antlers were desired, the optimum time for this would have been after the end of the antler growth period and before the antlers had been cast. In red deer, this would be in the autumn and winter months, the period when other food resources might be most stretched in an agricultural society.

If deer antler collection and perhaps even deer hunting were seasonal activities, then it is also possible that antler working may have been a seasonal activity. In the Iron Age, a tool that is very commonly made from antler rather than from bone is the so called weaving comb. These objects have been variously interpreted, but perhaps their most likely function is as combs for removing the naturally shed hairs from sheep. If the antlers were collected in the spring, then the tools too may have been made at this period so they would have been ready for the sheep's summer moult. Of course it is also possible that the antlers collected in the spring could have been stored for later use—this could be the explanation of the find of a pit with fourteen antlers in it at Casterly Camp (Cunnington 1913).

There also remains the possibility, especially when only small numbers of deer bones are found at a site, that the deer were killed because they were damaging crops. Deer are known to raid agricultural land, although this usually occurs when the deer, because of a reduction in their natural habitat, have not sufficient food for their needs in their natural environment. According to de Nahlik (1959) roe deer are especially fond of unripe corn and clover and may in some circumstances be an important enemy of agriculture. Deer remains may be indicative of a conflict between agriculture and the natural woodland environment of the wild animals, or even of periods of severe weather, when natural resources were stretched.

Finds of deer remains have not been given the attention they merit in the majority of excavation reports. Their presence or indeed their absence, may, as we have shown, give important indications of environmental conditions, agricultural efficiency, economic stability and organisation, social status and seasonal activities at occupation sites.

Bibliography

Burke, J. (1978) Life in the Villa in Roman Britain. London: Batsford.

Cantor, L. M. (1970) The Medieval Deer Parks of Leicestershire Transactions of the Leicestershire Archaeological and Historical Society 46 9-24.

Chapman, D. and Chapman, N. (1975) Fallow Deer. Suffolk: Terence Dalton Ltd.

Cornwall, I. (1953) The Animal Bones. In: Excavations at Clausentum, Southampton, 1951-1954 (Cotton, M.A. and Gathercole, P. W.) London: HMSO pp. 141-142.

Cunliffe, B. W. (1971) Excavations at Fishbourne 1961-1969. Volume I. London: Society of Antiquaries.

Cunnington, M. E. and Cunnington, B. H. (1913) Casterly Camp Wiltshire Archaeological Magazine 38 53-105.

Curwen, E. C. (1922) Blackpatch Flint Mine Excavation, 1922 Sussex Archaeological Collections 65 69-83.

de Nahlik, A. J. (1959) Wild Deer. London: Faber and Faber.

Eastham, A. (1972) Animal Remains. In: A Roman building at Chalk, near Gravesend (Johnson, D. E.) Britannia 3 141-142.

Evans, J. G. (1975) The environment of Early Man in the British Isles. London: Paul Elek.

Fraser, F. C. (1954) Animal Bones. In: The Stanwick Fortifications (Wheeler, R. E. M.) London: The Society of Antiquaries.

Frere, S. (1978) Britannia. London: Routledge and Kegan Paul.

Grant, A. (1971a) The Animal Bones. In: Excavations at the Deserted Medieval Settlement at Lyveden (Bryant, G. F. and Steane, J. M.) Northampton Museums and Art Gallery, Journal 9 90-93.

Grant, A. (1971b) The Animal Bones. In: Excavations at Fishbourne 1961-1969 (Cunliffe, B. W.) London: Society of Antiquaries pp. 337-387.

Grant, A. (1975a) The Animal Bones. In: Excavations at Portchester Castle. Volume I: Roman (Cunliffe, B. W.) London: Society of Antiquaries pp. 378-408.

Grant, A. (1975b) The Animal Bones. In: Excavations at Portchester Castle. Volume II; Saxon (Cunliffe, B. W.) London: Society of Antiquaries pp. 262-269.

Grant, A. (1978) The Animal Bones. In: Excavations at Portchester Castle. Volume III: Medieval, the Outer Bailey and its Defences (Cunliffe, B. W.) London: Society of Antiquaries pp. 213-238.

Grant, A. (1979) The Animal Bones from Bedford. In: Excavations in Bedford 1967-1977 (Baker, D., Baker, E., Hassall, J. and Simco, A.) Bedfordshire Archaeological Journal 13 286-288.

Grant, A. (forthcoming) The Animal Bones. In: Excavations at Harrold Pit. Odell, Bedfordshire 1974-1978 (Dix, B) Bedfordshire Archaeological Journal.

Grant, A. (forthcoming) The Animal Bones. In: Excavations at Burgh Castle by Charles Greene (Johnson, J. S.) East Anglian Archaeology.

Grant, A. (in preparation) The Animal Bones from the Iron Age Hillfort of Danebury.

Haglund-Cally, L. and Cornwall. I. (1963) The Animal Bones. In: Dinas Powys, An Iron Age, Dark Age and Early Medieval Settlement in Glamorgan (Alcock, L.) Cardiff: University of Wales Press pp. 191-193.

Halstead, L. B. (1974) Vertebrate Hard Tissues. London: Wykeham Publications.

Hamilton, R. (1971) Animal Remains. In: Latimer: Belgic, Roman, Dark Age and Early Modern Farm (Branigan, K.) Bristol: Chess Valley Archaeological and Historical Society pp. 163-166.

Harcourt, R. (1971) The Animal Bones. In: Durrington Walls: Excavations 1966-1968 (Wainwright, G. J. and Longworth, I. H.) London: Society of Antiquaries pp. 338-350.

Harcourt, R. (1979) The Animal Bones. In: Gussage All Saints, An Iron Age Settlement in Dorset (Wainwright, G. J.) London: HMSO pp. 150-160.

Jackson, J. W. (1943) Animal Bones. In: Maiden Castle, Dorset (Wheeler, R. E. M) Oxford: Society of Antiquaries pp. 367-371.

Megaw, J. V. S. and Simpson, D. D. A. (1979) Introduction to British Prehistory. Leicester: University Press.

Noddle, B. (1974) Report on the Animal Bones found at Dun Mor Vaul. In: Dun Mor Vaul: An Iron Age Broch on Tiree (Mackie, E.) Glasgow: University of Glasgow Press pp. 187-198.

Pilcher, (1974) Botanical Report. In: Dun Mor Vaul: An Iron Age Broch on Tiree (Mackie, E.) Glasgow: University of Glasgow Press.

Platt, M. I. (1948) Report on the Animal Remains. In: Report on the Excavation of a Broch at Skitten, in the Kilmster District of Caithness (Calder, C.) Proceedings of the Society of Antiquaries of Scotland 82 143-145.

Renfrew, J. (1974) Report on the Cereal Grains. In: Dun Mor Vaul: An Iron Age Broch on Tiree (Mackie, E.) Glasgow: University of Glasgow Press pp. 210-213.

Ritchie, J. (1920) The Influence of Man on the Animal Life of Scotland. Cambridge: University Press.

Ross, A. (1970) Everyday Life of the Pagan Celts. London: Batsford.

Turk, F. A. (1968) Report on the Animal Remains from Nour-nour, Isles of Scilly. Journal of the Royal Institution of Cornwall (n.s.) V iv 250-265.

Wilson, J. (1975) The Animal Remains. In: The Roman British Farmstead and its Cemetery at Lynch Farm, near Peterborough (Jones, R.) Northamptonshire Archaeology 10 94-137.

10. FLEECE CHANGES IN SHEEP

M. L. Ryder

Introduction

The last paper I presented in this department (Ryder 1980a) was concerned with medieval sheep and their wool types, so not only were the terms of reference wider, but more sources of evidence (including illustrations and records) were available. The main sources of evidence in the present account are wool in textile remains, and surviving primitive breeds of sheep.

It will be useful at the outset to say something about wool fibre types, fleeces and sheep breeds. The wild ancestor of domestic sheep has a short, hairy outer coat composed of bristly fibres known as kemps which obscure an even shorter, fine woolly undercoat (Fig. 10: 1). In the changes since domestication the kemps have been considerably reduced in thickness and number, and a second type of hairy fibre (named hair) has evolved, so that the fleeces of modern breeds contain kemp, hairs and wool in varying proportions.

The second point to note is that difference between the fleeces of different breeds are almost entirely genetic. It is true that better feed will make a short-woolled fleece grow somewhat longer, but this will not change it into a longwool type. Plate 10:I illustrates this distinction by showing that each type of fleece has a characteristic staple form which is brought about by the intermingling of varying proportions of different kinds of fibre. Difference in the size and appearnce of some main modern breed types are shown in Plate 10:II.

That fleece differences are genetic and not environmental might lead some to imagine that they lie outside the scope of the present conference. But one must remember that the environment will have influenced the way in which fleeces have evolved or been selectively bred (although the influence is rarely if ever obvious) and that man's livestock forms part of his environment. Indeed, it has been said that domestication was the greatest intervention made by man in his environment.

The domestication of sheep did not take place in Britain, so the first domestic sheep were introduced by neolithic settlers about 4000 BC. The prehistoric breed of Europe is thought to be represented today by the small, brown Soay sheep that survives in a feral state on St. Kilda off north west Scotland. Evidence for this comes from the similarity of neolithic bone remains with those of the Soay, and the similarity of wool in Bronze Age cloth with the fleece of the Soay (Ryder 1969).

The primitive features shared by the Soay with the wild ancestor are a short tail (all modern breeds have a long tail); a coloured fleece in which the belly is white (most modern breeds are completely white) and an annual moult (the wool of virtually all modern breeds grows continuously). To these

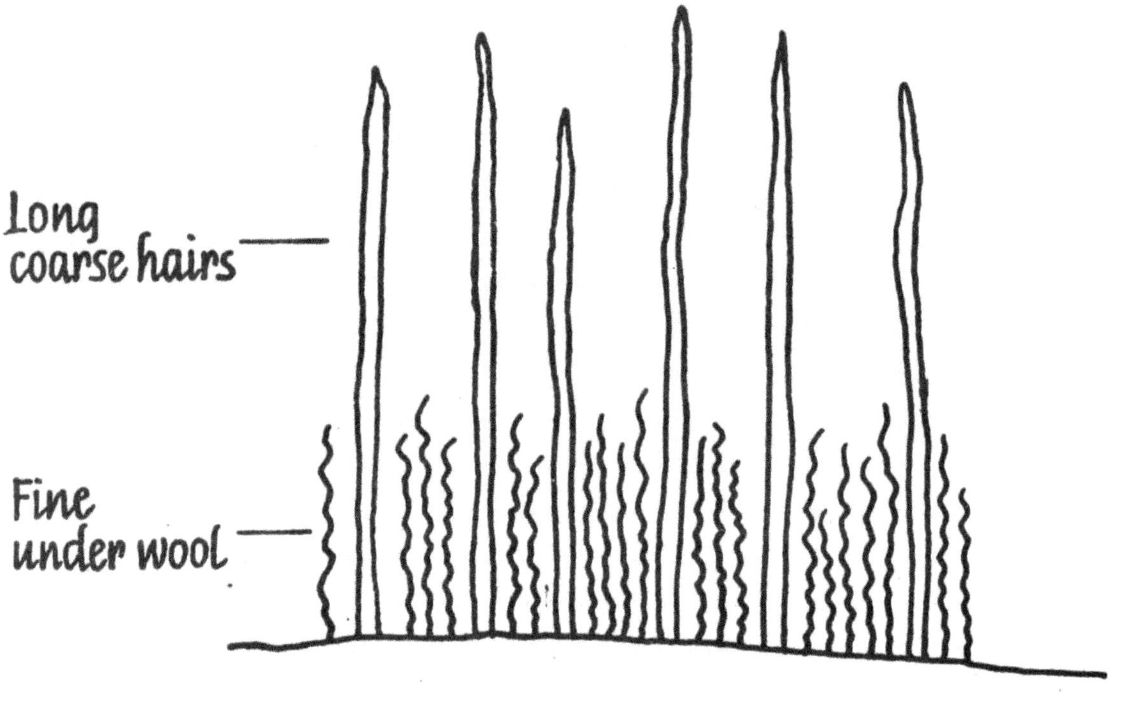

Fig. 10.1 Diagram of the structure of the double coat of the wild ancestor of domestic sheep (from Ryder and Stephenson 1968).

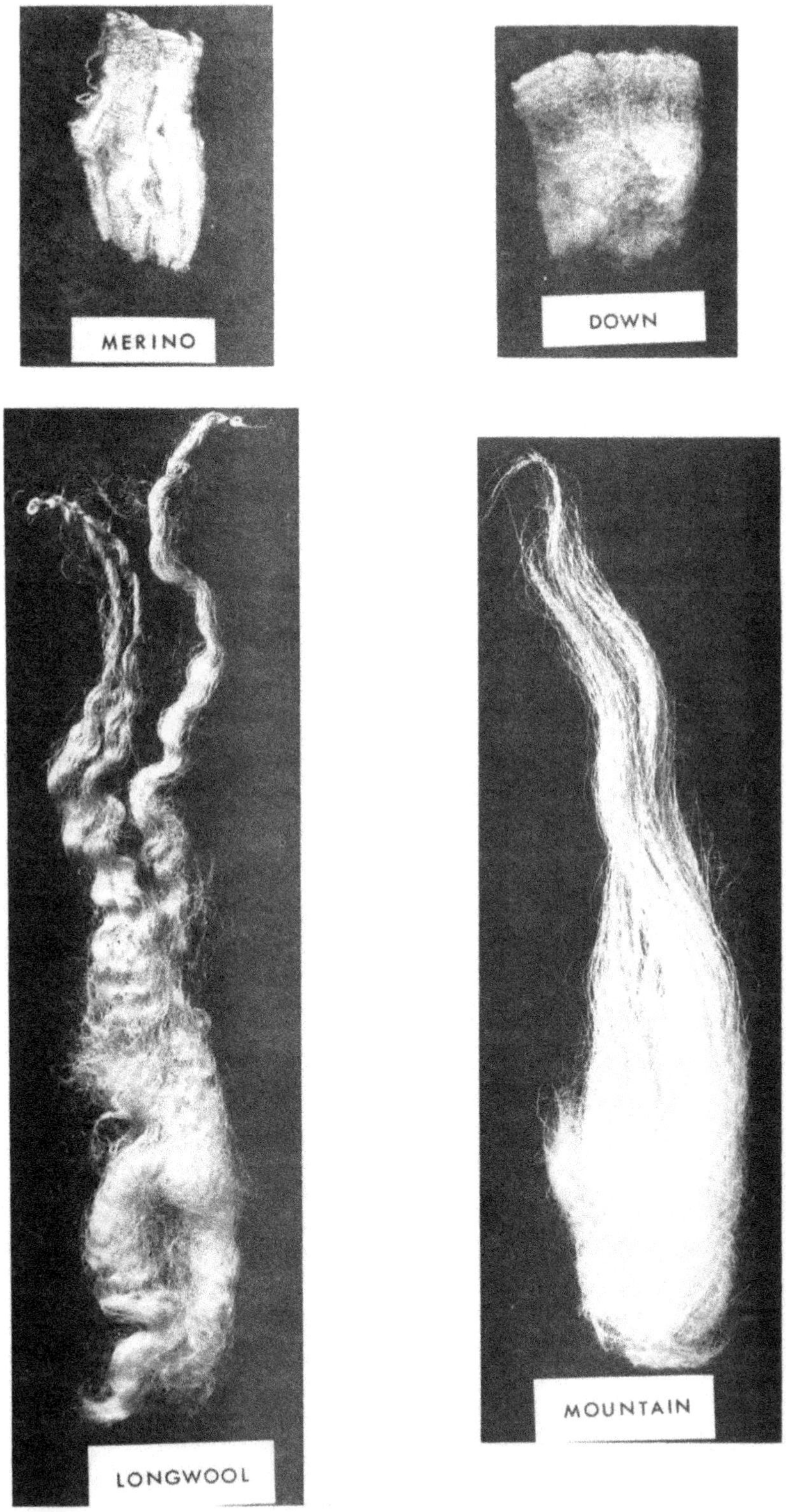

Plate 10.I Wool staples of different fleece types; the fine woolled Merino is not a native of Britain; the Down type is a modern representative of the shortwool; the longwool has a characteristic curl and lustre, and the fibres are coarse but not hairy; the mountain type is illustrated by a staple from the Scottish Blackface in which long hairy fibres obscure an undercoat of fine wool (from Ryder and Stephenson 1968).

major and obvious differences between wild sheep and modern domestic breeds can be added the change from a hairy coat to a woolly fleece, and the wool of the Soay is already much less hairy than the coat of the wild sheep, but there are hairy and woolly types.

This change is evident in Fig. 10:2 from Ryder (1969) which shows histograms of wool fibre diameter distributions in microns (one micron, μm, = 0.001 mm). The fibre diameter is shown along the horizontal axis, and the number of fibres on the vertical axis. The fibre diameter distribution of the wild sheep is shown at the top, the big difference in diameter between the coarse, hairy outer coat of fibres known as kemps, and the extremely fine woolly under coat being very striking. The first evolutionary change folloiwng domestication was a narrowing of the coarse, hairy outer coat kemps to produce the hairy medium fleece type of the hairy Soay. Further narrowing of these hairy fibres (fine kemps) presumably due to selective breeding by man, changed them into wool fibres of medium diameter, the result being the fleece type of the woolly Soay.

In evolutionary terms this is a generalised type forming an important link between more primitive hairy fleeces, and more highly evolved modern fleece types. The diameter distributions of surviving sheep have been shown in Fig. 10:2, but the lower five distributions have all been found in wool from ancient textiles. The hairy medium and generlised medium types appeared first in the Bronze Age, and were common until after the Middle Ages. The modern types on the bottom row began to appear in Roman times.

The Iron Age

Although there is evidence from illustrations and records of white sheep as early as 3000 BC in Mesopotamia and Egypt (Ryder and Stephenson 1968), there is no evidence of white wool in Asia and Europe before the Iron Age, where it occurs as part of an increase in the range of colour.

Very few Iron Age finds of wool are available for examination. Ryder (1969) described a single white specimen from central Asia dated about 400 BC which comprised a complete sheepskin with the wool intact from the Scythian burial at Pazirik in the Altai Mountains. This is of interest in having an outer coat of hairs instead of the kemps or medium fibres found in the Soay sheep. The hairs of modern true hairy breeds (such as the Scottish Blackface) are long, and are named heterotypes because they are intermediate in character between kemp and wool: they are thick and kemp-like in summer, but thinner in winter when they appear like true wool fibres. Whereas kemps cease to grown preparatory to moulting, hairs merely thin down and continue to grow.

The hairs of the Pazirik sample were longer and finer than kemps, but sections of the skin showed that they had stopped growing. This and a length comparable with only the shortest modern hairy fleeces (12 cm) suggest an early stage in the evolution towards modern heterotype hairs which usually grow continuously. The origin of hairs appears to have been associated with the change from a moulting fleece to one of continuous growth. Moulting is a disadvantage in leading to a loss of wool to the farmer. The normal way of harvesting wool from sheep that moult is by plucking, a method that was in use with the native breed of Shetland until recent times. Selective breeding against

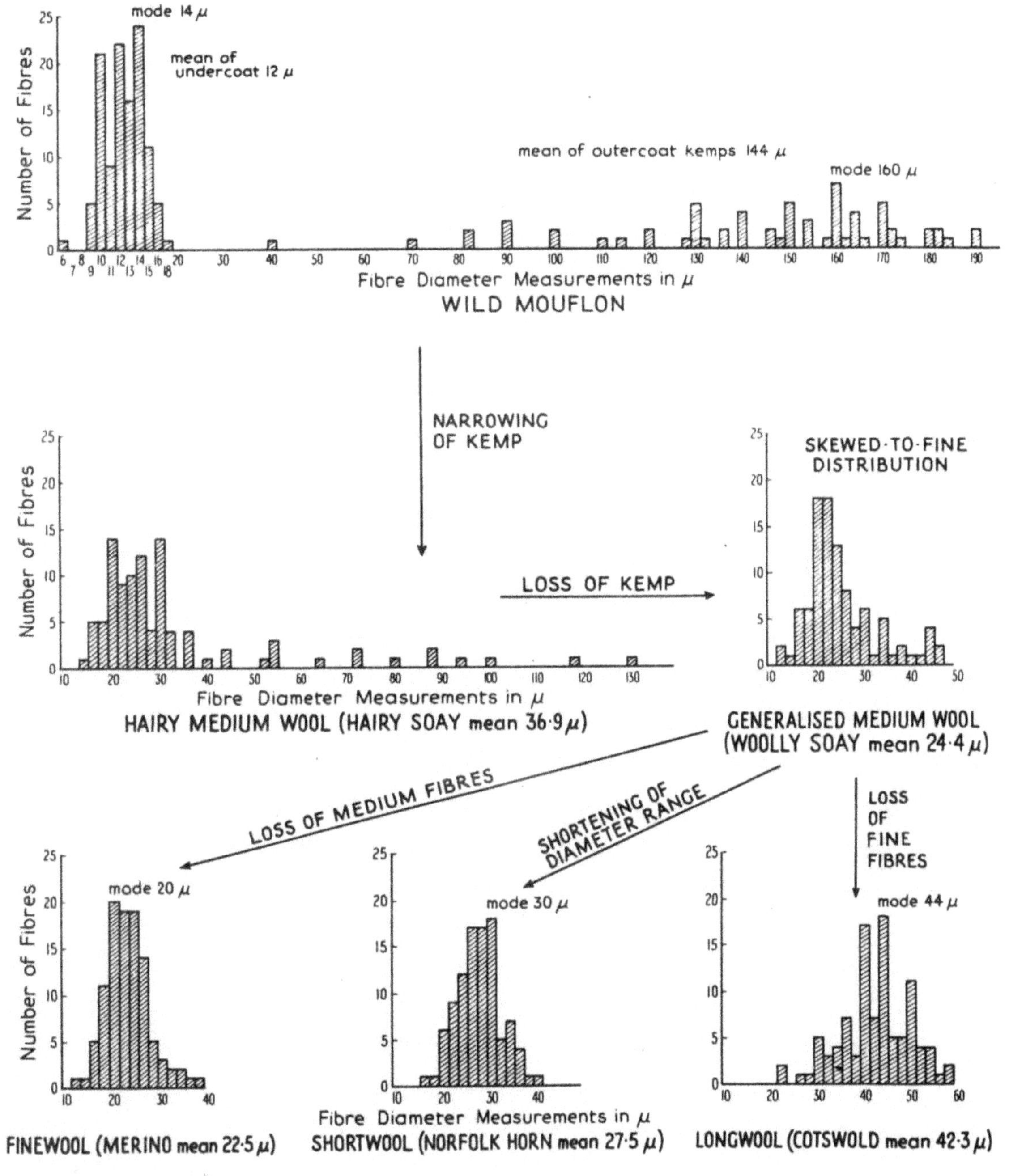

Fig. 10.2 Probable changes in wool fibre diameter distribution during fleece evolution; one μ (micron) = 0.001 mm (from Ryder, 1969).

moulting is unlikely to have begun until an alternative method of removing the fleece was available, and clipping became possible with the development of iron shears in the Middle East about 1000 BC. Shears of a size suitable for clipping a sheep are a common find on Iron Age and Roman sites in Europe, although Harding (1974) implies that they were introduced into Britain by the Romans.

A second Iron Age white fleece sample of somewhat earlier date from Hallstatt in Austria was described by Ryder (1969) as hairy, although the hairy fibres appeared more like kemps. A third example, again a sheepskin, from a bog dated about 500 BC in northern Germany was also white and hairy (Ryder 1977a). These three samples are of course inadequate to indicate the range of variation in Iron Age wool, and in fact true hairy fleeces of this type are rare until after the Middle Ages. Lacking direct textile evidence, one has to fall back on corroborative evidence from Roman textiles detailed below, and on surviving primitive breeds.

The breeds in question mostly have a short tail, and only the rams are horned, but the skeleton is otherwise similar to that of the Soay. There are still hairy medium as well as woolly fleeces, as in the Soay, and a spring moult, but the breeds are vari-coloured. In addition, to the brown of the Soay there are black, white, and grey, as well as piebald, animals. But the wild-pattern white belly is rare, and so this marks the Soay as truly unique among domestic sheep. This vari-coloured type appears to have been common until after the Middle Ages, and widespread throughout Europe, since small pockets of relic breeds survive in isolated areas (Ryder 1980b). Grey animals are a notable feature of the type, which could also be termed "European grey".

The main surviving breeds of this type in Britain are the Orkney and Shetland. During the 19th century these two breeds formed a single population, but the Shetland has since been selectively bred for white, woolly fleeces, and so only the Orkney retains the full range of colour and fleece type (Ryder 1968a and b). It has already been indicated that the tendency to shed meant that in the past the wool had to be obtained by plucking (Ryder 1966) but this tendency has largely been bred out today, so that most of the sheep are shorn.

It is pertinent to ask how this range of colour evolved from the brown Soay. First, all black and all brown animals are found in the breed, and so these can be increased in number by breeding the same colour together. Second, occasional white animals mutate from those with a white belly. Another way of obtaining white is by selective breeding for larger and larger areas of white on piebald sheep. Grey, which is brought about by a mixture of coloured and white fibres, must also have arisen by mutation, and black sheep have been observed to mutate to grey (Ryder 1980c).

Naturally-coloured wools were used until well after the Middle Ages, different colours often being woven together to produce patterns. It would be imagined that although white sheep would have been a novelty in themselves, the real stimulus to breed sheep with white wool may have come only after the development of dyes.

Plate 10. II Different breed types of British sheep: right, black-faced, horned hill type (Scottish Blackface; centre, white-faced hill sheep (Welsh Mountain); left, lustre longwool (Lincoln breed).

The Roman period

Roman textiles from Britain, Europe and the Near East show that the predominant wool type during that period, in addition to being white, was fine to the naked eye. But microscopic examination reveals that the wool contains medium fibres, and so is of generalised medium type. Hairy medium wools were still common, however, at Vindolanda on Hadrians wall for instance (Ryder 1977b and Table 1). More changes were taking place in fibre diameter distribution and fleece type at that time, and the new types, and how they evolved are summarised in Fig 10:2.

If the medium wool fibres of the generalised medium type had become narrower by selective breeding then the fibre diameter distribution of the true fine wool seen in the modern Merino would have been produced. The development of the true fine wool began in the Middle East probably soon after 1000 BC (Ryder 1969). But the emergence of the Merino as a distinct breed occurred in the late Middle Ages in Spain.

In the European breed survey already mentioned (Ryder 1980b) it was expected that remnants of an ancient sheep with a true fine fleece might be found in the Balkans. In the event the primitive breeds discovered had a fleece of generalised medium type. Thus although true fine wools undoubtedly existed in the Near East about the 1st century BC-AD (Ryder 1969) it is of immense interest that the most primitive sheep surviving in south east Europe have a fleece of the same type as the vari-coloured sheep in northern Europe.

Returning to Fig. 10:2, if on the other hand, the finer fibres had become coarser by breeding, then the true medium wool diameter distribution of the modern longwool would have been obtained. Thirdly, if both changes had taken place together, and the range of fibre diameter had become shortened, with a mean between the fine and medium values, the distribution would be comparable with that of the modern shortwool. Textile remains indicate that a few medium wools and shortwools had developed during Roman times, but neither became common in Britain until after the Middle Ages (Ryder 1969, 1974; 1977a; 1980d).

Using evidence from surviving "Iron Age" breeds again one should mention that in the primitive Orkney sheep the whole range of fleece types has been observed from hairy medium through generalised medium, to true fine, although the latter is rare (Ryder 1968a and b). In grey sheep of this breed it is usually the coarsest fibres that are black, and such mixtures of pigmented and non-pigmented fibres have been seen in textiles from as early as the Roman period (Ryder 1977a; 1980d).

Although there is virtually no direct evidence, it has always been assumed that the Romans introduced improved livestock into Britain, and this assumption is incorporated in the summary of breed type origin shown in Fig. 10:3. Following the Soay type, there are likely to have been vari-coloured sheep in the Iron Age, but to what extent these evolved locally or were introduced is not yet clear. It has been suggested that the Belgae introduced a white, fine-woolled sheep (Wild 1970).

Roman mosaics on the continent illustrate white-faced sheep in which the tail is no longer short, and there are horned as well as polled individuals

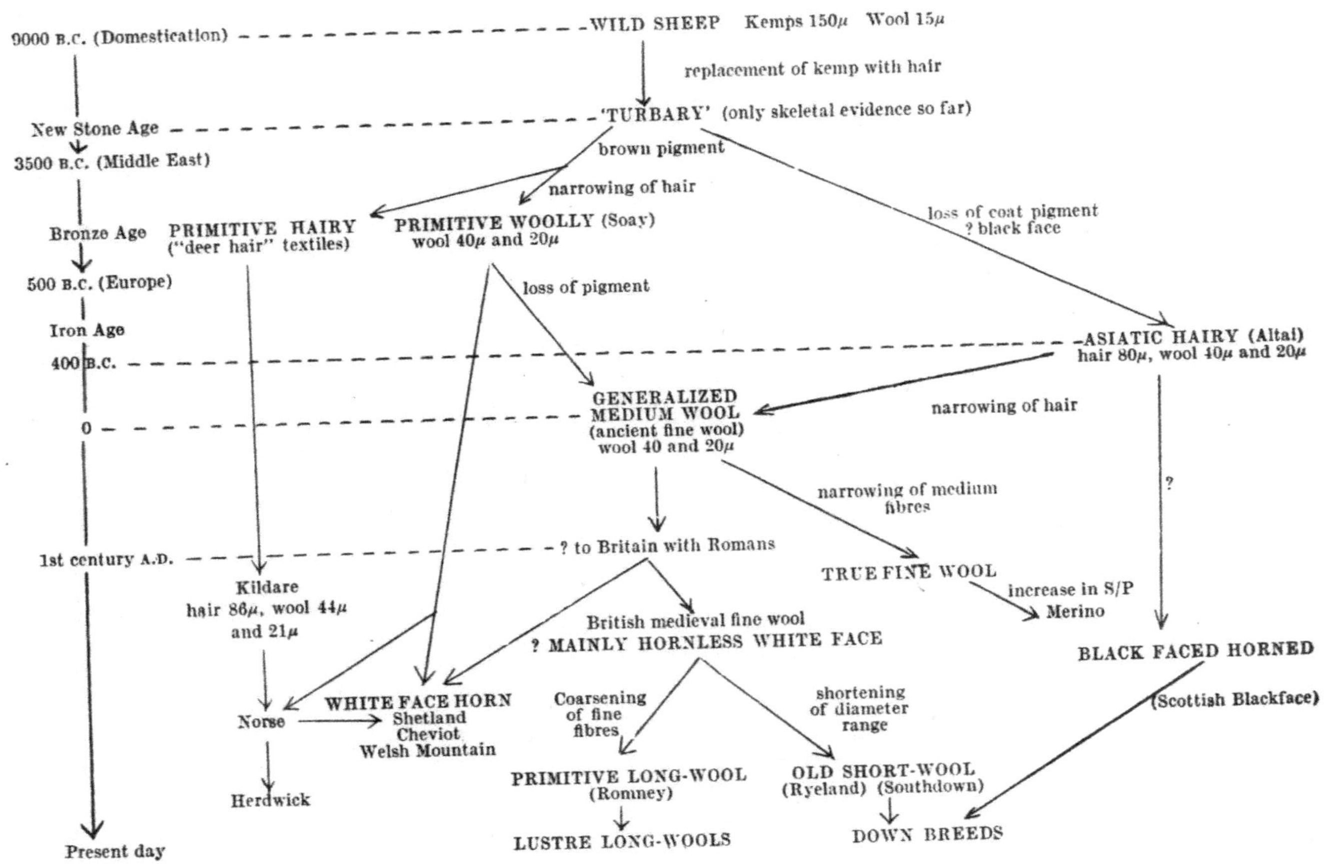

Fig. 10.3 Suggested lines of evolution of different fleece types in terms of breed types and historical periods (from Ryder 1969).

which could be rams and ewes, as in the Orkney-Shetland type and some modern breeds such as the Merino (Ryder 1980e p. 374). Fig. 10:3 assumes the introduction of such a sheep into Britain by the Romans. Crosses of this with the native Soay could have given rise to types that later emerged as breeds such as the white-faced Cheviot and Welsh Mountain in the North and the West. These are horned in the rams only, and any pigment tends to be brown rather than black.

Then the fleece changes outlined above could have produced a primitive longwool on the one hand, and a shortwool on the other. Both types have a white face and lack horns. Most medieval illustrations show white-faced, polled sheep with short wool.

Returning to Vindolanda, this was the first site to yield an adequate sample on which to base conclusions about the range of fleece types. Altogether 56 yarns and one unspun staple were measured. The latter was of interest because of the rarity of unspun wool, and it turned out to be the only true hairy sample from any Roman site. This was 4 cm long and comparable with the hairy Iron Age samples listed above, having a pointed tip not unlike that of staples in the modern Scottish Blackface breed.

The hairy sample has been omitted from the percentages in Table 1 (from Ryder 1980a) which show the proportion of different fleece types at Vindolanda and compares them with all other Roman wools examined, including samples from the Continent as well as all Saxon samples examined (Ryder 1969; 1974 and other minor reports). Hairy medium wools and generalised medium wools predominated at Vindolanda, and there was a moderate proportion of a finer type of generalised medium wool. It is of interest that there were more fleeces of the fine type that later emerged on the Continent as the Merino, than there were shortwools and true medium wools, which are thought to have developed in Britain. But these findings indicate evolution as early as Roman times, whereas it was previously thought that they appeared during the Middle Ages. Since Wild (1977) (see Ryder 1977b) considers that the Vindolanda wools are unlikely to have been imports, they probably indicate the range of fleece types found in northern Roman Britain in the 1st century AD. Comparison with the other Roman wools in the table, indicates that these have a greater proportion finer types (fine, generalised medium and true fine).

The pigment details given in Table 2 throw additional light on the breed types. The scarcity of completely pigmented wools indicates a move away from the prehistoric sheep exemplified by the Soay. Of particular interest is the large number of "grey" wools on this site. This accords with evidence from surviving breeds, that the predominant sheep in Europe from the Iron Age to the Middle Ages had a range of fleece colours (see above).

<u>The Saxon period</u>

Judging by the number of place names which embody a reference to sheep, they were found everywhere during the Saxon period. Excavations at West Stow, Suffolk, indicate well organised textile activity, and there are also indications that wool cloth was exported, as it had been during Roman times. Of course the export of wool, and later of cloth, during the Middle Ages was to become the major economic activity of the country. Saxon illustrations

TABLE 1: COMPARISON OF ROMAN AND SAXON WOOLS

	Vindolanda	All other Roman (incl. Europe)	Saxon (England only)
Hairy	(1)	-	-
Hairy medium	34%(19)	14%(12)	23%(10)
Generalised medium	34%(19)	15%(13)	23%(10)
Fine, generalised med.	18%(10)	39%(34)	30%(13)
True medium	2%(1)	3.5%(3)	5%(2)
Shortwool	4%(2)	1.5%(1)	-
Fine	9%(5)	27%(23)	19%(8)

(from Ryder, 1980a)

TABLE 2: EXTENT OF NATURAL PIGMENTATION IN THE VINDOLANDA WOOLS (ACTUAL NUMBER)

	100% pigmented	mixture ("grey")	no pigment	total
Hairy	-	-	1	1
Hairy medium	3	11	5	19
Gen. medium	1	12	6	19
Fine, gen. medium	1	3	6	10
Fine	-	2	3	5
Shortwool	-	-	2	2
Medium	-	1	-	1

(from Ryder, 1980d)

are virtually non-existent, although the Caedmon manuscript has an indication of horned rams and polled ewes (Ryder 1980e: 374).

Of fourteen yarns from the Saxon sites of Broomfield Barrow, Essex, Coombe, Kent, and Sutton Hoo, Suffolk, only five had any pigment. The same range of fleece types as those from Roman sites was found, but there were as many as seven fine wools, six generalised medium wools (half of which were the finer version) and only one hairy medium wool (Ryder 1969). The Sutton Hoo ship burial provided most of the fine wools, and these could have originated abroad.

Table 1 includes later samples from additional sites, even so the proportion of fine or relatively fine types is striking. One of the true medium samples was a sheepskin from Durham dated about AD 1000 and described by Ryder (1977a).

Three main possible introductions of sheep into Britain have been mentioned, first, the brown Soay; second, the Iron Age vari-coloured, which still, however, had a short tail, and third a Roman white sheep with a long tail. A fourth main type is the black-faced, horned type with a true hairy fleece, which persists today in such breeds as the Scottish Blackface. Few textile remains of this fleece type have been found, possibly because the wool is too hairy for clothing, being used today mainly in carpets, so there is less evidence on when it arrived. It will be remembered that all the Iron Age remains (skins) and one Roman staple, were of this type (see above). Because these sheep are superficially similar to many Asiatic breeds, it has long been suggested that their ultimate origin is Asia. It is therefore of interest that the Pazirik sample was hairy.

The area occupied by the Danes before the Norman Conquest corresponds very roughly with the area in which the black-faced, horned type later emerged into history, and this has led to the suggestion that this fourth main stock was introduced by the Danes (Ryder 1972). Support for this can be gained from the existence of a similar sheep in the past in Denmark. The Dutch, German, and Polish "Heath" breeds were once thought to be of the same general type, but these have grey individuals and tend to have a short tail, and so appear to belong to the vari-coloured group. Some Danish textiles excavated in York in fact contained true hairy wool (Ryder 1974). It is just possible that the black-faced horned type evolved in Britain.

Many northern areas were occupied by the Vikings, and most Norse wools measured were from Scotland (Ryder 1968b). The general impression was that they were more hairy (i.e. hairy medium) but had less pigmentation than Roman and Saxon wools. Orkney and Shetland were occupied by Norsemen until the 15th century, and so the vari-coloured sheep persisting in these island groups may contain Scandinavian influence instead of being simple survivals from the Iron Age.

This paper began by setting the scene in the periods prior to those under consideration. By the same token it should end by summarising the fleece types of the succeeding medieval period. Despite the vital importance of wool to England during the Middle Ages, only during the last decade have appreciable textile remains become available for study. The findings from these are

TABLE 3: SUMMARY OF FLEECE TYPES IN BRITISH MEDIEVAL WOOLS

		Hairy	Hairy medium	Generalised medium	Fine gen. medium	True medium	Short	Fine
Winchester	11th cent.	14%(1)	57%(4)	29%(2)	-	-	-	-
London, Baynard's Castle	1200	-	25%(2)	63%(5)	-	-	-	12%(1)
York	12th-13th cent.	27%(3)	37%(4)	27%(3)	-	9%(1)	-	-
Southampton	13th-14th cent.	-	11%(2)	58%(11)	-	5%(1)	5%(1)	21%(4)
Baynard's Castle	14th cent.	13%(3)	8%(2)	8%(2)	38%(9)	4%(1)	13%(3)	16%(4)
Baynard's Castle	15th cent.	7.5%(2)	11%(3)	15%(4)	18%(5)	11%(3)	30%(8)	7.5%(2)
Yorkshire	15th cent.	-	33%(2)	66%(4)	-	-	-	-
Perth	12th-14th cent.	19%(17)	44%(39)	18%(16)	6%(5)	8%(7)	5%(7)	-
Aberdeen	13th-14th cent.	12.5%(2)	19%(3)	44%(7)	6%(1)	12.5%(2)	-	6%(1)
	Average	10%	27%	36%	8%	6%	6%	7%

(from Ryder, 1980a)

summarised in Table 3 from Ryder (1980a) from which it can be seen that the generalised medium fleece type predominated.

Summary

The wool in the earliest textile remains (from the Bronze Age) indicates a fleece type like that of the brown Soay sheep (hairy medium and generalised medium). By the Iron Age it appears that there were black, white and grey sheep, in addition to brown fleeces, but they were still hairy medium and generalised medium in type. The greatest number of finds come from the Roman period, and these suggest that there were more hairy types in Britain than on the continent. The modern true medium and shortwool fleeces were also evolving. The few Saxon remains available have included a high proportion of fine wools, but it could have been about this time that the true hairy type (now used in carpets) reached Britain or evolved locally.

Bibliography

Harding, D. W. (1974) The Iron Age in Lowland Britain.

Ryder, M. L. (1966) Shetland sheep and wool. Journal of the Bradford Textile Society 1965-66 91-99.

Ryder, M. L. (1968a) Fleece structure in some native and unimproved breeds of sheep. Zeitschrift für Tierzüchtung und Zuchtungsbiologie 85 143-170.

Ryder, M. L. (1968b) The evolution of Scottish breeds of sheep Scottish Studies 12 127-167.

Ryder, M. L. (1969) Changes in the fleece of sheep following domestication (with a note on the coat of cattle). In: The Domestication and Exploitation of Plants and Animals. (eds Ucko, P. J. and Dimbleby, G. W.) London: Duckworth pp. 495-521.

Ryder, M. L. (1972) The wools of Britain Chap. 4 In: The Wool Textile Industry of Great Britain (ed Jenkins, J. G.) London: Routledge & Kegan Paul.

Ryder, M. L. (1974) Wools from Antiquity Textile History 5 100-110.

Ryder, M. L. (1977a) Some miscellaneous ancient fleece remains. Journal of Archaeological Science 4 177-181.

Ryder, M. L. (1977b) The Vindolanda wools In: Vindolanda III The Textiles (Wild, J. P.) Vindolanda Trust, Hexham pp. 34-41.

Ryder, M. L. (1980a) Medieval sheep and their wool types. Council for British Archaeology (in press).

Ryder, M. L. (1980b) The primitive breeds of domestic sheep of Europe. Proceedings of the 3rd international Archaeozoological Conference, Poland 1978 (in press).

Ryder, M. L. (1980c) Fleece colour in sheep and its inheritance Animal Breeding Abstracts 48 305-324.

Ryder, M. L. (1980d) Wools from Vindolanda Journal of Archaeological Science (submitted).

Ryder, M. L. (1980e) "Livestock" pp. 301-410 in Piggott, S. (Ed.). The Agrarian History of England & Wales vol. 1 Part I Cambridge University Press (in press).

Wild, J. P. (1970) Textile Manufacture in the Northern Roman Provinces. Cambridge University Press.

11. DISEASE AS AN ENVIRONMENTAL PARAMETER

Don Brothwell

It is to the credit of the organisers of this conference that disease was considered to be an aspect of the environment worthy of particular consideration. This indicates not only an expanding range of subjects being investigated within environmental archaeology, but also the fact that a better balance, a more ecologically complete view of past human environments, is being achieved.

The study of disease is both a specialised biological field, and yet at the same time has universal interest for us all. Every one of us has direct experience of disease, and if we include trauma (or accidental damage), then disease in some form or another is obviously responsible for every death. This is so for all vertebrates, and it is therefore something which can be investigated not only in earlier human populations, but also in relation to the more general field of zooarchaeology.

It is not my intention here to give a general account of this vast subject—time doesn't permit this. For anyone wishing to read further on general vertebrate palaeopathology, one can be referred to a new book called Animal Diseases in Archaeology (Baker and Brothwell 1980). In contrast to this wide-ranging study, my concern here is to look in particular at the evidence for disease in the livestock of early British farmers from post-Bronze Age to medieval times. This was a period when an expanding human population was probably making increasing demands of farming economies, and yet this was also a time when standards of livestock maintenance were likely to have been elementary and perhaps rather experimental.

There is no question that the early farmers had many problems to resolve, and even in advanced farming today, disease, premature death (and therefore serious economic wastage) can occur. An example of this is seen in the 1957 report (Project No. 345) of the O.E.E.C. concerned with losses just in young farm animals. Pre-weaning mortality alone varied in cattle from different European countries, from one to as many as thirty percent. The chief causes of mortality were genetic factors, infectious disease from micro-organisms, and the diseases produced by larger parasites. In pigs, pre-weaning mortality was between eighteen and over thirty percent, and in sheep average losses were about fifteen percent. It seems highly unlikely that early farming communities would have experienced percentages which were less than these figures, for although intensive breeding may have boosted the frequency of some factors, hygiene generally is likely to be significantly better today. So even prior to weaning, early farmers would be presented with the common and puzzling fact that numbers of young animals died. Added to this, of course, are all the problems of keeping the adult animal healthy.

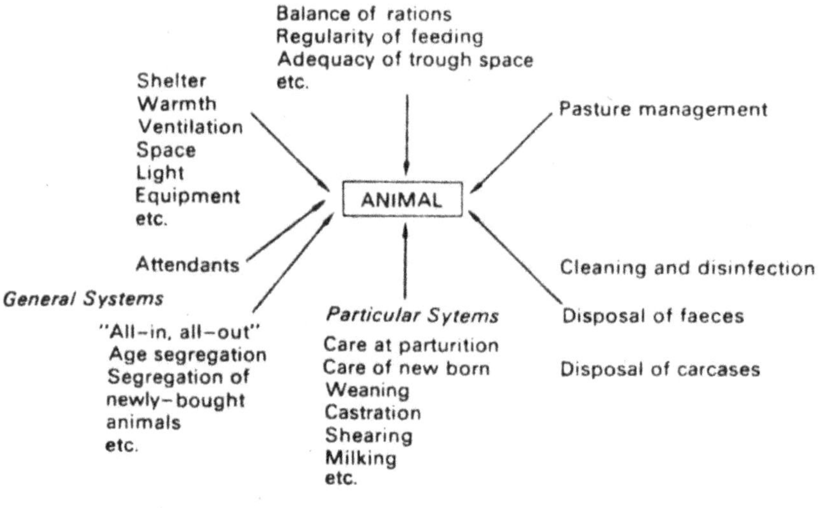

Fig. 11.1 The complex of husbandry factors which may affect the ability of mammals to withstand disease (After Halpin 1975).

Morbidity and overall wastage, as exemplified by modern dairy herds is in the region of twenty percent, and this is with reasonable veterinary care. So what could it have been in the past? Moreover, how much notice was taken of disease, how much was caused by man, and what happened to the animals showing abnormality? Some of the answers are certainly to be found in their remains, even though soft tissue is absent. Also, it is not simply a matter of attempting to establish some kind of crude estimate of overall morbidity, for we can hope to separate some of the environmental factors influencing or producing disease. I dont mean that the complex mesh of factors in disease ecology, some of which are shown in fig. 11: 1, can be precisely revealed, but that some aspects of feeding, housing, handling, type of parasite, and so on, might be identifiable. The range of skeletal pathology which may occur is in fact considerable and the nature of the bone changes is also highly variable, as I hope to demonstrate in a brief way.

Clearly, with this sort of information one might go in for that popular archaeological pastime of model making, this time in terms of reconstructing possible mortality, or disease occurrence, but my concern here is to review some of the actual evidence of disease in the livestock of these early British farming communities and urban samples.

During the past few years, a significant start has been made regarding the collection of evidence for livestock disease as shown by changes to the bones and teeth. This, of course, only provides restricted information, but nevertheless it is possible to demonstrate quite a range of environmentally-linked pathology. Rather than attempt any detailed listing of all reported abnormality, I want to concentrate on particular specimens, disease categories, and samples which demonstrate well the extent of the evidence so far, and the potential of this field.

Congenital Abnormality

We should begin by considering inborn errors of development. These can indicate genetic variation or in some conditions growth changes linked to an intra-uterine environment. Variations in the number of teeth may well be genetically programmed and thus noticeable frequency differences may indicate inbreeding or the maintenance of specific kinds of stock which carry these variants. (See Andrews and Noddle 1975 for a general discussion). It is therefore interesting to note that five cases of absence of second mandibular premolars in cattle appear in both Roman and medieval Exeter (Maltby 1979), sixteen cases of mainly premolar absence occur at Iron Age Gussage All Saints (Harcourt 1980), and in Saxon cattle from 'Hamwih' (Southampton) 10.9% of the lower second premolars are absent (Bourdillon and Coy 1980). Although these frequencies are too few as yet, they certainly suggest that the collection of further sample data may provide evidence of both temporal and spatial variation linked with the micro-evolution of cattle stocks in relation to these early British populations. Other congential traits may similarly deserve more detailed consideration. John Baker has kindly given me unpublished information on the animal pathology from Roman to medieval Winchester, and this included 32 more noticeable congenital variants, including abnormal perforations through the occipital bone of cattle skulls, and the non-traumatic clefting of the articular surfaces of some bones in cattle and sheep. This clefting of the articular surface

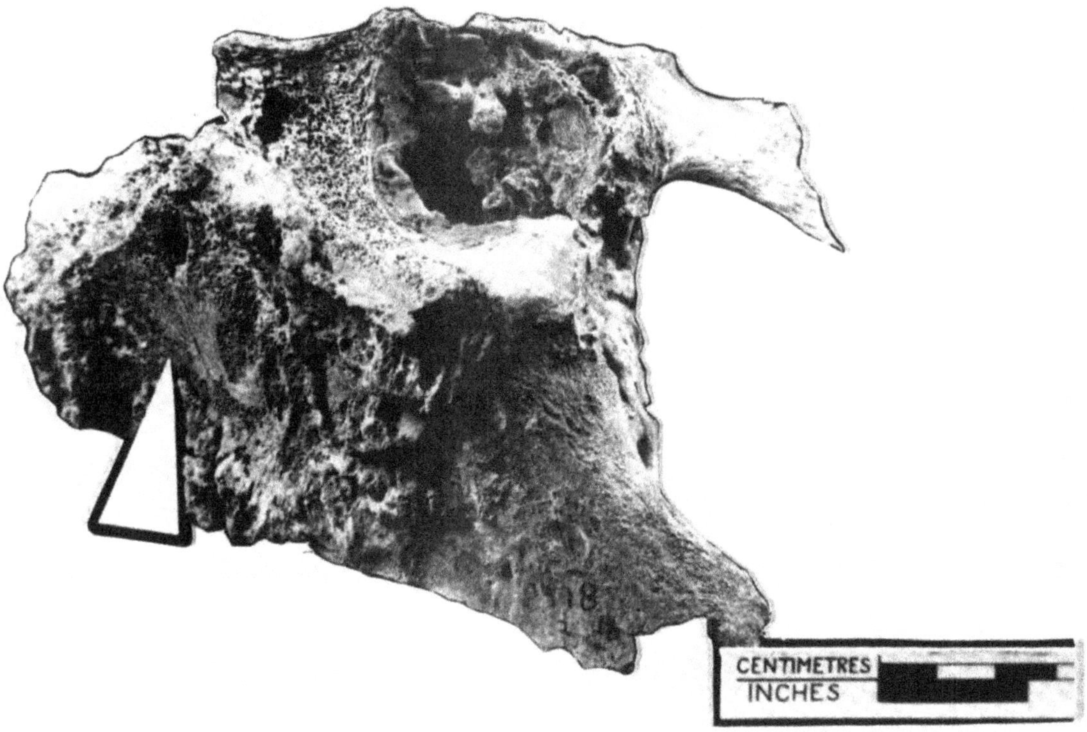

Plate 11.I Fragment of bovid pelvis showing evidence of dislocation and a false joint (with eburnation arrowed). Danebury.

probably has not been fully reported in the past, although its frequency at Norton Priory was 13% at one period. Syndactilism is known in various mammals (Gruneberg 1963) and archaeological examples of the fusion of digits of the hands or feet are known. A sheep at Gussage All Saints displayed syndactyly in two united third phalanges (one abnormal in shape), and similar instances are known from the Osborne House site, Romsey (Coy, in Siegel 1976).

As in other forms of pathology, there is value at times in attempting to construct an overall frequency estimate, but for some abnormalities even one or two specimens may tell us a lot. From the Roman site of Uley, for instance, domestic fowl skulls included one with a genetically determined severe anomaly called cerebral hernia (Brothwell 1979). This occurs only in certain breeds of so-called crested fowl, and could indicate the occurrence of this variety by Roman times. A final example of an unusual possible congenital variant, which poses some provocative questions, is from Portchester Castle. The material has been kindly handed to me for study by Annie Grant (with the agreement of Barry Cunliffe). This post-Roman specimen (Plate 11:I) is part of a cow pelvis, and displays an acetabular socket modified by disuse, with changes and 'polishing' on the bone above the acetabulum indicative of a false joint with the femoral head. The history of this case may well be that a congenital dysplasia of the hip eventually gave rise to a full dislocation of the femur some time after birth. Once the dislocation occurred, movement would have been difficult and clumsy. But the evidence suggests that the animal survived into adulthood and the Portchester community did not kill it as noticeably defective stock. In a similar way, the Iron Age lumbar vetebrae from Danebury seen in Plate 11:II (again by courtesy of Barry Cunliffe and Annie Grant) show deformities probably indicative of congenital abnormality, which is further evidence of another community tolerant to noticeable abnormality in its livestock.

Growth and Nutrition

Congenital defects must be considered as only phase 1 of a series of problems which early communities became acquainted with in their livestock. The growing animal may experience climatic stress, variable 'housing' conditions and hygiene, and especially differing nutritional levels. The effect on skeletal growth can be modest, producing small scale size differences, or enormous, as in severe rickets. Most signs of feeding problems fall between these extremes. The incidents can be isolated, as in the poor bone and tooth calcification in an immature Iron Age dog from Maiden Castle (Siegel 1976) or the bovine long bones from the Goss Street site in Chester where growth seemed to have been inhibited for a period some time prior to death (Baker and Brothwell 1980). Similarly, a bovine skull from Southampton displays a form of osteoporosis which may indicate protein and mineral deficiency (Noddle 1975). Growth at the epiphyseal plate may be varyingly affected, a probable example being seen in an Iron Age pig forelimb from Danebury (Plate 11:III). Rib and long bone fragments from a number of Scottish and English sites show changes suggestive of rickets (Baker and Brothwell 1980).

But by far the commonest evidence possibly indicative of different nutritional environments is to be seen in sheep horn cores and in the oral health of sheep.

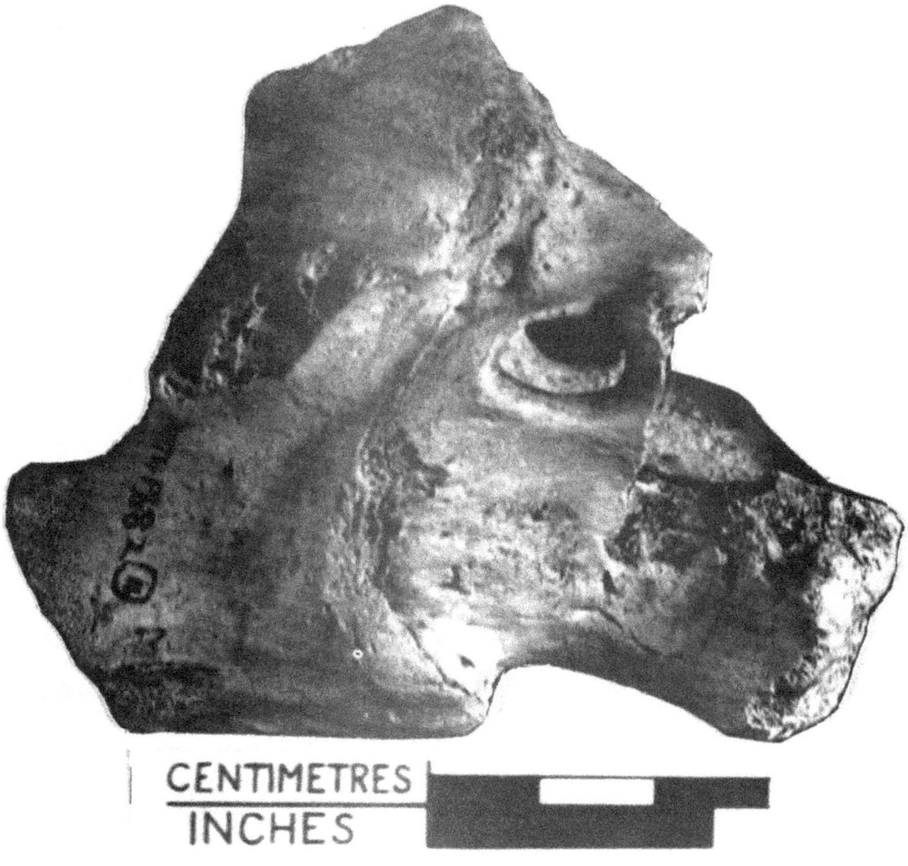

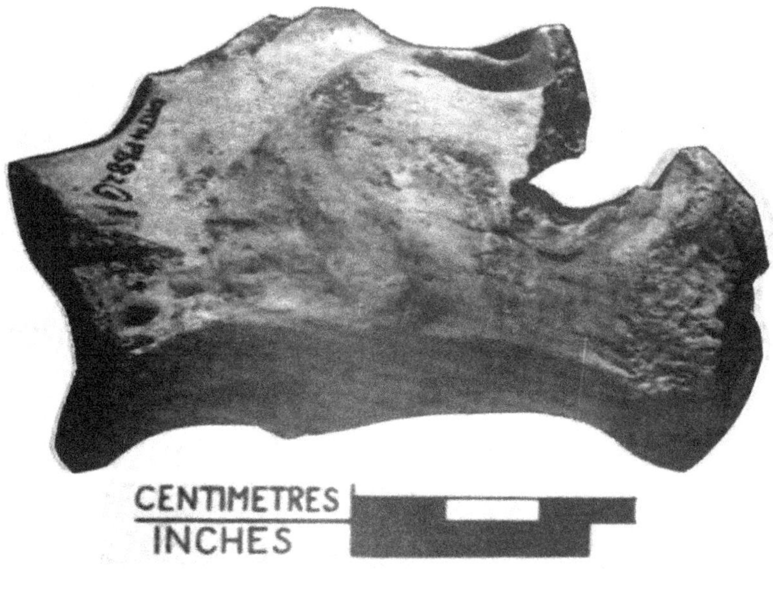

Plate 11. II Lumbar vertebral bodies partly united and deformed, suggesting congenital deformity. Danebury. Two views of same specimen.

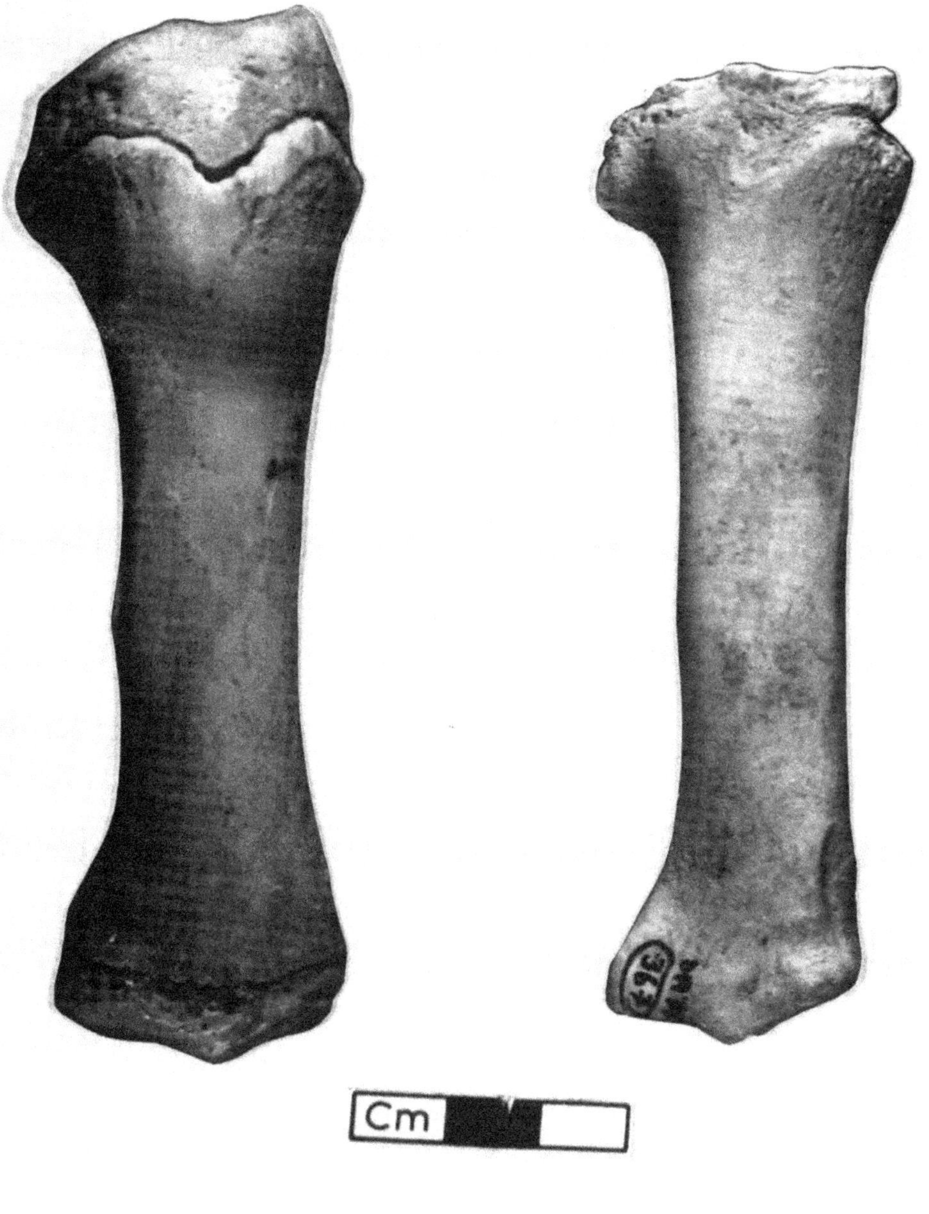

Plate 11. III Evidence of retarded growth (especially at the epiphyseal plates) in a Danebury pig forelimb bone (right), seen in comparison with a healthy modern specimen (left).

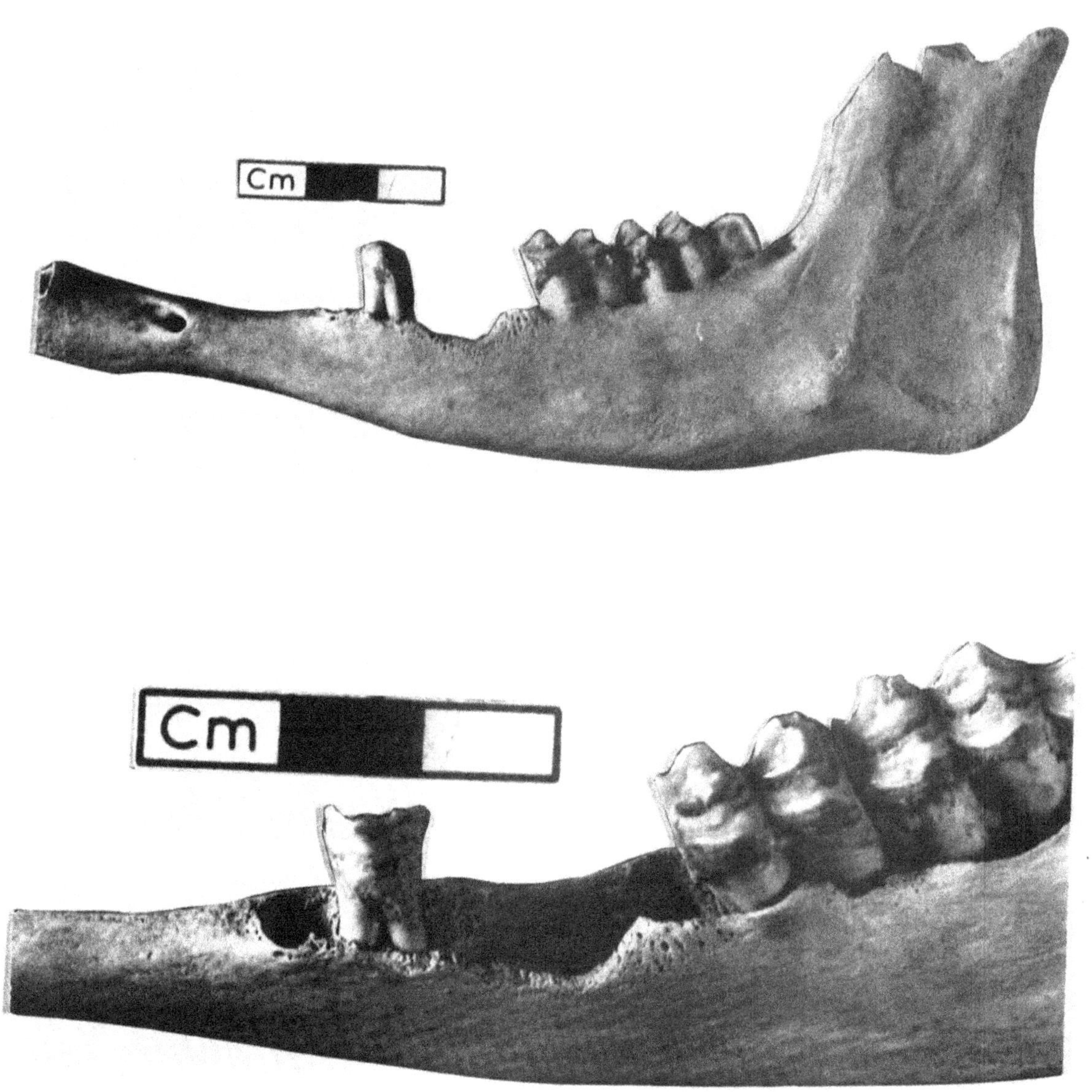

Plate 11.IV Typical evidence of Danebury sheep oral pathology (calculus, periodontal disease and tooth loss). Two views of same specimen.

Some years ago, Hatting (1975) suggested that indentations on sheep horn cores might be related to severe winter conditions and malnutrition. When we look for this defect in early British material, it can be seen at various sites, Iron Age Danebury, Roman Portchester, Winchester and Canterbury. At Saxon Southampton, nearly a quarter of the horn cores show this feature (Bourdillon and Coy 1980), and clearly further studies—including those done on modern sheep—need to be undertaken to clarify this potentially useful environmental marker.

In the case of the pathology of the mouth, the situation is complex, with not only a number of distinct disease entities, but with such factors as food hardness and composition, extent of calculus (tartar) deposits, alignment of teeth, and the nature of the oral micro-organisms of the mouth, all producing different effects. What is impressive is the amount of oral pathology in some samples. Moreover, there is some modern clinical evidence to suggest that the health of the gums and alveolar bone in the region of the teeth is related to some extent to dietary health. Harcourt (1980) suggests this link in his interpretation of sheep periodontal disease in Iron Age material from Gussage All Saints. My own investigation of Danebury sheep jaw pathology (exemplified in Plate 11:IV) suggests a food link also. Annie Grant (1975) similarly reports that Roman Portchester Castle sheep were prone to diet-related periodontal disease. In a study of a large sample of Saxon sheep from Wicken Bonhunt, Levitan (1977) showed that a very large number had periodontal disease, especially in the region of the last premolar and first molar. In contrast to these findings, and in fact in conflict with the evidence of the horn core indentations, the Saxon Hamwih sheep (Bourdillon and Coy 1980) seemed to be remarkably free of this oral pathology. Incidentally, some of the degrees of gum inflammation and alveolar bone swelling (at times with tooth loss) could have deterred feeding and dragged down the general health of the sheep, but clearly the shepherds were either not aware of this, or ignored the fact in some localities.

Trauma and Its Consequencies

Throughout growth and into mature life, animals may suffer trauma, or accidental damage. This may reflect an aspect of survivorship within an environment or even provide information on the treatment of animals by Man. There are, of course, many archaeological examples of fractures and other injuries to bone, but I want to consider here only those which are meaningful environmentally. One category is the traction injury, resulting from the use of animals for tractive purposes over long periods, leading to degenerative joint changes. In Baker's (1980) analysis of the Winchester mammals of Roman to medieval date, he found evidence of use injury in 82 bovids and 3 horses. In the majority of cases, this was in the form of pelvic joint changes (eburnation within the acetabulum), the result of repeated over-rotation of the femur head, perhaps in an attempt on the part of the animal to move heavily laden carts on bad surfaces or ploughs in heavy soil.

'Penning elbow' is a traumatic joint condition of sheep, which is recognised by shepherds as being linked with the use of pens for management purposes. Like cattle traction injuries, there is a need for a systematic investigation of this condition in archaeological samples, and one can only say at present that

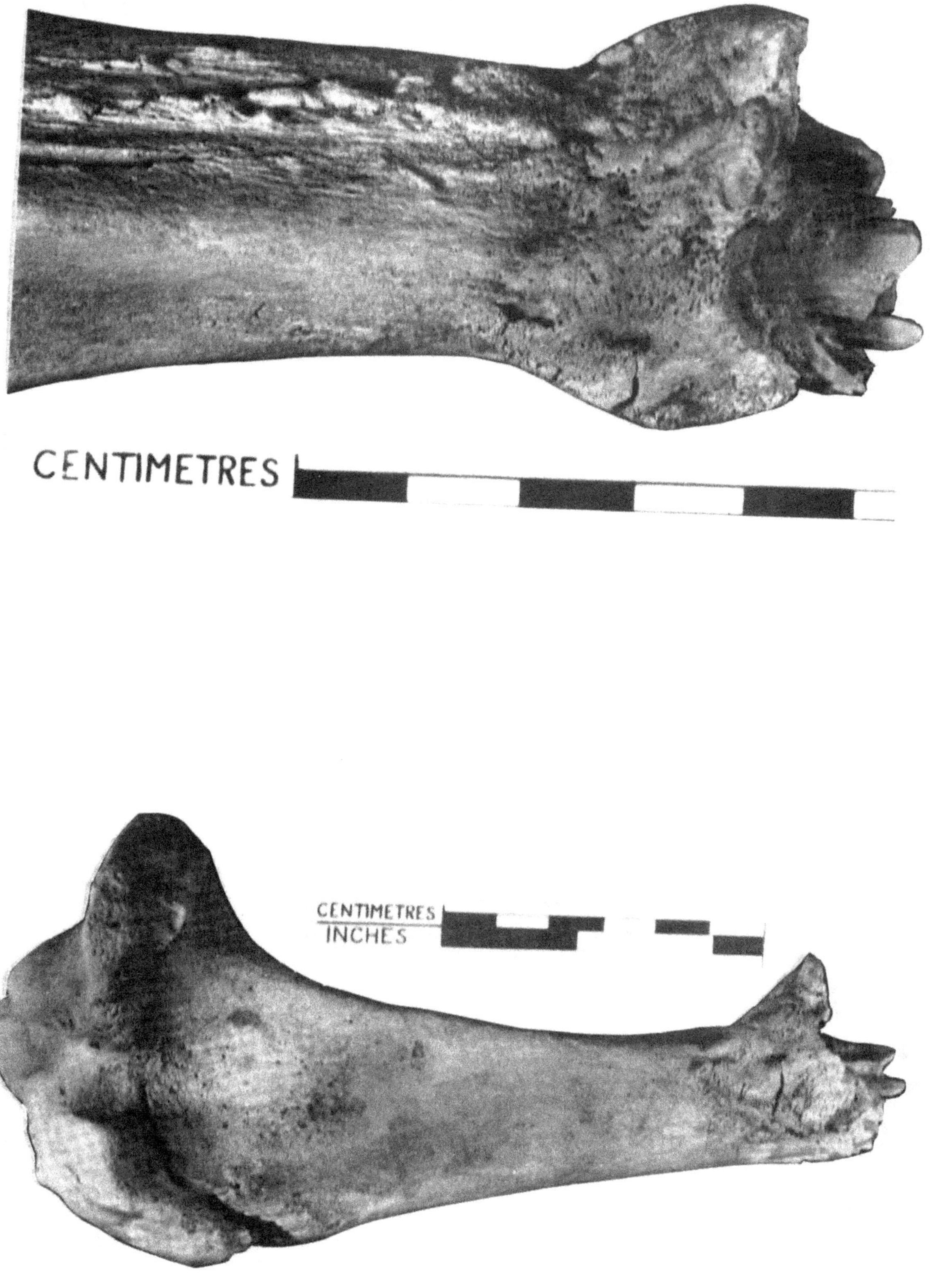

Plate II: V Posterior (above) and lateral (below) views of a bovid tibia showing fracture and some healing. Danebury.

it occurred at some sites. Bourdillon and Coy (1980) note ten bones of the sheep elbow from Saxon Hamwih with arthritic changes suggestive of this condition. By far the greatest evidence comes from Roman to medieval sheep at Winchester, where the large number of cases indicates that penning was used over a long period of time (Baker 1980).

A final aspect of trauma which should be mentioned is the question of injury in relation to serious manhandling. Examples come readily from the Iron Age Danebury pathology which I am at present working on. One bovid tibia (Plate 11: V) shows a shaft fracture with some healing. It is difficult to know what caused this injury to such a large strong bone, and one can only question the motives of the community in keeping alive an animal with such major leg damage. In the same way, what incident, what impact, produced the deep injury to the thoracic neural spine (Plate 11: VI), or the broken immature caprovid pelvis (Plate 11: VII) or the fractures deep in the well muscled scapulae of pig and bovid from Danebury? Most surprising of all is the Danebury dog evidence: in one specimen, injury had caused tooth loss and damage to the snout and orbital area; in another, a blow to the pelvis had broken the femur head off in the socket, producing aseptic necrosis (Plate 11: VIII), after which the dog had limped and survived on a false joint. The surprise is that both dogs show butchery marks, and were presumably eaten—hardly the treatment for a family pet.

<u>Infection and Inflammation</u>

Inflammatory conditions in bones can be in the form of fine surface reactions or massive changes—destroying or expanding bone shape. Roman Portchester Castle provides an example of extremes, with the knee end of an ox tibia abscessed and greatly swollen (Grant 1975). The swelling and pain would have been obvious (not to mention a pus-draining sinus), but the animal was not slaughtered for a considerable time. Similarly, the Danebury Iron Age community kept a horse (Plate 11: IX) with severe long-term osteomyelitis of the mandible, and with this condition it is unlikely to have been a working horse, and even as a prize stallion its performance would probably have left much to be desired.

More important than these general pathologies of uncertain aetiology are of course the diseases where one can make a guess at the micro-organisms which are probably involved. This not only helps to establish the antiquity of a specific parasite and disease in relation to early agricultural developments, but in some instances may throw light on the possibility of health changes in the associated human community. The list of specific infectious disease is at present short but variable. It seems likely that if greater attention were given to evaluating disease, more conditions and more specimens would be found.

Of the disease entities caused by the micro-organism <u>Fusiformis necrophorus</u>, one condition called 'foul-in-the-foot' has occurred at a number of sites. Osteomyelitis may develop from the initial soft tissue lesion at the digits. It has now been identified in Saxon and medieval cows from Hereford, two cases from Saxon and medieval Winchester, and at least five other instances, including two from Iron Age Dragonby. The condition is not uncommon today, and the question arises as to whether the evidence has been missed or the disease has become commoner in relation to modern breeds of cattle?

Plate 11. VI Fractured and semi-healed bovid neural spine from Danebury.

Plate 11. VII Broken and semi-healed immature caprovid pelvis.

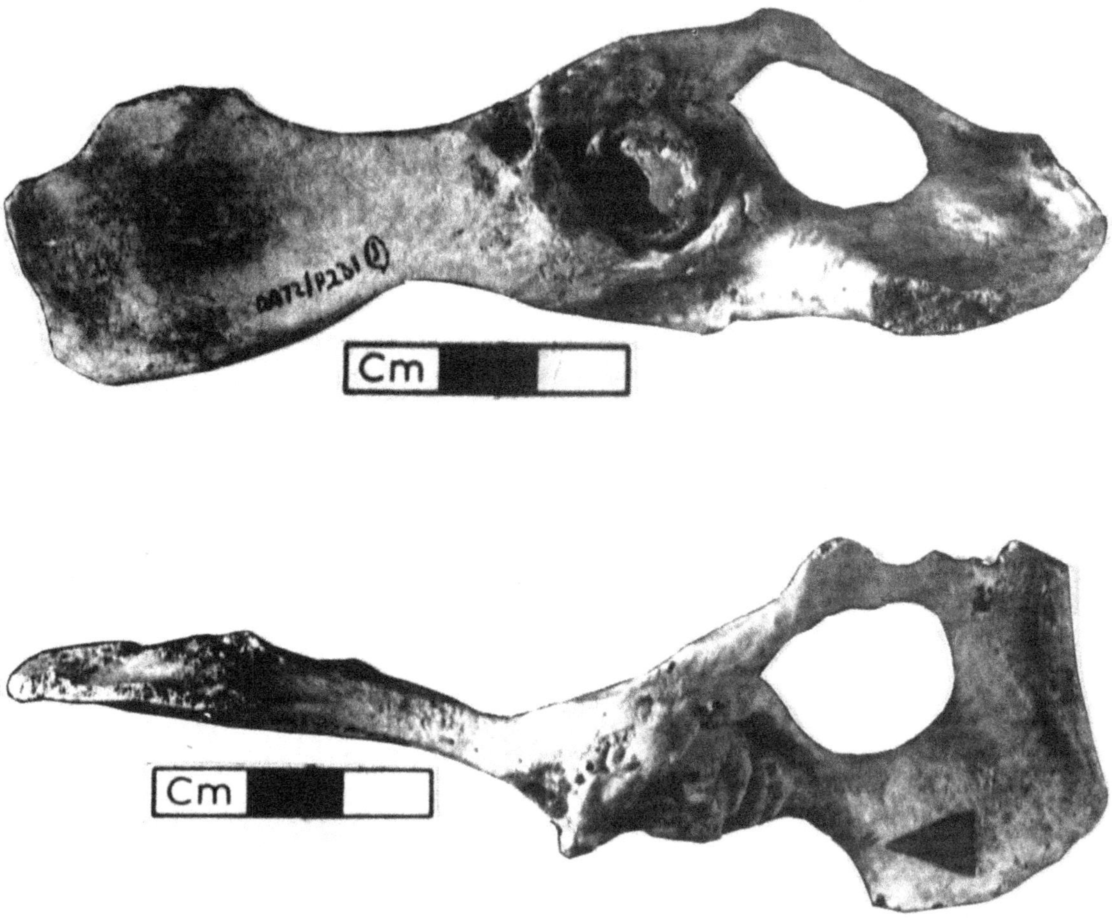

Plate 11. VIII Two views of a Danebury dog pelvis with the femoral head broken off ante-mortem and remaining free within the acetabulum. Butchery marks are also to be seen (arrowed).

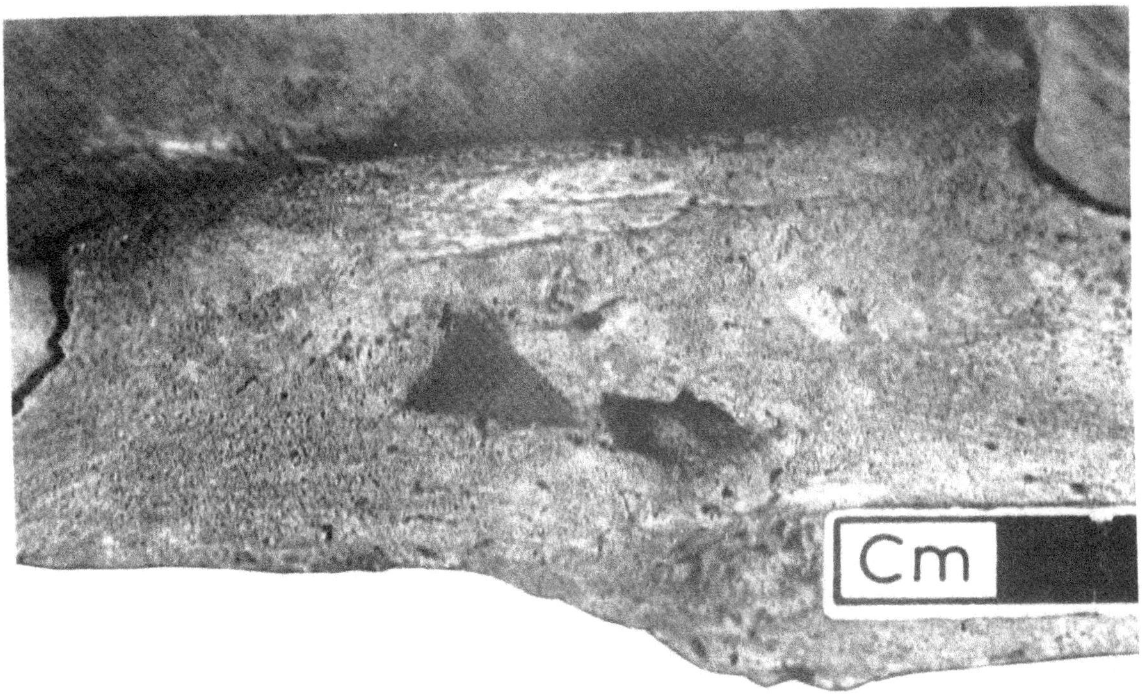

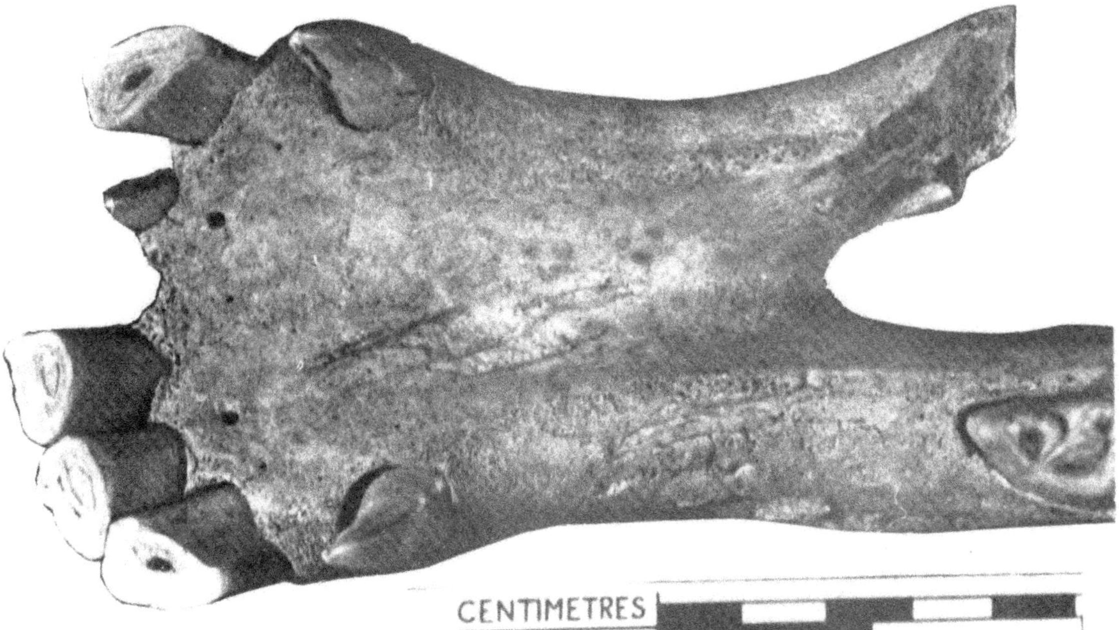

Plate 11. IX Danebury horse mandible (two views) showing an extensive inflammatory reaction in the region of the diastema and mental foramen on the left (arrowed).

Lumpy jaw is also quite common today. Actinomyces bovis can invade the lower jaw and cause massive changes, but there is again surprisingly little evidence of the disease so far. There is a possible case from Saxon Southampton (Siegel 1976). Harcourt (1971a) notes two other possible instances, and the earliest specimen is from Roman Winchester (Baker 1980). Because of the problems of differentiating between the early stages of lumpy jaw and jaw swellings resulting from other types of infection, a more detailed and careful evaluation of much oral pathology needs to be undertaken.

The organism Brucella abortus primarily affects cattle and man, but also to a lesser extent, horses. Surprisingly, the only possible evidence so far is in a vertebral specimen from Dragonby (Baker 1980b). If this is indeed a case, then we should begin to look for brucellosis in contemporary human and cattle groups. At present, we know nothing of brucellosis in cattle before 1805 (according to Dalrymple-Champneys 1960). Finally, one viral disease which is specific to chickens is known as avian osteopetrosis (Baker 1965). The virus stimulates the formation of much extra bone, and in the later stages this is easy to identify in long bones. As a result of archaeological material, we can now establish that the disease may have affected early fowls of Thistleton and Wicken Bonhunt (Harcourt 1971b; Baker and Brothwell 1980). Surprisingly, no other cases are known, and it is possible that the disease frequency may have increased noticeably with the development of modern husbandry methods.

As I said at the beginning, this can only be a brief summary of a vast research field. Diseases are as yet little considered in the field of environmental archaeology. My hope is that, by reference to some early British examples, one can see that it has much information to give and a significant place within the subject as a whole.

Acknowledgements

I greatly appreciate the considerable help of Annie Grant in providing from her intensive zooarchaeological study at Danebury, material for this paper. Thanks are similarly extended to Barry Cunliffe, the excavator of the Danebury site, and John Baker who provided various unpublished information from his own recent studies on mammal bone pathology. Noel Syers kindly provided both photographic and radiographic assistance.

Bibliography

Andrews, A. H. and Noddle, B. A. (1975). Absence of premolar teeth from ruminant mandibles found at archaeological sites. Journal of Archaeological Science 2 137-144.

Baker, J. (1965). Avian Osteopetrosis. Ph.D. thesis, University of Bristol.

Baker, J. (1980a). Manuscript on the palaeopathology of Winchester mammals. Unpublished, but submitted for publication.

Baker, J. (1980b). Dragonby. Diseased bones. Manuscript submitted for publication.

Baker, J. and Brothwell, D. (1980). Animal Diseases in Archaeology. London: Academic Press.

Bourdillon, J. and Coy, J. (1980). The animal bones. In: Excavations at Melbourne Street, Southampton, 1971-76. (ed Holdsworth, P.) Published for S.A.R.C. by Council for British Archaeology, London. pp. 79-121.

Brothwell, D. (1979). Roman evidence of a crested form of domestic fowl, as indicated by a skull showing associated cerebral hernia. Journal of Archaeological Science 6 291-293.

Dalrymple-Champneys, W. (1960). Brucella Infection and Undulant Fever in Man. London: Oxford University Press.

Grant, A. (1975). The animal bones. In: Excavations at Portchester Castle, Vol. 1. Roman. (ed Cunliffe, B.) Reports Research Comm. Soc. Antiq., 32, London. pp. 378-406.

Grünenberg, H. (1963). The Pathology of Development. Oxford: Blackwell.

Harcourt, R. A. (1971a). The palaeopathology of animal skeletal remains. Veterinary Record 89 267-272.

Harcourt, R. A. (1971b). Animal bones from Thistleton. Report on file at DoE, Ancient Monuments Laboratory, London.

Harcourt, R. A. (1980). The animal bones. In: Gussage All Saints. An Iron Age Settlement in Dorset. (ed Wainwright, G. J.) H.M.S.O., London. pp. 150-160.

Hatting, T. (1975). The influence of castration on sheep horns. In: Archaeological Studies. (ed Clason, A. T.) North-Holland Publishing Co., Amsterdam. pp. 345-351.

Levitan, B. (1977). Pathological Anomalies in Sheep Mandibles: a Methodological Approach. B.Sc. thesis, University of London.

Maltby, M. (1979). Faunal Studies on Urban Sites. The Animal Bones from Exeter 1971-1975. Exeter Archaeological Reports, Vol. 2. Department of Prehistory and Archaeology, Sheffield.

Noddle, B. A. (1975). The animal bones. In: Excavations in Medieval Southampton, 1953-1969. (ed Platt, C. and Coleman-Smith, R.) Leicester: University Press. pp. 332-335.

ORGANISATION FOR EUROPEAN ECONOMIC COOPERATION (1957). Control of Losses in Young Farm Animals. Project No. 345, OEEC, Paris.

Siegel, J. (1976). Animal palaeopathology: possibilities and problems. Journal of Archaeological Science 3 349-384.

THE ECOLOGY OF SETTLEMENT

12. THE IRON AGE TO EARLY SAXON ENVIRONMENT OF THE UPPER THAMES TERRACES

Mark Robinson

INTRODUCTION

This paper is an attempt to trace the environmental history of the area under consideration using a broad spectrum of evidence, mainly biological, from sites excavated by the Oxfordshire Archaeological Unit. The principal biological groups used were: pollen, waterlogged macroscopic plant remains, insects and molluscs.

The Upper Thames Valley itself is a coherent geographical division, more or less contained by the present day county of Oxfordshire. From the junction of a small Cotswold River Thames with the Rivers Churn, Leach, Coln and Cole in the region of Cricklade and Lechlade, to the Goring Gap, where the Thames forces its way through the chalk scarp of the Chilterns and Berkshire Downs, the lower part of the valley has a consistent character: broad expanses of calcareous gravel terraces and short clay slopes stepping down to a floodplain that is still sometimes inundated. Above the terraces is a hinterland of wider clay slopes, sands higher up and rather brashy limestone hilltops while the chalk downs are ever present further to the south.

On the gravel terraces and to a lesser extent the floodplain exist numerous archaeological sites of Iron Age to early Saxon date. On the terraces, ploughing has usually removed any ancient soils but on the floodplain some sites survive in good condition under alluvial coverings. Anaerobic waterlogged deposits are frequently present on these terrace or floodplain sites in well and ditch bottoms while old stream beds can be encountered on the floodplain as well.

Before this investigation, no work had been done on plants or insect remains from waterlogged post glacial contexts on the gravel terraces. Most of the work on other biological remains of Iron Age to early Saxon date has been piecemeal identifications of excavated material. Recently, however, with the creation of the Oxfordshire Archaeological Unit, a systematic programme of environmental investigations has been undertaken, with work by M. Jones on charred seeds and R. Wilson on animal bones running in parallel with the work on waterlogged material. Their results so far have been presented site by site in Hinchliffe and Thomas (1980), Lambrick and Robinson (1979), Miles (1981) and Parrington (1978).

GEOLOGY, DRAINAGE AND SOILS

The landscape of the area has resulted largely from the erosion of Jurassic and Cretaceous sedimentary rocks which gently slope to the southeast. Where clays were exposed, valleys formed while on the higher ground their protective covering of limestone or chalk survive (Ordnance Survey one inch drift geological

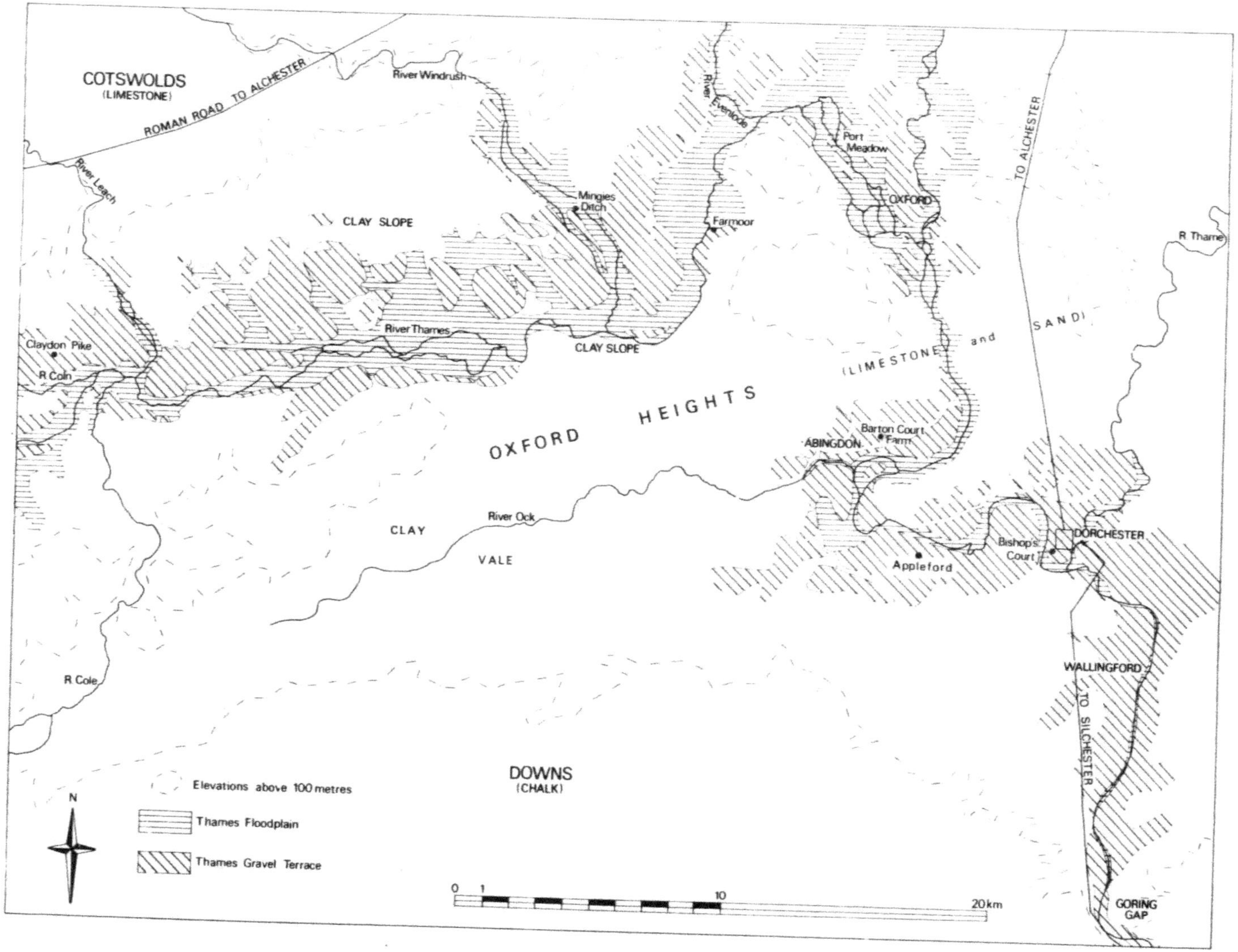

Fig. 12.1 The Upper Thames Valley showing sites mentioned in text and principal Roman Roads.

maps). Over the region the River Thames drops about 30 m before the Goring Gap, which forms the base level of the Upper Thames (Wood 1968: 15).

Throughout the Pleistocene there have been periods when the Thames or a proto-Thames has accreted an extensive limestone gravel terrace over the clay vales only to start downcutting, remove much of the gravel and then form a new terrace. The sequence of terraces was established by Sandford (1924). The youngest and lowest of the terraces is the first terrace, which rises 2 to 3 m above the present river level. Above the first terrace is the second terrace at 8.5 m above modern river level, the third terrace at 18 m and the fourth terrace at 30 m (although not all the terraces have the same gradient as the modern river). The first and second terraces are the most extensive and it is on them, as well as the modern floodplain, that the sites investigated were located. In some parts of the region the two terraces are 5 km or more wide. The present floodplain of the upper Thames and its tributaries can be over 2½ km wide and is covered with a usually calcareous alluvium.

The gravel terraces are well drained although wet flushes sometimes occur on the clay slopes between terraces. The River Thames in this region is well known for its extensive winter floods and at one time there was much marshy land on the Thames floodplain (Emery 1974: 81). Except where it is protected by floodbanks, it is still usual for the Thames to inundate part of its floodplain almost every year. In about one year out of three most of the unprotected floodplain floods almost to the edge of the gravel terrace and in about one year in twelve, serious flooding occurs, which extends onto some parts of the first gravel terrace (Lt.-Col. D. Williams, pers. comm.).

On the floodplain are poorly drained alluvial clays which are neutral to alkaline (Jarvis 1973: 181-2). A variety of soils is to be found on the first gravel terrace, the most important being: the Sutton series, a well drained neutral sandy loam; the Badsey series, a calcareous and sometimes more clayey series; and the Caswell series, neutral to calcareous imperfectly drained clay or clay loam. On the higher gravel terraces is only the Sutton series (Jarvis 1973: 116-17, 179-81). The Sutton series are <u>sols lessivés</u> transitional to the brown calcareous group but their drainage is not impeded (Jarvis 1973, 28, 181). The changes in these soils through time are discussed on p. 270.

The soil of the floodplain still presents problems for arable agriculture even with extensive modern drainage. The soils of the higher gravel terraces which are free draining are easily cultivated but especially on the higher terraces crops can suffer from drought and the Sutton series is deficient in potassium (Jarvis 1973: 181), but these factors probably tended to result in reduced, but still acceptable, yields rather than discouraging arable completely.

THE ARCHAEOLOGY OF THE REGION

The archaeology of the gravel terraces of the Upper Thames Valley is particularly well known because their soils are very suitable for the occurrence of cropmarks and arable is the main use of the gravels. After 40 years of intensive aerial photography all gravel cropmarks known to date were plotted for Oxfordshire (Benson and Miles 1974) and further upstream to the headwaters (Leech 1977). Recent work by R. Hingley has increased by 30% the number of sites given in Benson and Miles between Oxford and Lechlade. Gravel extraction initially provided an easy way of obtaining finds from these sites but an increasing effort has been put into rescue archaeology. The floodplain has received rather less attention because, until recently, most of it was permanent grassland. However, a few sites have been discovered from the air in very dry years as parch-marks in grass and others still survive as earthworks.

The Upper Thames gravels had already been extensively occupied by late neolithic times and large ritual sites, such as the Dorchester henge and cursus complex had been built on them (Atkinson et al. 1951). Few undoubted settlement sites from the Bronze Age are known but ring-ditches are ubiquitous on the gravel terraces and they are also present on the floodplain.

The earliest acceptable radiocarbon date for an Iron Age context in the area is 520 $\pm$ 70 bc but unfortunately the dates obtained from the other two samples in the series were highly suspect (Parrington 1978: 39). Settlements of the early Iron Age on the Thames gravels remain relatively enigmatic. Only one enclosure of this date has been excavated but it was incompletely published and its dimensions are unknown (Harding 1972: 15-16). However, pit clusters from this period have frequently been encountered on the gravel terraces, for instance at Ashville, Abingdon (Parrington 1978: 3) and Farmoor (Lambrick and Robinson 1979: 65-6). Some were probably for grain storage. On the Berkshire Downs a series of hillforts were constructed and they were integrated with Bronze Age land divisions (Bradley and Richards 1978). No major land divisions or field systems are known from the gravel terraces at this date.

For the middle Iron Age, that is from about 350 bc onwards, a much greater amount of information has been recovered for settlement sites on the gravel terraces and floodplain. Recent excavations of them have been summarised by Lambrick (1978: 112-18). They fall into two types. Firstly, the relatively uncommon enclosed settlements, where all the huts and occupation areas were surrounded by a relatively substantial ditch. Only one such site has been excavated which was at Mingie's Ditch, on the floodplain (p. 257). The other type is Harding's category of open settlements (Harding 1972: 14-21). They are characterised by hut circle gullies, sometimes with small annexes or farmyards attached, and rectangular to irregular shaped enclosures up to about 50 m across, clusters of pits and seemingly random lengths of gully. These settlements range from what amounted to a hamlet or village on the second terrace, as at Ashville, with many hut circles and storage pits (Parrington 1978), to single hut complexes, as on the floodplain at Farmoor The earliest buildings identified on the gravels b elong to this period and they were usually in the form of post-built circular huts up to about 8 m in diameter, usually surrounded by an open hut circle gully. At Mount Farm, Dorchester, Lambrick (1979) discovered the beginnings of a system of sub-rectangular, small ditched fields dating from the middle Iron Age and continuing in use into the Roman period. On present evidence, however, such a system does not seem to have been usual on the gravel terraces at this date.

With the coming of the Belgic period in the 1st century BC, some major changes seem to have occurred in the Thames Valley. At the beginning of this period the region seems to have been on the boundaries between the historic tribes of the Dobunni to the west, the Atrebates to the southeast and the Catuvellauni to the northeast but by the early years of the 1st century AD the Catuvellauni had expanded over much of the area (Cunliffe 1974: 76-105). The hillforts seem to have fallen out of use in the Belgic period (Harding 1972: 45) and their possible successors were Dyke Hills at Dorchester where a double rampart cut off a loop of the Thames, and the Big Ring at Cassington Mill by the River Evenlode. Their dating is by no means secure (Harding 1972: 54-6).

The development taking place on the gravel terraces during this period was the appearance of relatively large (100 m across or more) rectangular fields or paddocks. At Appleford, on the first terrace, a hut circle with an attached compound was incorporated into a larger rectilinear system of gullies with a trackway (Lambrick 1978: 116; Hinchliffe and Thomas 1980). At Ashville rectangular enclosures replaced the middle Iron Age hamlet. However, such enclosures only occur in the immediate vicinity of settlements and do not seem to represent part of extensive field systems.

The Thames Valley was within the initial area of Roman conquest after the invasion of 43 AD. The two ultimately walled Roman towns of Oxfordshire, Dorchester and Alchester, probably succeeded conquest period forts (Rodwell 1975: 41, 101). Another ten or so unwalled small town-like settlements developed in the region, including perhaps ones at Abingdon and Wallingford on possible Thames crossing points (Rodwell 1975: 13). Thus, with Dorchester, there were probably three Roman towns on the gravels. The major roads built by the Romans in the area were Akeman Street, running east-west from Verulamium to Cirencester via Alchester and one running north from Silchester through Dorchester and Alchester to Towcester. The important part played by long-distance communications during this period is emphasised by the presence of marine oyster shells on even the smallest of Roman rural sites in Oxfordshire. The major non-agricultural activity of the region in the Roman period was the pottery industry situated on the Dorchester to Alchester road east of Oxford. The industry began in the 2nd century AD and by the 4th century was one of the most important pottery industries in Roman Britain (Young 1977).

Many villa estates were established on the limestone of the Cotswold slopes including in the area which became the medieval Wychwood forest. The villa at Ditchley began life as a rectangular timber building in about 70 AD and in the early years of the 2nd century was replaced by a stone winged-corridor villa. Some of these Cotswold villas were very large, for example North Leigh (Frere 1967: 268-9).

While the Cotswold villas seem to have been the wealthiest farms in the region, the Thames gravels had the most intensive agricultural settlement. The Roman sites on them tend to consist of a ditched trackway with small rectangular enclosures laid out alongside them (Miles 1978; Robinson 1978). Such sites can be regarded as a proliferation and enlargement of the Belgic type of settlement on the gravels. Within the small fields occur various occupation features such as pits and wells but the buildings themselves have rarely been discovered. As in the earlier periods the enclosures do not represent extensive field systems and seem to be related to the occupation areas themselves. The droveway sites occur on all the gravel terraces and very many are known from aerial photographs but they do not extend onto the floodplain. On the higher gravel terraces a few of these sites developed into stone-built villas, as at Barton Court Farm (p.266) but at a later date than the Cotswold villas, perhaps in the late 3rd century and not on such a grand scale. Until recently, no villa sites were known on the first gravel terrace but excavations by D. Miles (1980) at Claydon Pike revealed stone Roman buildings on this terrace.

At the end of the 4th century there seems to have been a breakdown in communications of the region resulting in the decline of the pottery industry and the villa system. Metalwork finds from the town of Dorchester and two burials in the rampart of Dyke Hills are thought to suggest that a Germanic Roman military garrison was present in the town in the very early years of the 5th century (Brown 1974: 16; Rowley 1974: 42). By the middle of the 5th century Saxon colonisation of the Upper Thames Valley was taking place and before the end of the century had extended upstream to beyond Lechlade. The Saxons occupied Dorchester from at least the later 5th and throughout the 6th centuries (Rowley 1974: 44) but it cannot be regarded as functioning as a town in the Roman sense. On the gravel terraces the Pagan Saxons are best known from their cemetery sites, rich in grave goods, for example at Berinsfield (Miles 1975). There is little evidence for a continution of use of the Roman enclosure systems themselves, although Saxon occupation sometimes occurred on such sites as at Mount Farm (Lambrick 1979). An interesting exception was provided by the Barton Court Farm villa where a 5th century extension was made to a Roman enclosure. Pagan Saxon settlements of the 5th and 6th centuries on the gravels can be characterised by the presence of several to many sunken huts, perhaps a few post-built hall houses, some pits and shallow wattle-lined wells (e.g. Hawkes and Gray 1969).

METHODS

Waterlogged and dry samples were taken from contexts which could be related to the archaeology of the sites investigated and processed to extract plant and invertebrate remains. These were identified and used to build up ecological information about the sites. Full details of sampling procedure, extraction techniques, reference collections used, identification methods, sources of ecological information and aspects of interpretation are given in Lambrick and Robinson (1979: 77-111).

Interpretation of coleopteran remains is further discussed in Appendix I

THE IRON AGE

Major investigations of two floodplain sites belonging to this period, Mingie's Ditch and Farmoor (which also extended onto the first terrace) were undertaken. In addition a small sample from the first terrace site of Claydon Pike was examined so that its macroscopic plant remains could be compared with Farmoor.

Mingie's Ditch, Hardwich with Yelford (SP 391059) Floodplain (Interim reports, Robinson and Allen 1978; Lambrick 1978; Allen and Robinson 1979).

The site, on the floodplain between the two courses of the River Windrush, consisted of a circular double-ditched enclosure about 80 m in diameter. Occupation features namely huts, four post structures and pits, were confined entirely to the area within the inner enclosure ditch. A thin alluvial covering over the site had resulted in the survival of occupation layers and floors for some of the huts. Initially the enclosure stood in isolation, with a pair of 'antennae ditches' running from a break in the outer ditch and a corresponding gap in the inner ditch provided an entrance to the interior. Access to the area between the ditches (about 10 m wide) was provided by a break in the outer ditch where a meander of a stream, now silted up, approached it. Subsequently a small field was added to the enclosure, by extending the line of one of the 'antennae ditches'. One side of this field was formed by the side of the main enclosure, the second by the old stream and the other two by new ditches. Radiocarbon dates are awaited for the site but on the basis of the pottery a likely date for its use would be between 350 bc and 250 bc.

Beneath the gravel banks of the Iron Age enclosure was about 0.1-0.2 m of alluvial clay. It had been worm sorted, concentrating all the riverine Mollusca at the bottom of the profile and the upper part of the profile had become decalcified. This suggests that there was a period (possibly lasting more than a millennium) when the site was not regularly flooded by the calcium carbonate-rich waters of the river. Cutting this alluvial layer were a series of pits, the deeper of which sometimes contained humified peat, a few very rotten fragments of *Alnus* wood and *Alnus* charcoal. In one of these pits, preserved by a spill of gravel, were woodland Mollusca. These pits probably represented a pre-Iron Age phase of clearance of alder wood.

There was no evidence from any of the deposits that the Iron Age settlement suffered flooding while it was occupied. Pollen analysis (by G. W. Dimbleby) of peat from the bottom of the enclosure ditches suggested a landscape substantially cleared of woodland although willow trees may have been conspicuously present. Tree pollen formed 4% of the total tree pollen while shrub pollen, mostly *Salix* comprised another 13% of the total. This picture is confirmed by the Coleoptera. Wood and tree dependent species (excluding *Anobium punctatum* and *Lyctus* sp.) made up only 1% of the total terrestrial Coleoptera identified from Mingie's Ditch compared with 17% from Buscot Lock (Robinson unpublished), a riverine deposit probably surrounded by woodland, and 5% from Shustoke B (Kelly and Osborne 1964-5), which was regarded as having an open environment although with some secondary woodland and scrub present.

The pollen spectra suggest that species-rich grassland was important with Gramineae pollen averaging 34%, *Plantago lanceolata* 6% and Compositae-Liguliflorae 7% of the pollen from the four samples. The grassland was likely to have been wet, grading into marsh along the margin of the River Windrush and the ancient stream while such vegetation probably covered the bed of various minor streams traversing the area which perhaps only contained water in the winter. Seeds of wet grassland or marsh species included *Ranunculus* cf. *flammula*, *Lychnis flos-cuculi*, *Filipendula ulmaria*, *Mentha* sp. *Pulicaria* sp., and various rushes and sedges. Grassland species were well represented amongst the Coleoptera, especially *Phyllopertha horticola*, which has larvae that feed on roots in permanent grassland (Raw 1951-2: 6445). The Coleoptera provide overwhelming evidence for the importance of pasture in the vicinity of the site. Dung beetles from the genera Geotrupes, Colobopterus, Aphodius and Onthophagus made up 16% of the total terrestrial Coleoptera from the site, some of which are restricted to dung of large herbivores in the field (rather than in manure heaps).

Alongside the River Windrush peat deposits with abundant remains of Schoenoplectus lacustris and Menyanthes trifoliata as well as brightly coloured fragments of donaciine Chrysomelidae were present. These probably represent a reedswamp community along the edge of the river, perhaps with the cattle grazing the aquatic vegetation in shallow water enabling the Menyanthes to compete against the taller growing Schoenoplectus (Hewett 1964: 729).

The discovery of two pairs of saddle querns and the presence of cereal remains, both waterlogged and charred, show that grain must have been used on the site. There was no direct evidence for cultivation either inside or outside of the small field to the south of the occupation enclosure. There was no trace of mixing of gravel into the Iron Age soil profile even though it was only 10 cm deep in places (it had not been truncated) and worm-sorting had not taken small gravel spills from the upcast of Iron Age features to the bottom of the profile. A preliminary examination by M. Jones (pers. comm.) of an assemblage of carbonised cereals from the site (mostly Hordeum vulgare but also Triticum spelta) suggested that they had been grown on dry ground, and probably not on the floodplain.

Amongst the macroscopic plant remains were abundant seeds, wood and leaves of many shrubs and small trees:

Acer campestre
Corylus avellana
Crataegus cf. monogyna
Prunus spinosa
Rhamnus catharticus
Rosa sp.
Rubus fruticosus agg.
Salix sp.
Sambucus nigra
Thelycrania sanguinea
Viburnum opulus

The pollen, however, did not give even a slight indication of their presence apart from Salix and Corylus, but mostly they are entomophilous species. Excluding Salix and Corylus, pollen from these species has usually only been recorded at very low levels (for example Acer, see Godwin 1975: 170) and tends not to be preserved well. In addition to the shrubs there were also seeds of various herbs of scrub and shady places including Moehringia trinerva, Stachys sylvatica and the climber Humulus lupulus. In one sample from the inner enclosure ditch seeds from light-demanding terrestrial plants were almost absent while various species which can grow in woodland were well represented.

The strong presence of scrub suggested by the macroscopic plant remains contrasts with the poor evidence for woody species but substantial indication of grassland from the Coleoptera. This probably reflects the much more local origin of the plant remains compared with the insects, and the scrub was localised to the vicinity of the settlement. The diversity of woody species suggests either that the scrub had been deliberately planted in the form of hedges around the enclosure or the scrub was of ancient origin, perhaps even a relic of the woodland which had been cleared from the site prior to the Iron Age. Acer campestre (represented by fruits, wood and leaves) is a species of woods, hedgerows and scrub preferring a high base status but it is not a pioneer species in the invasion of grassland. It is present in scrub only during the later seral stages (Jones 1944-5: 241-3). Pollard (1973: 343-52) has shown that A. campestre is common in old woodland on appropriate soils but has little ability to colonise hedges, even if growing in woodland nearby. It has been suggested that hedges were created by selective scrub clearance in parts of the Netherlands during the Neolithic (Groenman van Waateringe 1978: 138-40).

Some of the pre-Iron Age tree pits were deeper than the deepest Iron Age features yet preservation of organic material in them was very poor, the only identifiable remains being very rotten fragments of wood. By the time of the Iron Age occupation of the site, the water table had risen and there was standing water in the bottom of the enclosure ditches, resulting in good preservation of organic remains. As the ditches began to silt up so the quality of preservation deteriorated and in some ditch profiles a non-organic deposit built up. Then a gravel trackway belonging to the final phase of occupation was constructed through the entrance. Either just before or just after the abandonment of the site, but stratigraphically later than the gravel road, there was a very distinct recurrence horizon for peat development in the enclosure ditches. It is possible that the construction of the road surface was a response to the rising water table and the abandonment of the site may also have been related to this problem. Perhaps there was a resumption of flooding. In the Roman period drainage ditches over the site probably represented an attempt to lower the water table of the floodplain, but they fell out of use by the early Saxon period and the alluvium which formed a thin covering over the site was deposited.

Farmoor (SP443057) First Terrace and Floodplain
(Full report, Lambrick and Robinson 1979)

The site extended from the floodplain onto the first terrace, a cut-off channel of the River Thames separating the floodplain from the terrace. On the floodplain were three middle Iron Age farmsteads, each comprising a small ditched enclosure about 12 m in diameter for a roundhouse, with one or more ditched subsidiary yards or stock pens attached. Other small enclosures, one with an adjoining post-built semicircular building were made on the gravel terrace. Peaty deposits had accumulated in the ditches of the floodplain enclosures and in a sump of one of the enclosures on the gravel terrace. The dating of the middle Iron Age settlement has proved difficult. On the basis of the pottery it is certainly pre-Belgic and likely to post-date Mingie's Ditch, being perhaps some time between 250 bc and 100 bc. Two of the three radiocarbon dates obtained fall within the above date range but a third was 460 $\pm$ 100 bc.

The earliest sample examined was from a thin non-gleyed sandy silt covering the gravel underlying the occupation layer of one of the enclosure complexes (3) on the floodplain. All of the few molluscan remains were from terrestrial species. The occupation layer of Enclosure Complex 3, however, was a gleyed clay and along with charcoal flecks and a little pottery, also contained riverine aquatic molluscs. These species were also present in some of the ditches of the floodplain enclosures. Bithynia tentaculata was present in the ditches of Enclosure Complexes 2 and 3. It has been included by Sparks in his flowing-water group of molluscs (Sparks 1959-60: 76) and Boycott (1936: 139-40) states that the freshwater operculates live almost exclusively in running water, never occurring in small, closed ponds. B. tentaculata would not have been able to breed in water in the bottom of these ditches (which had been traced in their entirety and formed a closed system). Flooding is regarded as the only likely explanation of its introduction, implying that the sites were seasonally occupied.

The pollen analyses (by G. W. Dimbleby) gave an even lower value for tree pollen than the samples from Mingie's Ditch, 1.4% of the total pollen. Shrub pollen was also down to a level of 1.4% of the total pollen. These values suggest that clearance had been very thorough in the region of Farmoor by this date and unlike Mingie's Ditch there was no evidence to the contrary from macroscopic plant remains for a significant preserve of woody species. Only a single beetle out of over 1,000 terrestrial individuals identified fell into the group of tree and wood dependent species (excluding Anobium punctatum and Lyctus linearis). The major biotype indicated by the pollen seems to be grassland.

The waterlogged seeds were mostly from plants of disturbed ground but their origin was probably extremely local but there were seeds of about 35 species which could have grown in grassland, including some which are restricted to wet habitats. Damp pastureland is suggested by Eleocharis S. Palustres sp., which was well represented. These species (E. palustris and E. uniglumis) are restricted to places where competition from tall growing species is slight or absent and the water level is at or above soil level in the spring, when rhizome growth takes place (but can tolerate drier conditions during the rest of the year) (Walters 1949: 196-7, 204). E. palustris is an abundant plant on the wetter areas of Port Meadow, heavily grazed grassland on the floodplain near Oxford which has been pasture for many centuries. Another species which grows on the wetter areas of Port Meadow, Potentilla anserina, was also well represented amongst the seeds. The importance of pasture at Farmoor in the Iron Age is confirmed by dung beetles of the genera Geotrupes, Colobopterus, Aphodius and Onthophagus. They made up almost 28% of the total terrestrial Coleoptera, a higher proportion than for any other Thames Valley site so far investigated.

There was a wide range of seeds from plants of disturbed ground which probably grew around the settlement, but two species which are particularly common in nutrient-rich ground around settlements and whose seeds are normally found in large numbers from waterlogged archaeological deposits, Urtica dioica and Sambucus nigra, were almost absent. One or both of these species were well represented from all the other sites investigated for this project, including the Iron Age site of Mingie's Ditch which was on the floodplain and Claydon Pike which had somewhat similar enclosures to Farmoor. U. dioica and S. nigra are both capable of growing on the floodplain today, sometimes in great abundance, but neither normally grow in grassland (though once given the opportunity to invade, U. dioica can survive for many years) (Greig-Smith 1948: 345). S. nigra must be several years old before flowering and U. dioica does not flower in its first year (Greig-Smith 1948: 349).

A comparison for Farmoor between the proportion of seeds of some of the more common annual weeds with the proportion of seeds of the following perennial weeds which do not flower in their first year: U. dioica, S. nigra and Rubus sp., shows the perennials to have been much better represented in the Roman deposits than in the Iron Age. (The annual species used in the comparison were:

Atriplex sp.	Polygonum aviculare agg.
Chenopodium album	Sonchus asper
Chenopodiaceae indet.	Stellaria media gp.
Papaver spp.	Urtica urens

All these species, none of which normally occur in undisturbed grassland, were present in both Iron Age and Roman samples. In the 4 Iron Age samples with a large number of these seeds, the three perennials ranged between 7.5% and 0.4% of the total for both the annuals and the perennials, whereas in the 6 sizeable Roman samples these perennials ranged from 29% to over 99% of the total. A possible reason for this difference is that the Iron Age settlements at Farmoor were temporary whilst the Roman ones were permanent. The annual weeds would be able to build up large populations rapidly on the freshly disturbed grassland in the vicinity of the sites but by the time the perennials had become established and were flowering in profusion, the settle ments had been abandoned and their ditches allowed to silt up.

A small number of charred grains of Triticum sp. and Hordeum from the Iron Age features on the floodplain were the only crop remains identified. They certainly indicate usage of cereals but in view of the evidence for flooding it would seem most unlikely that the floodplain was cultivated. If the inhabitants of the Iron Age enclosure complexes were growing their own grain in the vicinity, it would be most likely that their fields were on the first gravel terrace, in which case the siting of the settlement on the floodplain, where presumably they could only be seasonally occupied, seems surprising.

Pteridium aquilinum frond fragments were present in Iron Age samples from both Mingie's Ditch and Farmoor. It requires somewhat acidic conditions for sporeling establishment although the adult plants themselves can grow in the presence of calcium salts (Conway and Stephens 1957). Limestone heath might have developed somewhere over the gravels in a stone-free soil where there was not an input of calcium carbonate (for instance from the lime-rich floodwaters of the rivers) but it is more likely that the bracken had been brought to the sites from the acidic sands on the hillsides. There is evidence of prehistoric man bringing bracken to sites elsewhere, perhaps as litter for animals (Dimbleby and Evans 1974: 132).

Some time after the Iron Age sites on the floodplain had ceased to be occupied, the nature of the flooding changed. Rather than being the sort of flooding which occurs on Port Meadow today, depositing mostly flotsam with very little mineral material, a much greater mineral load was deposited, causing a build up of alluvium over the Iron Age enclosure complexes, the filling in of the old river channel and a levelling up of undulations in the floodplain. Perhaps the type of flooding was already beginning to change when Complex 3 was in use because its occupation layer was a similar clay to the alluvium above it and had the same proportion of aquatic molluscs. Evidence from the Roman site at Farmoor showed that the main aggredation of alluvium must have been substantially complete by the 4th century AD.

Claydon Pike, Fairford (SP193996) First Terrace
(Interim report, Miles 1980)

Trial excavations on this very large crop-mark site have shown the presence of many middle Iron Age hut circles and other small enclosures as well as Roman stone buildings and trackways. The bottom layer from the ditch of a small middle Iron Age sub-rectangular enclosure was the only waterlogged deposit on the site. Urtica dioica made up about 20% of the weed seeds using the same categories as at Farmoor plus Chenopodium rubrum. Excluding C. rubrum, U. dioica out-numbered all the other annuals put together. These results provide a useful contrast to those from the Iron Age enclosures at Farmoor, although they do not, of course, show that the enclosure at Claydon Pike was in long-term use, the nettles could, for example, have been growing amongst scrub.

THE ROMAN PERIOD

Substantial waterlogged samples from three Roman sites have been investigated, 'native droveway settlements' on the first terrace at Farmoor and Appleford, and a small villa on the second terrace at Barton Court Farm.

Farmoor (see above)

There was a break in the occupation of the site during the Belgic period and beginning of the Roman period but the site was re-occupied from the later 2nd century AD until the end of the 4th century (dated by pottery and coins). The settlement was confined to the gravel terrace and based around a ditched droveway which crossed the terrace along its edge with the floodplain. The old river channel had been filled in by this date but a stream probably still ran along the edge of the gravel terrace and a gravel layer at the end of the droveway is likely to have represented a ford crossing onto the floodplain. Laid out alongside the droveway were a series of small fields or paddocks within which occurred pits, wells and two corn-driers. The site had clearly been inhabited but no structures that could definitely be identified as buildings were located. Waterlogged samples were obtained from pits, wells, the droveway ditch and peat over the possible ford.

The pollen analyses (by G. W. Dimbleby) show that, as in the Iron Age, the landscape was predominantly unwooded. The highest value for tree/shrub pollen in any sample was 12% of the total pollen but the average from the other samples was about 3%. The proportion of tree and wood-feeding Coleoptera (excluding Anobium punctatum and Lyctus linearis) out of the total number of terrestrial individuals, at 1.4%, is very low when compared with woodland sites but more than ten times greater than the comparable value for the Iron Age features at Farmoor.

Grazed pasture is again suggested by the presence of Eleocharis S. Palustres sp. but seeds of species of wet grassland were only present in features which were close to the edge of the gravel terrace. The large group of dung beetles from the super family Scarabaeoidea was almost as important as it was for the Iron Age at Farmoor, Geotrupes, Aphodius and Onthophagus spp. comprising 21% of the terrestrial fauna. Their occurrence in the various Roman samples was, however, by no means uniform. Perhaps this reflects different usage of the various small fields on the site, some having animals concentrated in them for particular purposes, others only occupied by the number of animals the grazing can permanently support or not being under grass at all.

Although the cut-off river channel between the gravel terrace and the floodplain had become filled in, there was still a marsh along the terrace edge. Seeds from the peat area at the end of the droveway included:

Alisma sp.	Mentha sp.
Apium nodiflorum	Menyanthes trifoliata
Carex spp.	Potamogeton sp.
Iris pseudacorus	Ranunculus S. Batrachium sp.

There was also a wide range of Coleoptera from marsh or bankside habitats in the Roman features close to the edge of the gravel terrace including:

Agonum viduum	Notaris acridulus
Apthona coerulea	Pterostichus vernalis
Blethisa multipunctata	Prasocuris phellandrii

The seeds from plants of disturbed ground have already been mentioned. Most of the annual weeds were probably not from plants which grew in arable fields but from weeds growing on disturbed ground in the immediate vicinity of the deposits. Some of the species identified, however, show a particular association with arable agriculture: Ranunculus arvensis, Agrostemma githago and perhaps Anthemis cotula. A. cotula, although being abundant as a cereal weed and not being found on waste ground, is sometimes common growing in farmyards and occasionally occurs at roadsides (Kay 1971: 625). Unlike the Iron Age samples from Farmoor, some of the Roman samples contained a significant quantity of cereal remains. In all of these samples the remains represented the debris from the threshing of spelt wheat. T. spelta does not have free-threshing grains and requires parching over a corn-drier or fire to render the glumes brittle, unless a sophisticated mill is to be used (Helbaek 1953: 233). It has been assumed that the carbonised remains from one of the corn-driers (examined by M. Jones), which contained five times as many glume bases as cereal grains, were from a crop which had been subjected to this process, threshed and then the winnowings perhaps thrown onto the corn-drier fire. A pit contained a substantial quantity of waterlogged spelt chaff and charred spelt chaff was well represented in one of the water holes along with a little Hordeum vulgare. A few seeds of another arable crop, Linum usitatissimum were identified from one of the wells.

Three other species which were present stand out as crops: Prunus domestica, Anethum graveolens and Coriandrum sativum. Two varieties of plum were found in the Roman samples, one with stones the size and shape of a modern variety of damson, the other from a larger fruited variety. Dill has been identified from several Roman sites in Britain (Willcox 1977: 272-8). as has coriander (Robinson in Young 1975: 167). Another species which might have been cultivated was a non-British umbellifer, seeds of which were present in Roman samples from both Barton Court and Farmoor. They somewhat resembled seeds of Chaerophyllum aromaticum.

It has already been said that there was a low frequency of tree pollen in the Roman samples from Farmoor. Macroscopic remains of woody species, however, were not uncommon, and combined with evidence from the arthropod remains suggest that hedges might have been present. All the Roman deposits contained thorny twigs of Prunus spinosa/Crataegus type, Prunus wood and most contained leaf abscission pads and bud scales. They were almost absent from the Iron Age samples while the proportion of tree and wood-feeding Coleoptera (excluding species which can invade structural timbers) were ten times more frequent than from Iron Age Farmoor. The most numerous species of ant was Stenamma westwoodi, which, with a total of 46 individuals occurred in all but one of the Roman samples. It nests under large stones or amongst the roots of trees in shady woodland or in hedgerows (Boulton and Collingwood 1975: 22). Several of the other ants which were represented by workers nest in old tree stumps, hedges etc. (large stones being notably absent from the site). Similarly, many of the mites (identified by S. Denford) were species associated with trees. (Both the worker ants and the mites would be expected to have much more local origin than the Coleoptera or pollen, probably more comparable with that of the macroscopic plant remains.)

Unlike Iron Age Farmoor, there seems to be abundant macroscopic evidence for trees or shrubs, especially thorny species from the family Rosaceae growing on the site in the Roman period in the vicinity of some of the waterlogged features. It has already been shown for Mingie's Ditch how insect pollinated shrubs and trees can be under-represented or absent from the pollen record. Had Farmoor become overgrown with scrub, much of it would probably have been P. spinosa and Crataegus monogyna initially but other species would probably have established themselves after a while such as Salix spp., Rhamnus catharticus and eventually Corylus avellana. If such a natural scrub had been present on the site, as at Hardwick, some of the species such as Salix spp. ought to have been detected by the pollen analyses and the others by a diversity of macroscopic remains. Neither is there any archaeological evidence which would suggest that the site had been abandoned to scrub, indeed its occupation seems to have been at its most intensive during the late Roman period. Therefore it is postulated that the way in which the results from the various biological groups can be explained is by the presenee of thorn hedges, especially P. spinosa.

Evidence for hedges of another sort came from the discovery of box leaves in two of the Roman features, in one instance along with Buxus sempervirens fruit fragments. Box is a plant of chalk and limestone scarps which is regarded as native to Britain (Pigot and Walters 1953). It seems a most unlikely plant to be growing wild at Farmoor. It is one of the few native evergreen hedging plants and Pliny (XVI: xxviii) describes its use for ornamental hedges.

In the Roman period at Farmoor there was a considerable increase in the number of beetles which live in relatively intimate association with man compared with any of the earlier sites investigated. The first group to be considered are those beetles which are capable of invading structural timbers (although they can also inhabit naturally occurring dead wood). Anobium punctatum, the woodworm beetle, made up 2.5% of the total terrestrial Coleoptera from the Roman samples at Farmoor compared with 0.5% for A. punctatum plus Lyctus linearis, a beetle of not too dissimilar habitats from the Farmoor Iron Age deposits and 0.9% from Mingie's Ditch.

The following species from Roman Farmoor can be regarded as synanthropic:

Mycetophagus quadriguttatus	Tenebrio molitor
Ptinus fur	Tipnus unicolor
Stegobium paniceum	Typhaea stercorea

All but S. paniceum certainly have habitats in Britain away from human influence but some of them are considerably more esoteric than the natural habitat of the woodworm beetle, birds' nests being a particular favourite of some of these species. All these species can occur in granary refuse. They are often associated with hay waste or straw and can perhaps live in thatch. These species made up 3.3% of the total terrestrial Coleoptera compared with a complete absence from Iron Age Farmoor and 0.1% (a single individual of P. fur) from Mingie's Ditch.

Another group of species likely to be related to human habitation, the Lathridiidae, were about five times as abundant at Roman Farmoor, (where they comprised 9.6% of the total terrestrial Coleoptera) compared with either of the Iron Age sites. They tend to occur in "sweet" compost such as haystack bottoms and thatch.

Appleford (SU523936) First Terrace
(Full report, Hinchliffe and Thomas 1980 forthcoming)

In many ways the Roman site at Appleford seems to have resembled a larger version of Farmoor. The Roman features excavated consisted of part of a ditched droveway with small fields laid out against it. Within the small fields were the usual pits and water holes but no definite evidence of buildings was discovered. The period of use of the site seems to have ranged from the late 1st century AD until the end of the 4th century (on the basis of pottery and coins). Cropmarks showed that the droveway converged with two others about 150 m away from the excavated area. The area of floodplain between the first terrace and the River Thames was considerably less than at Farmoor.

Examination of samples from the water holes showed that as with the other sites discussed so far, grassland was important. Interestingly the dung beetles from the super-family Scarabaeoidea comprised just under 10% of the terrestrial Coleoptera, a lower value than for any of the other Iron Age and Roman sites but indicators of meadowland were much better represented than from any of those sites. Appleford water hole 32 was especially rich in species which tend to be characteristic of flower-rich meadows such as Cantharis spp. and Olibrus sp. but they are by no means confined to it. Water hole 32 also had, for an archaeological sample, an extraordinary diversity of Coccinellidae (ladybirds) with five species present. Although a wide range of Coccinellidae can be collected in old haymeadows, again, they are not indicative of them. Perhaps more useful are weevils of the genus Apion (excluding the identified mallow and nettle feeding species) and the genus Sitona. This group made up about 11% of the total terrestrial Coleoptera in the Appleford samples. Their species are largely confined to feeding on members of the family Papilionaceae and from personal experience can be caught in very high numbers is meadows, certainly more than on pastureland. Some individuals, however, would be expected even if meadowland was completely absent.

A couple of seeds of *Agrostemma githago*, as well as a small quantity of both charred and waterlogged cereal remains, were found. A few flax seeds and plum stones were also present but not the culinary herbs which were identified from Farmoor.

Even though the landscape was predominantly open some scrub seems to have been present at Appleford. The evidence included wood of *Prunus* sp. and *Crataegus* type along with prickly stems of *Rosa* sp. and seeds of *Bryonia dioica*. It is uncertain whether or not the scrub was in the form of hedges.

Beetles from the synanthropic category made up only 0.7% of the total terrestrial Coleoptera but this was probably a result of the excavated area being some distance from any buildings at Appleford rather than that occupation was necessarily less intensive than at Farmoor or Barton Court.

Barton Court Farm, Abingdon and Radley (SU510978) Second Terrace
(Full report, Miles 1981, forthcoming)

This multiperiod site was on the terrace edge, overlooking the clay slope down to the first terrace. In the later 1st century AD a Roman rectangular timber farmhouse within a rectilinear enclosure replaced a Belgic farmstead on the site. There was then perhaps a break in occupation of the site until about AD 270 when a small stone corridor-type villa was built. The villa enclosure was set in a rectangular farmyard with several sub-divisions and a droveway leading into it. In the farmyard was also a second building, a corn drier and two stone-lined wells, both of which filled up during the 4th century (dated by pottery and metalwork). One of the wells was 7 m deep and the massive stones of its lining had been bedded into thick layers of moss, perhaps to filter the water. Waterlogged samples were examined from both wells but no waterlogged features were encountered from the pre-villa phase.

J. Greig's pollen analyses showed that the villa was surrounded by an open, relatively unwooded environment. This impression is confirmed by the Coleoptera, a few individuals of *Phymatodes testaceus* being the only representatives of the Cerambycidae, and Scolytidae were entirely absent. At 0.2%, the proportion of the total Coleoptera feeding on wood or trees was lower than either of the other two Roman sites investigated. Grassland and disturbed ground were the major ecotypes indicated by the waterlogged plant and insect remains. Dung beetles from the genera *Geotrupes*, *Aphodius* and *Onthophagus* comprised 16% of the total terrestrial Coleoptera. At 0.07% the proportion of chrysomelid beetles and weevils that feed on aquatic or marsh plants out of the total terrestrial Coleoptera was lower than for any of the other sites. Seeds of species from wet grassland were also relatively low. This result is what would be expected considering the situation of the various sites.

Remains of three species of arable crops which had been brought to the villa farmyard for processing were found waterlogged: spelt wheat, barley and flax. By far the most numerous of these was flax (*Linum usitatissimum*), represented by seeds and capsule fragments from all the waterlogged samples (except the moss from the well lining). It must be remembered that flax can be grown for linseed as well as to make linen. Flax retting requires water-filled ditches, pits or tanks, for which there was no evidence, and it is a disgustingly smelly process which would not be pleasant for the occupants of the site. It seems more likely that the capsules were harvested (rippled) and brought to the villa for pressing or consumption of the seeds while their stems were retted elsewhere, perhaps at the place of cultivation. A most interesting discovery was made along with the flax remains in one of the samples, silicula fragments of Camelina alyssum. This weed is particularly associated with flax cultivation and the presence in a flax field of its close relative, C. sativa, seriously affects the yield of the crop due to the secretion of growth inhibitors (Dimbleby 1978: 84). However, seeds of C. alyssum harvested with the linseed would themselves contribute to the oil production on pressing and it is even possible that C. alyssum itself was cultivated as an oil crop.

Spelt wheat and barley were identified from waterlogged chaff fragments. In addition, a couple of waterlogged cereal caryopses were found and 'cereal' pollen was present. However, the evidence from the carbonised seeds was much more extensive. Work by M. Jones on the corn drier and other samples showed spelt wheat followed by hulled six-row barley to be the most abundant charred cereals. This apparently contradicts the results from the waterlogged samples, in which flax was the most important. The processes which led to the incorporation of the remains and their preservation, however, would have been biased; the parching of cereals tends to result in accidental charring and cereal remains do not preserve well under waterlogged conditions.

The insects provide a little information which might suggest that there was a greater area of arable land in proximity to the villa than at either the Farmoor or Appleford Roman sites. Some species of the carabid genus Amara, in particular A. apricara, A. bifrons and A. tibialis live on open, rather sandy, bare or disturbed ground (Lindroth 1974: 91-3). They made up 1.3% of the terrestrial Coleoptera from the Barton Court Villa samples (34 individuals) but were absent from Farmoor and present at Appleford in reduced numbers. Their abundance would, however, be related to soil type and bare ground in the vicinity of the settlements as well as arable.

Excluding trees grown for their wood, there were seeds of three more species of cultivated plant: Anethum graveolens, Coriandrum sativum and Prunus domestica. Another species, the opium poppy (Papaver somniferum), represented by 71 seeds in one sample, could have grown as a weed, been cultivated in small garden plots or harvested as a field crop. Alternatively the seeds could have been imported. P. somniferum had been introduced into Britain by the Iron Age (Godwin 1975: 129) and seeds of it have been recorded from several Romano-British sites including Silchester. Reid (1905: 377) suggested that at Silchester the seeds were used sprinkled or bread. It was almost certainly cultivated in Britain during the Roman period and is useful as an oil, as well as a medicinal crop, the crushed stems being used as animal fodder (Crawford 1973: 230-1). Contrary to popular mythology, it does seem that opium poppy can be grown as a drug plant under English climatic conditions (Godwin 1978: 157-8). Crawford (1973: 232) cites several Roman (though Eastern) examples of medicinal usage.

There was no good evidence to suggest the presence of hedges on the site. The biological remains suggested that woodland was absent from the immediate vicinity of the site, although as is only to be expected, wood was used by the villa including substantial oak timbers and hazel rods. An unusual insight on a piece of woodland exploited by the villa, but not its location, was given by the moss which had been packed between the stone lining of one of the wells. The mosses (worked on by J. Dixon) were all woodland species. The most abundant, *Hylocomium brevirostre* occurs on calcareous soils in deep shade and is at present uncommon. Many other biological remains had been transported with the moss from the woodland floor. They included insects which overwinter in moss, for example the bug *Sehirus bicolor* (Southwood and Leston 1959: 28) and the woodland snail *Azeca goodalli*. 62% of the pollen from the moss was tree/shrub pollen, 50% being *Corylus*. The seeds confirm the evidence from the moss for the predominantly calcareous substrate of the wood for they included *Viburnum lantana* and *Fragaria vesca*. Oak galls of parasitic *Hymenoptera* from the sub-family Cynipinae were present, including leaf spangle galls of *Neuroterus* cf. *albipes* and bud galls of *Andricus* cf. *albopunctatus* (Robinson 1980). There was a low value for oak pollen, only 6% of the total from the moss. The 'old woodland' bryophyte flora might suggest that the wood had an ancient origin but the high value of hazel pollen and the low value for forest trees show that by the late 3rd century AD there had been human interference with the wood. Perhaps it was an oak/hazel coppice and the gall-infested oaks had not reached flowering size. The wood may have been situated on the coral limestone to the north of the site as is the present day wood of Bagley, with which its flora and fauna have some affinities. However, there are many other areas with calcareous substrates where it could have grown, in particular the gravel terraces which have now been so thoroughly cleared of trees that the type of forest it would have supported is uncertain.

Anobium punctatum made up 1.6% of the total terrestrial Coleoptera, not as much as for Roman Farmoor. The proportion of the synanthropic species, the Lathridiidae and species of foul, somewhat anaerobic, decaying organic remains including manure heaps, such as the smaller Sphaeridiinae and many of the Staphylinidae were no better represented than at Farmoor. Another species has been added to the synanthropic group for Barton Court, *Mycetaea hirta*. It is a fungal feeder which most often occurs in buildings and seems to be particularly partial to the dry rot fungus (Horion 1961: 116). Only four individuals of *Stegobium paniceum* from Barton Court can be regarded as true pests of stored farinaceous material. The paucity of grain beetles was perhaps only to be expected for Farmoor and Appleford but it is interesting that the major grain pests, which can infest granaries in their millions (*Cryptolestes*, *Oryzaephilus*, *Sitophilus* and *Palorus* spp.) should have been absent from a true villa which probably had a greater involvement in grain production. These species were all present in Britain by the date of the wells and serious infestations had already occurred in towns (Kenward and Williams 1979; Buckland 1978). Some of these species have been recorded from villas elsewhere, for example *Oryzaephilus surinamensis* at Barnsley Park, Gloucestershire (Coope and Osborne 1978) and *Sitophilus granarius* at Winterton, Lincolnshire (Robinson, unpublished).

THE SAXON PERIOD

The discovery of an early Saxon well at Barton Court and also a well on the first terrace Saxon site of Bishop's Court provided sufficient waterlogged material to extent the study beyond the end of the Roman period, although not on the scale of the investigation of the Iron Age and Roman environment.

Barton Court (see above)

At the end of the 4th century or beginning of the 5th century the villa went out of use, although a coin hoard from the other building suggests it may have stood longer. The site was occupied by the Saxons from at least the mid 5th century (on the basis of decorated pottery) and continued into the 6th century when two burials were inserted into the rubble of the villa and two more into the remains of the other building (dated by their accompanying metalwork). An irregular enclosure was added to the south of the villa farmyard and in the vicinity several sunken huts and post-built hall-houses were constructed. The Saxons also built a wattle-lined well cut from the bottom of a partly filled in Roman ditch, a very meagre structure compared with the Roman wells.

Superficially there seems to have been very little difference between the general environment around the site in Roman and Saxon times. As for the Roman period, the pollen (examined by J. Greig) suggested an open relatively unwooded environment. It is of particular interest to know whether there was any increase in scrub showing abandoned land after the end of the Roman period but the slight rise in tree and shrub pollen from 4.5% to 10% was not sufficient to indicate any change on the basis of only one Saxon deposit. Indeed the Coleoptera show the reverse trend with no wood or tree dependent species (even Anobium punctatum) being present.

The same type of grassland was probably present as in the Roman period although dung beetles of the genera Geotrupes, Colobopterus, Aphodius and Onthophagus had fallen to a level of 8% of the total terrestrial Coleoptera.

The evidence from the waterlogged cereal remains takes on a special importance because the presence of residual Roman pottery in the Saxon features would suggest that some of the carbonised seeds could be residual too. Roman material which had not been charred would have decayed in the soil. The ony waterlogged cereal was Hordeum vulgare, identified from its rachis nodes. It is disappointing that no waterlogged wheat was found because it would be most interesting to know whether spelt continued in use into the Saxon period here. However, bread wheat chaff is not as robust as spelt chaff, so it is possible that it is not as easily preserved in a recognisable form. The other arable crop indicated was flax, again in the company of Camelina alyssum silicula fragments. The range of exotic species of Umbelliferae used by the Romans was absent as were the fruit stones.

Coleoptera associated with accumulations of plant debris and manure in various stages of decay were present though in lower proportions than in the Roman period. The species particularly associated with buildings were at a much reduced level or absent. The only representative of the synanthropic group was a single specimen of Ptinus fur and Anobium punctatum was not found.

Bishop's Court, Dorchester (SU571939) First Terrace

Gravel extraction about 500 m to the west of the Roman town of Dorchester in an area relatively free of cropmarks exposed an early Saxon sunken hut and some Saxon wells. A sample was obtained from one of these wells, which was wattle lined and dated to the 5th-6th century AD by pottery.

Pollen analysis was not undertaken for this deposit, but the Coleoptera did not include any tree or wood feeding species. Dung beetles of the genera *Geotrupes*, *Colobopterus*, *Aphodius* and *Onthophagus* made up 10% of the total terrestrial Coleoptera, sufficiently high to indicate the presence of at least some pasture nearby. Remains of crops were sparse, but apart from a single charred grain of bread/club wheat, they were the same two species as for the Saxon well at Barton Court, flax and barley.

The representation of species of decaying plant remains and those associated with human habitation were also similar to Saxon Barton Court with *Anobium punctatum* absent, a low level of Lathridiidae and a single individual of *Typhaea stercorea* being the only member of the synanthropic grouping.

PATTERNS OF CHANGE AND CONTINUITY

There was no evidence for any climatic change during the period investigated or a difference in climatic conditions from those of the present. All the changes detected were probably the result of human activity, intentional or otherwise.

Soil change and Flooding

The undisturbed natural soils on the gravel terraces mostly seem to have been stone-free silt loams (Robinson in Palmer 1980, forthcoming). The lack of limestone gravel in these soils would have meant that decalcification by rainwater leaching and the formation of extreme *sols lessivés* was a possibility once forest cover had been removed. The modern soils on the terraces are likely to have resulted from ploughing deep enough to mix the natural soils with the underlying sandy limestone gravel. Any *sols lessivés* would have been destroyed on shallow soils and probably have been made more permeable as a result of earthworm activity over deeper hollows. When this occurred on an extensive scale is uncertain. Late Bronze Age/early Iron Age ploughing at Mount Farm, near Dorchester, Oxon. took place in a stony soil and plough marks cut into the gravel (G. Lambrick, pers. comm.). Soil depth would have been an important factor and the presumably early Saxon soil under Oxford is relatively stone-free.

The pre-Iron Age soil on the floodplain tends to be less gleyed and have a higher silt content than the present alluvial clays (Robinson in Palmer 1980, forthcoming). The decalcified upper part of their profile, lacking in aquatic molluscs suggests that they were not regularly flooded by the mollusc-rich calcareous waters of the Thames. Even before the founding of the Iron Age settlement at Mingie's Ditch (in circa 350-300 bc) the water table had risen somewhat on the floodplain although the site probably did not flood. At about the time of abandonment there was a further rise in the water table and possibly an onset of flooding, although there was no definite evidence for alluvial deposition on the site until much later. At Farmoor, flooding of the floodplain enclosures was taking place at a date later than the abandonment of the Mingie's Ditch site, perhaps around 200 bc. After the abandonment of the floodplain enclosures, extensive deposition of up to about 0.5 m of alluvium took place over parts of the floodplain. The water table rose to above what had been the Iron Age ground surface of one enclosure. The bulk of the alluvium had been deposited by the 4th century AD but seasonal flooding continued until the present day. There seems no reason why alluvium covering some Bronze Age sites on the floodplain (Robinson in Palmer 1980, fo thcoming) should not date to the main period of deposition at Farmoor, although the main aggradation at Mingie's Ditch took place no earlier than the early Saxon period.

It has been suggested that soil erosion following large-scale clearance, arable agriculture or the introduction of winter cereals have caused aggradation of alluvium and a rise in the water table in river valleys (Limbrey 1978: 23, 25; Shotton 1978). By the date of the Farmoor deposit the gravel terraces had been extensively farmed and species of cereal suitable for winter sowing had been cultivated for over 500 years. Also, the gravel terraces are almost flat and their drainage tends to be by water percolating through the soil and then flowing relatively freely through the gravel. They would not seem very susceptible to erosion. A possible, but by no means definite, explanation for the source of the deposits would be middle and late Iron Age agricultural expansion onto the limestone slopes of the Cotswolds.

Exploitation of the Man-made Environment

By the beginning of the Iron Age woodland was probably only a background feature to the agricultural landscape of the gravel terraces. Areas of scrub, however, probably existed on the floodplain, especially up the tributaries of the Thames. In the early Iron Age, Triticum spelta, which is suitable for winter sowing, was being cultivated at the second terrace site of Ashville and a normal range of domestic mammals was kept (Jones and Wilson in Parrington 1978).

The most complete range of evidence for land usage in the Thames Valley during the Iron Age comes from the middle Iron Age. Throughout this period the 'village' at Ashville was occupied. With its group of huts, small enclosures and numerous storage pits it probably represents well the type of agricultural settlement which exploited the terraces. Both the pastoral and arable aspects of its economy seems to have been important but there was no evidence as to how its agricultural landscape was divided into its various components. All that can be said is that the area under cereal cultivation had a variety of soil types including wet ground which would not be expected on the gravel terrace (on the basis of charred weed seeds). Not only was T. spelta present but also charred seeds of weed species which would suggest at least some autumn sowing (Jones in Parrington 1978, 105-6).

The Mingie's Ditch site on the floodplain was very different from Ashville and other gravel terrace sites. It can perhaps be regarded as a pioneer site bringing an area of scrubby pasture into more intensive use. The layout of the double ditched enclosure, with its 'antennae' ditches at the entrance and access to a watering place on the old stream help to confirm the evidence from the plant and invertebrate remains for its primarily pastoral function. Whether or not it was a subsidiary farm of one of the nearby agricultural settlements on the second gravel terrace, it would have been dependent upon a source of grain from elsewhere. The site can be seen as a hedged enclosure set in an expanse of grassland and marshy grazing, perhaps with some scrub still present and bounded only by the River Windrush and the old stream, though crossed by a mesh of small seasonal streams.

The Farmoor floodplain settlements, seasonal self-contained farmsteads set in a treeless expanse of unenclosed grassland, were perhaps only occupied for a few summers until flood damage was not worth repairing. The idea of seasonal occupation of the Thames floodplain to exploit rich alluvial grazing during the Iron Age was first expounded by Case (1963: 51). The minding of grazing herds was probably virtually the only concern of these farmsteads. From the animal bones, sheep appeared to have been relatively less important than cattle compared to Ashville on the second terrace (Wilson in Lambrick and Robinson 1979: 133). Horses were also kept. The Farmoor settlements were probably dependent upon a permanent community of the Ashville-type to which the transhumant herdsmen returned for winter grazing and where the arable plots were situated. It is possible that the Farmoor enclosures represented a response to the onset of inundations of the floodplain which perhaps caused the abandonment of the Mingie's Ditch site.

It seems likely that a major change in the way in which land was managed occurred in the late (Belgic) Iron Age. Occupation (though presumably not usage) of the floodplain ceased. Forerunners of the Roman type of droveway settlements appeared on several sites but waterlogged biological evidence is lacking from them at present.

In the Roman period, the droveway settlements were greatly expanded on the gravel terraces, a few of them developing into villas on the higher terraces. At Farmoor, once again the settlement was concerned with the grazing of cattle, but it was managed in a different way and, unlike the Iron Age, other agricultural activities were probably important as well. The droveway and enclosed fields, possibly with thorn hedges are clear indications of a more careful control of land usage. The discovery of a large scythe in the peat at the end of the droveway serves as a reminder that a crop of hay could have been taken off the floodplain before animals were allowed to graze on it. Various models can be postulated for the management regime of the grassland at Farmoor and it seems most likely that the clay hillsides and limestone hilltops should be taken into consideration as well as the gravel terrace and floodplain. Floodplain grassland would probably have been the most productive during the height of the summer and the least useful in winter. The gravel terrace would probably have been the most resilient to winter grazing and provide the earliest hay crop, but be the most susceptible to summer drought. The gravel and limestone soils would also have been the most suitable for arable which may have resulted in the overwintering of some animals on the less suitable clay slopes. There is considerably more evidence for the involvement of the settlement in arable agriculture than for the Iron Age but its location is problematic. Parching prior to the final threshing of grain may have been undertaken at the place of consumption and unknown economic factors would have determined whether it was better for the inhabitants of the site to use the gravel terrace for pastoral or arable farming. (The use of corn driers for parching grain would require them to be run at a higher temperature and take a much smaller load than if they were used to dry the entire crop at harvest time prior to storage). Perhaps cereals were only grown for use on the site.

The Roman site is perhaps best envisaged as a collection of thatched timber houses and barns set in small hedged fields along the droveway with farm yards, piles of domestic rubbish and compost in which filth beetles lived and neglected corners where plants such as Chelidonium majus and Sambucus nigra grew. Some of the houses may have had vegetable and herb gardens enclosed by box hedges. Other of the small fields were perhaps cultivated for minor crops (i.e. other than cereals) or orchards or were special pastures. The main agricultural land of the site, however, must surely have been beyond these small fields: lush unenclosed grassland on the floodplain; elsewhere on the terrace was perhaps winter pasture and arable, either in open or very large fields.

The Appleford site probably had a similar function to Farmoor although perhaps with less dependence on floodplain grazing and the settlement itself was larger. The Barton Court Farm villa had somewhat better evidence for the proximity of arable fields which is what might be expected with its situation on the second gravel terrace. The site was also involved in pastoral farming and there was a rather higher proportion of sheep bones than for Farmoor (Wilson in Miles 1981). The area exploited by the villa, however, probably extended down the clay slope, across the first terrace and onto the floodplain (Miles 1981), which would tend to minimise the bone evidence for any differences between the usage of the first and second terraces.

Cereals were not the only important crop; flax was especially well represented from Barton Court. Flax seeds were identified from all the Roman sites investigated but not found waterlogged from any of the earlier deposits. An increase in the cultivation of flax might well have resulted in the particular areas under cultivation being slightly different from those in the Iron Age to include more damp ground.

The entomological evidence shows an increase in species often associated with structural timbers and various synanthropic species from all the Roman sites. How much this difference was due to an increase in the number and size of buildings per capita (perhaps there was even indoor housing of some domestic animals) and how much it was due to an increase in population of the occupation sites is uncertain. It was probably related to some sort of intensification of agricultural exploitation of the area. Grain storage was also changed in this period from below ground to above ground. Although grain beetles were only present in Roman samples, they do not seem to have reached serious pest proportions on the sites investigated. Perhaps grain surplus to local needs was stored in the towns.

It might have been expected that the biological remains identified from Barton Court would reflect the higher economic status of the villa in relation to the other two Roman sites. They did not, and it was Farmoor which had evidence for ornamental plants with Buxus sempervirens present. Apart from Papaver somniferum, all the culinary herbs or fruits identified from Barton Court also occurred at Farmoor. Exotic species which have been identified from some Roman towns and could not be grown in Britain (Wilcox 1977) were absent.

It seems curious that relatively few villas are known from the gravels in Oxfordshire despite the soils of the gravels being agriculturally useful and resilient to damage. In contrast, many large Roman villas occur on the limestone of the Cotswolds. It has been argued that the system of land inheritance may have kept Romano-British holdings small in the Thames Valley (Stevens

1966: 125). If the Roman conquest had not resulted in dispossession of the land held by the existing populace, the technological innovations of the Roman period alone may not have been enough to enable many holdings to become sufficiently profitable to support villas. Agriculture would already have been intensive on the gravel in the Iron Age and population levels relatively high. In areas where populations were smaller, individual holdings perhaps less efficient and larger, or clearance incomplete, there would have been a much greater potential for consolidation or improvement and increased profits on a single holding to support a villa. Interestingly, after the breakdown of Roman administration, some of the Cotswold villas were abandoned to woodland (p. 256) but the 'great estates' of post-medieval Oxfordshire tended to be in this area (Hoskins 1955: fig. 13).

It is not known how much of the Roman system of agricultural organisation continued into the Saxon period on the gravels. At Mingie's Ditch the Roman drainage system on the floodplain fell into disuse and even at Barton Court, the 5th century addition to the villa enclosure did not last long into the Saxon occupation of the site. The intensiveness of human occupation as reflected by synanthropic beetles fell back to the level of the Iron Age. Another Iron Age practice may have been re-introduced. There is documentary evidence that can be interpreted as showing that transhumance was undertaken to exploit floodplain grazing (Sturdey 1963). There was no evidence for an agricultural decline but there may have been a general decrease in efficiency. There were no remains of the orchard or garden crops but flax cultivation continued.

Perhaps the most important factor that all the sites from the Iron Age to the Saxon period have in common (other than the agricultural exploitation of the Thames Valley) is a lack of evidence for any main field boundaries. Extensive areas of arable 'celtic fields' and 'ranch boundaries' dividing up the pasture are known to have existed on the chalk of the Berkshire/Oxfordshire Downs from the middle Bronze Age onwards (Bradley and Richards 1978). The gravel terraces are flat enough not to need ploughing on the celtic field pattern to prevent erosion and are mostly sufficiently well drained for drainage ditches not to be necessary. It might be argued that any boundaries were in the form of hedges alone but it would then seem curious that the small fields which surrounded the Roman droveway settlements were ditched (i.e. most hedged boundaries would have been ditched and therefore identifiable). The individual areas of presumed pasture enclosed by ditches on the Berkshire Downs are to be measured in terms of many square kilometres. Perhaps both the areas of arable and pasture on the gravels were of a similar size to those on the Downs and their boundaries were the few streams which cross the terraces. Marker posts, trees or stones may have been all that served to indicate corners of different plots within the arable fields. Some of the prehistoric earthworks on the Downs were used by the Saxons as estate boundaries (Gelling 1976: 629).

Some Aspects of Species Differences between Sites

The ecological reasons for many of the differences have already been covered in the discussion of the environments of the various sites but some interesting points remain. The Iron Age weed seeds illustrate several. Firstly, some species which have no pre-Roman records given by Godwin

(1975) and might perhaps have been thought of as Roman (or for some, later) introductions such as *Valerianella rimosa* were present in Britain at least by the Iron Age. Secondly, *Hyoscyamus niger*, a denizen of farmyards etc. which is now uncommon (Bowen 1968: 229; Salisbury 1964: 48) was present on all the Iron Age, Roman and early Saxon sites. Finally, a couple of species which are not particularly associated with human settlement and are not common in the Thames Valley persistently occurred in the samples: *Anthriscus caucalis* and *Onopordum acanthium*. The reasons for their past abundance remain obscure, *O. acanthium* might even have been regarded as a useful plant. Several other weeds from the Iron Age and Roman sites, in particular *Papaver argemone*, *Valerianella dentata* and *V. rimosa* also seem to have been rather more abundant then than they are in the area at present, probably as a result of changing agricultural practice. The recent decline of *Agrostemma githago* (which was present in the Roman samples) due to improved seed cleaning is well known (Salisbury 1964: 36).

Anethum graveolens, *Coriandrum sativum*, *Papaver somniferum*, *Linum usitatissimum* and at least two varieties of *Prunus domestica* were identified from the Roman sites but were absent from the prehistoric samples. All of these species seem to be relatively frequent finds from Roman waterlogged deposits elsewhere in Britain (e.g. Willcox 1977). The Roman liking of Umbelliferae seeds for flavouring has been commented upon by Godwin (1975: 222) and they were an important component of Roman dishes in Italy itself (Apicius). Dill and coriander were possibly Roman introductions. The other aspect of Roman cookery which would previously have been alien to the British Isles was the use of olive oil. It may have been imported although as yet no evidence for its use in Britain has been discovered. It would not, however, have become commonly used throughout the Romano-British populace unlike, perhaps coriander and dill. Were opium poppy and linseed grown as oil-seed crops and did the Roman taste for oily as well as spicy foods become widespread in Britain? Both of these species, however, have been recorded from pre-Roman deposits in Britain (Godwin 1975: 129, 167). Also did the Romans introduce the first varieties of 'dessert plum' into Britain? This fruit had apparently been present in the country since neolithic times (Godwin 1975: 197) but plum stones are rarely encountered in pre-Roman deposits. One piece of evidence which suggests that the Romans introduced culinary change to Britain was the extensive use of the *mortarium* (mortar), a type of pottery vessel unknown from the Iron Age.

Agrostemma githago was not identified from any of the pre-Roman waterlogged samples. Until recently it was regarded as a probably Roman introduction but the discovery of 248 carbonised seeds of *A. githago* at the bottom of a late Iron Age pit at Barton Court Farm (M. Jones, pers. comm.) serves as a warning that many presumed Roman introductions could in fact have occurred in the Belgic period just before the Roman Conquest. *Anthemis cotula* is another weed common in the Roman samples which might well have been a Belgic introduction but at present no pre-Roman records of it in reliable quantities are known (see Jones, this volume, p. 111).

Grain beetles were not present in any of the pre-Roman samples, which is in agreement with evidence from elsewhere. Only two species, *Stegobium paniceum* and *Tenebrio molitor* were present in the Roman samples and there

TABLE I: SPECIES PRESENCE

Species	Iron Age			Roman			Early Saxon	
	Mingie's Ditch	Farmoor	Claydon Pike	Farmoor	Appleford	Barton Court	Barton Court	Bishop's Court
Valerianella rimosa Bast.	+	+	-	-	-	+	-	-
Hyoscyamus niger L.	+	+	+	+	+	++	+	+
Anthriscus caucalis Bieb.	+	+	+	+	+	+	+	+
Onopordum acanthium L.	-	+	+	-	-	+	-	+
Anethum graveolens L.	-	-	-	+	-	+	-	-
Coriandrum sativum L.	-	-	-	+	-	+	-	-
Linum usitatissimum L.	-	-	-	+	+	+	+	+
Prunus domestica L.	-	-	-	+	+	+	-	-
Agrostemma githago L.	-	-	-	+	+	+	-	-
Anthemis cotula L.	-	-	-	+	+	+	+	+
Stegobium paniceum (L.)	-	-	-	+	+	+	-	-
Tenebrio molitor L.	-	-		+	-	-	-	-
Apion urticarium (Hbst.)	+	-		+	+	+	+	+

were none from the Saxon samples. It would be interesting to establish which species of grain beetles survived the rigours of the early Saxon period. T. molitor, with its alternative habitat (p.265) would probably have been one of them.

Finally, the weevil Apion urticarium no longer occurs in Oxfordshire but was present on all the Roman and Saxon sites we well as in an Iron Age sample from Mingie's Ditch. Its host plants, Urtica dioica and U. urens are still common. There is no obvious reason for its present absence. It has also been identified from 12th century deposits in Oxford (Robinson in almer 1980, forthcoming) and other Roman records are given by Robinson (in Young 1975: 167).

Acknowledgements

I would like to thank the directors of the various O.A.U. excavations for co-operation when sampling their sites. The samples were investigated in the Hope Entomological Collections University Museum, Oxford and at the Institute of Archaeology, London. I am grateful to Dr. M. W. R. deV. Graham and Professor G. W. Dimbleby for the provision of working facilities and access to the collections under their care. I am indebted to Dr. J. H. Dickson for provision of a report on the moss from Barton Court, Professor G. W. Dimbleby and Mr. J. R. A. Greig for pollen analyses, Mr. M. Jones for use of his carbonised seed results and Mrs. S. Denford for working on the Farmoor mites. Mr. C. S. Elton has provided me with access to unpublished information in the Wytham Ecological Survey and I have had many useful discussions on interpretation with Mr. J. R. A. Greig, Dr. A. R. Hall and Mr. H. K. Kenward. Dr. K. D. Thomas made many helpful comments on a preliminary version of the text while Mr. T. G. Hassall, Mr. G. H. Lambrick and Mr. R. Wilson read through the final draft. Almost all their suggestions have been incorporated. Finally I would like to thank Mrs. Wendy Page for drawing Fig. 12:1 and Mrs. Lindsay Donaghy for efficient typing.

APPENDIX 1

THE USE OF ECOLOGICAL GROUPINGS OF COLEOPTERA FOR COMPARING SITES

The 'New Approach' of Kenward (1978) has proved useful in the interpretation of urban insect death-assemblages but does not exploit the full potential of rural deposits. The aim with urban samples tends to be attempting to establish precisely very local conditions, whereas with rural deposits the aim is usually to elucidate general information about the surrounding landscape. Attempting to define useful ecological groupings was suggested by Kenward (1978: 40-43) as another method likely to be of use in the interpretation of archaeological insect faunas and it has been found so. The particular groupings used must be regarded as preliminary but when expressed as a percentage of the total number of terrestrial individuals from the samples the results seem to be corroborated by other lines of evidence. The reason that aquatic species have been eliminated is because all the deposits accumulated under water and any differences due to the environment of the deposit itself can thus be eliminated. Water beetles (defined as any species which can spend part of its adult life under water) ranged from 1.1% of the total terrestrial Coleoptera in the Roman wells at Barton Court Farm, which were probably unsuitable habitats, to 54% from the ditches of Mingie's Ditch, in which they apparently flourished. Several samples were grouped together for the calculations partly to eliminate differences due to very local conditions. Some of the results have been encouragingly consistent between sites, for example Lathridiidae ranged between 8.8% and 9.6% on the three Roman sites while on the two Iron Age sites they were 1.3% and 2%.

When more sites have been examined and a larger corpus of results becomes available for comparative purposes, it ought to become possible to 'calibrate' the percentage values for the various species groups so that the importance of a particular habitat can be assessed. For instance the results from Buscot Lock (Robinson, unpublished) and Shustoke (Kelly and Osborne 1964) suggest that wood and tree-dependent species (not including *Anobium punctatum* and *Lyctus* sp.) perhaps make 20% of the total terrestrial Coleoptera in a deposit which formed surrounded by closed woodland (possibly even higher if the woodland was dying) but perhaps fall to only 1% from a cleared landscape with shrubs and trees only present in hedges. More work could also be done on improving the species composition of the groupings and establishing more rigorously the ecological conditions they indicate. However, it must be expected that many of the groups can only be used to show trends rather than absolute importance or absence. For example some individuals of *Apion* spp. (excluding mallow and nettle feeding species) plus *Sitona* spp. would be expected even if no meadowland was present while a high proportion of these species can only hint that meadowland was important.

TABLE II: PERCENTAGES OF TERRESTRIAL COLEOPTERA

Species Grouping	Iron Age		Roman			Early Saxon	
	Mingie's Ditch	Farmoor	Farmoor	Appleford	Barton Court	Barton Court	Bishop's Court
1. Aquatic	53.8	18.3	7.6	5.9	1.1	30	3
2. Pasture/Dung	16.5	27.9	21.0	9.8	16.1	8	10
3. ?Meadowland	4.1	2.5	5.2	10.8	3.0	3	4
4. Wood and Trees	1.0	0.09	1.4	1.2	0.2	0	0
5. Marsh/Aquatic Plants	0.5	1.6	1.3	0.1	0.07	2	2
6. Bare Ground/Arable	0.0	0.0	0.0	0.8	1.3	0	1
7. Dung/Foul Organic Material	12.6	11.7	15.4	9.1	9.0	10	7
8. Lathridiidae	2.0	1.3	9.6	9.4	8.8	4	1
9. Synanthropic	0.1	0.0	3.3	0.7	2.1	0.4	0.7
10. Esp. Structural Timbers	0.9	0.5	2.4	1.3	1.6	0	0
Total Number of Terrestrial Individuals	1330	1105	2217	952	2715	257	147

Samples used: Mingie's Ditch, 10/9, 212/7; Farmoor Iron Age with further identifications, all except 1172; Farmoor Roman with further identifications, all except 1072 and 1169; Appleford, all; Barton Court Roman, all except 832 spit 4 moss; Barton Court Saxon, all; Bishop's Court, all.

Species in groupings: 1. see p. 279; 2. see p. 258; 3. see p. 265; 4. wood and tree feeding Coleoptera (plus fungus feeders and predators which are strictly associated with wood e.g. *Cis* sp. and *Pediaecus* sp. but not e.g. *Pterostichus oblongopunctatus*) *Anobium punctatum* and *Lyctus* sp. excluded; 5. Chrysomelidae and Curculionoidea which feed on marsh or aquatic plants; 6. see p. 267; 7. *Cercyon* spp., *Megasternum* sp. *Cryptopleurum* sp., *Anotylus rugosus*, *A. sculpturatus* and *Platystethus arenarius*; 8. Lathridiidae; 9. see p. 265, 268; and 10. *Anobium punctatum* and *Lyctus linearis*.

This approach could be extended to other types of deposit if autochthonous Coleoptera were excluded but they might prove rather harder to define. The method will have its problems, for instance if insect-rich compost had been dumped in a well or a long-distance component became concentrated in a deposit. Some samples will have to be eliminated from the consideration, as was done with the moss from the deep Barton Court Well. Insects living circumjacent to the deposit can also create problems. For instance marshy ground around the top of the Barton Court Saxon well (because it had been cut in an old Roman ditch) probably resulted in beetles that feed on marsh and aquatic plants being over-represented. Despite the various problems this method does seem to provide a useful way to interpret archaeological insects quantitatively.

Bibliography

Atkinson, R. J. C., Piggott, C. M. and Sandars, N. K. (1951). Excavations at Dorchester, Oxon. Oxford: Ashmolean Museum.

Allen, T. G. and Robinson, M. A. (1979). 'Hardwick with Yelford, Mingie's Ditch' Council for British Archaeology Group 9 Newsletter 9 115-17.

Apicius, The art of cooking (Translated by Flower, B. and Rosenbaum, E. 1958, London: Harrap).

Benson, D. and Miles, D. (1974). The Upper Thames Valley: an archaeological survey of the river gravels Oxfordshire Archaeological Unit Survey 2.

Boulton, B. and Collingwood, C. A. (1975). Hymenoptera: Formicidae Royal Entomological Society Handbook 6, pt 3(c).

Bowen, H. J. M. (1968). Flora of Berkshire. Reading: Privately Published.

Boycott, A. E. (1934). 'The habitats of land Mollusca in Britain' Journal of Ecology 22 1-38.

Bradley, R. and Richards, J. (1978). 'Prehistoric fields and boundaries on the Berkshire downs' In: Early land allotment in the British Isles, (eds Bowen, H. C. and Fowler, P. J.) B.A.R. 48 pp. 53-60.

Brown, P. D. C. (1974). 'Problems of continuity' In: Anglo-Saxon settlement and landscape, (ed Rowley, R. T.) B.A.R. 6 pp. 16-19.

Buckland, P. C. (1978). 'Cereal production, storage and population: a caveat' In: The effect of man on the landscape: the Lowland Zone, (eds Limbrey, S. and Evans, J. G.) C.B.A. Research Report 21. pp. 43-45.

Case, H. (1963). 'Notes of finds and ring-ditches in the Oxford Region' Oxoniensia 28 19-52.

Conway, E. and Stephens, R. (1957). 'Sporeling establishment in Pteridium aquilinum: effects of mineral nutrients' Journal of Ecology 45 389-99.

Coope, G. R. and Osborne, P. J. (1967). 'Report on the coleopterous fauna of the Roman well at Barnsley Park, Gloucestershire' Transactions of the Bristol and Gloucester Archaeological Society 86 84-87.

Cunliffe, B. W. (1974). Iron Age communities in Britain. London: Routledge and Kegan Paul.

Dimbleby, G. W. (1978). Plants and archaeology. 2nd Ed. London: Baker.

Dimbleby, G. W. and Evans, J. G. (1974). 'Pollen and land-snail analysis of calcareous soils' Journal of Archaeological Science 1 117-133.

Emery, F. (1974). The Oxfordshire landscape. London: Hodder and Stoughton.

Frere, S. S. (1967). Britannia. London: Routledge and Kegan Paul.

Gelling, M. (1976). The place-names of Berkshire. Eng. Place Name Soc. 51.

Godwin, H. (1975). The History of the British flora. 2nd Edn. Cambridge: University Press.

Godwin, H. (1978). Fenland: its ancient past and uncertain future. Cambridge: University Press.

Greig-Smith, P. (1948). 'Urtica L.' Journal of Ecology 36 339-355.

Groenman-van Waateringe, W. (1978). 'The impact of Neolithic man on the landscape in the Netherlands' In: The effect of man on the landscape: the Lowland Zone, (eds Limbrey, S. and Evans, J. G.) C.B.A. Research Report 21. pp. 135-146.

Harding, D. W. (1972). The Iron Age in the Upper Thames Basin. Oxford: University Press.

Hawkes, S. C. and Gray, M. (1969). 'Preliminary note on the early Anglo-Saxon settlement at New Wintles Farm, Eynsham' Oxoniensia 34 1-4.

Hewett, D. G. (1964). 'Menyanthes trifoliata L.' Journal of Ecology 52 723-35.

Hinchliffe, J. and Thomas, R. (1980). 'Excavations at Appleford' Oxoniensia 45, forthcoming.

Horion, A. D. (1961). Faunistik der Mitteleuropaischen Käfer 8. Uberlingen-Bodensee: Kommissionverlag Buchdruckerei.

Hoskins, W. G. (1955). The making of the England landscape. London: Hodder and Stoughton.

Jarvis, M. G. (1973). Soils of the Wantage and Abingdon District Memoirs of the Soil Survey of Great Britain: England and Wales.

Jones, E. W. (1944-45). 'Acer campestre L.' Journal of Ecology 32 239-52.

Kay, Q. O. N. (1971). 'Anthemis cotula L.' Journal of Ecology 59 623-36.

Kelly, M. and Osborne, P. J. (1963-64). 'Two faunas and floras from the alluvium at Shustoke, Warwickshire Proceedings of the Linnean Society of London 176 37-65.

Kenward, H. K. (1978). The analysis of archaeological insect assemblages: a new approach The Archaeology of York 19/1.

Kenward, H. K. and Williams, D. (1979). Biological evidence from the Roman warehouses in Coney Street The Archaeology of York 14/2.

Lambrick, G. H. (1978). 'Iron Age settlements in the Upper Thames Valley' In: Lowland Iron Age communities in Europe, (eds Cunliffe, B. W. and Rowley, R. T.) B.A.R. S48.

Lambrick, G. H. (1979). 'Berinsfield, Mount Farm' Council for British Archaeology Group 9 Newsletter 9 113-15.

Lambrick, G. H. and Robinson, M. A. (1979). Iron Age and Roman settlements at Farmoor, Oxfordshire C.B.A. Research Report 32.

Limbrey, S. (1978). 'Changes in quality and distribution of the soils of Lowland Britain' In: The effect of man on the landscape: the Lowland Zone, (eds. Limbrey, S. and Evans, J. G.) C.B.A. Research Report 21.

Lindroth, C. H. (1974). Coleoptera: Carabidae Royal Entomological Society Handbook 4, pt. 2.

Miles, D. (1975). 'Berinsfield, Wally Corner' Council for British Archaeology Group 9 Newsletter 5 40-41.

Miles, D. (1978). 'The Upper Thames Valley' In: Early land allotment in the British Isles, (eds Bowen, H. C. and Fowler, P. J.) B.A.R. 48 pp. 81-88.

Miles, D. (1980). 'Lechlade/Fairford, Claydon Pike' Council for British Archaeology Group 9 Newsletter 10 160-64.

Miles, D. (1981). Archaeology at Barton Court Farm, Oxon. Forthcoming.

Palmer, N. J. (1980). 'Excavations at the Hamel, Oxford' Oxoniensia 45, Forthcoming.

Parrington, M. (1978). The excavation of an Iron Age settlement, Bronze Age ring-ditches and Roman features at Ashville Trading Estate, Abingdon (Oxfordshire) 1974-76. C.B.A. Research Report 28.

Pigott, C. D. and Walters, S. M. (1953). 'Is the box tree a native of England?' In: The changing flora of Britain, (ed Lousley, J. E.) Oxford: Botanical Society of the British Isles.

Pliny, Natural History. XII-XVI (translated by Rackham, H. 1968, London: Heinemann).

Pollard, E. (1973). 'Hedges VII: woodland relic hedges in Huntingdonshire' Journal of Ecology 61 343-52.

Raw, F. (1951-52). 'The ecology of the garden chafer, Phyllopertha horticola (L.) with preliminary observations on control measures' Bulletin of Entomological Research 42 605-46.

Reid, G. (1905). in Hope, W. H. St. J. and Fox, G. E. 'Excavations on the site of the Roman City at Silchester, Hants. in 1903 and 1904' Archaeologia 59 pt 2 367-8.

Robinson, M. A. (1978). 'A comparison between the effects of man on the environment of the first gravel terrace and floodplain of the Thames Valley during the Iron Age and Roman period' In: The effect of Man on the landscape: the Lowland Zone, (eds Limbrey, S. and Evans, J. G.) C.B.A. Research Report 21. pp. 35-43.

Robinson, M. A. (1980). 'Archaeological finds of wasp galls' Journal of Archaeological Science 7 93-5.

Robinson, M. A. and Allen, T. G. (1978). 'Hardwick with Yelford, Mingie's Ditch' Council for British Archaeology Group 9 Newsletter 8 17-19.

Rodwell, K. A. (1978). Historic Towns in Oxfordshire: a survey of the new county Oxfordshire Archaeological Unit Survey 3.

Rowley, R. T. (1974). 'Early Saxon settlements in Dorchester on Thames' In: Rowley, R. T. ed. Anglo-Saxon settlement and landscape 42-50, B.A.R. 6.

Salisbury, E. (1964). Weeds and aliens. 2nd Edn. London: Collins.

Sandford, K. S. (1924). 'The river gravels of the Oxford district' Quarterly Journal of the Geological Society of London 80 113-179.

Shotton, E. W. (1978). 'Archaeological inferences from the study of alluvium in the lower Severn-Avon Valleys' In: The effect of man on the landscape: the Lowland Zone, (eds Limbrey, S. and Evans, J. G.) C.B.A. Research Report 21 pp. 27-32.

Southwood, T. R. E. and Leston, D. (1959). Land and water bugs of the British Isles. London: Warne.

Sparks, B. W. (1959-60). 'The ecological interpretation of non-marine Mollusca' Proceedings of the Linnean Society of London 172 71-80.

Stevens, C. E. (1966). 'The social and economic aspects of rural settlement' In: Rural settlement in Roman Britain, (ed Thomas, C.) C.B.A. Research Report 7 pp. 108-128.

Sturdy, D. (1963). 'Traces of Saxon nomadic life near Oxford' Oxoniensia 28 95-8.

Walters, S. M. (1949). 'Eleocharis L.' Journal of Ecology 37 192-206.

Willcox, G. H. (1977). 'Exotic plants from Roman waterlogged site in London' Journal of Archaeological Science 4 269-82.

Wood, P. D. (1968). The Oxford and Newbury area British Landscapes Through Maps 11. Sheffield: The Geographical Association.

Young, C. J. (1975). 'The defences of Roman Alchester' Oxoniensia 40 136-70.

Young, C. J. (1977). The Roman pottery industry of the Oxford Region B.A.R. 43.

13. CROP HUSBANDRY AND ENVIRONMENTAL CHANGE IN THE COASTAL AREA OF THE FEDDERSEN WIERDE, NEAR BREMERHAVEN, NORTHWEST GERMANY

Udelgard Körber-Grohne

Preface

A wierde or wurt is a mound built up by the hands of the humans who wanted to live upon it, safe from the storm floods, before the marshes were protected by dikes. (The plural of wurt is wurten). Such dwelling mounds are found everywhere in the marshes of the southern North Sea and along the lower courses of the rivers that flow through the marshes. The wurten were built over a time span of approximately 1500 years from the 1st century AD to the late Middle Ages. They did not go out of use immediately the construction of dikes began, as the early dikes were too weak to ensure permanent protection. The wurten were not in general erected in single operation. They rather began as small low mounds that were systematically broadened and heightened in the course of several hundred years. The applied construction layers may be dated archaeologically on the basis of structural features and remains of prehistoric man within them. In the Feddersen Wierde, seven consecutive settlement layers were detected during excavation. These were deposited over about 500 years.

The archaeological investigation of wurten allows the way of life of their occupants to be determined. The concurrent geobotanical treatment of plant remains in the various wurt layers and in the original soil surface beneath the wurten serves several functions. It acquaints one with the environment of the people of those times and provides information about the details of their agriculture, but also helps accumulate important information about the evolution of the North Sea coastal region itself.

The excavation of the Feddersen Wierde is the largest wurt excavation in the German coastal area to date. The following is an account of the geobotanical investigations carried out in connection with the wurt, which have been published in full by Körber-Grohne (1967).

The setting of the Feddersen Wierde

The Feddersen Wierde lies in the coastal region of the southeast part of the North Sea (Fig. 13:1). The wierde has a diameter of 200 m and is 4 m high. It was, like several others in the area, settled from the end of the 1st century AD to the 5th century AD. If one stands today upon the Feddersen Wierde, a low mound amidst pasture, one sees only marshes in the vicinity. The present coastline runs at about 4 km to the west of the site. To the east the flat ridge of low hills may be seen, about 4 km away. They are pleistocene sands of the last glaciation. If the coastline was not protected today by a sea dike against tides and storm floods, these would transform the approximately

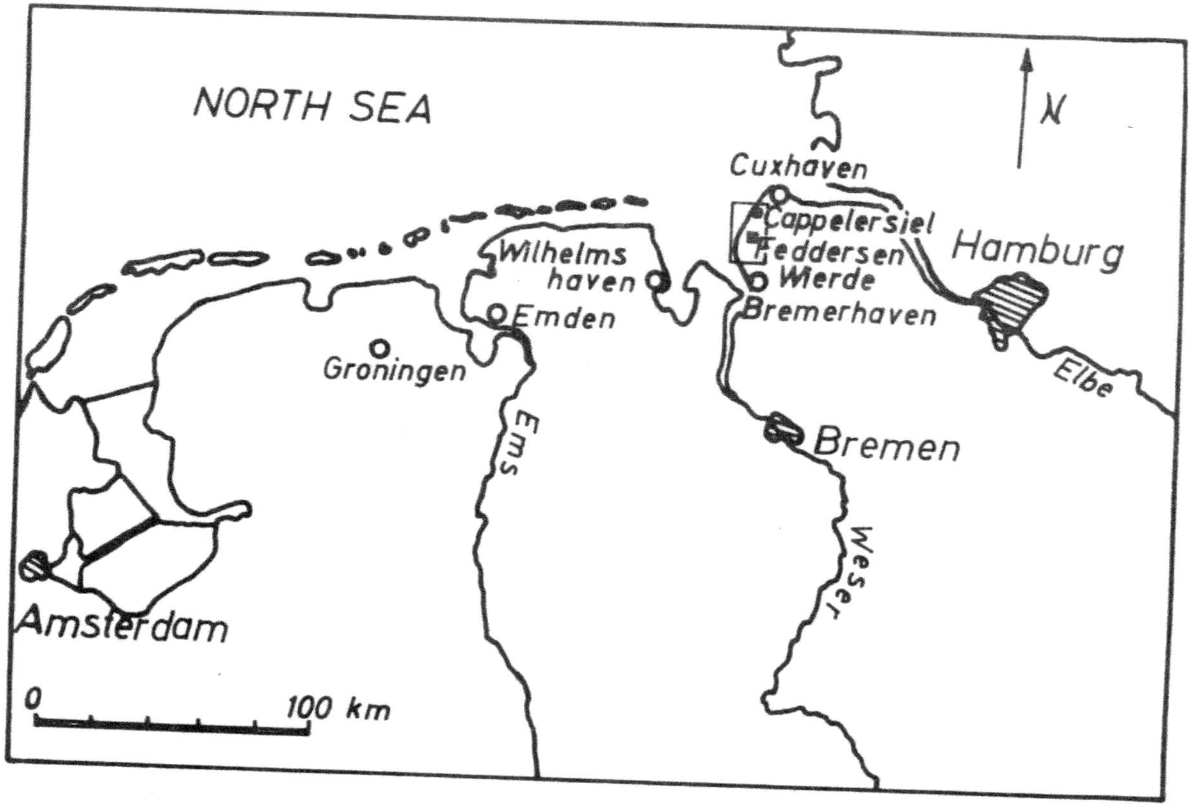

Fig. 13.1 Map of the North Sea showing the positions of the Feddersen Wierde, and of Cappelersiel (site of the agricultural trial discussed on p. 306).

8 km wide strip of marshland in front of these low hills known as "the geest" into a halophytic zone of lagoons.

What provisions were necessary in prehistory in order that people could live on the site for five centuries?, and what was the appearance of the landscape at that time? The archaeological, geobotanical, and geological investigations of the wurt and its surroundings go some way to answering these questions.

The selection of the Feddersen Wierde as an object of research

The excavation began in 1955 after a 10 year programme of archaeological investigation of other wurten and geological investigation of the marshes along the section of coast between the Ems and Elbe rivers, carried out by the Institute for Marsh and Wurt Research in Wilhelmshaven under the direction of Prof. Dr. W. Haarnagel. The close co-ordination of archaeological, and geological and botanical investigations is vital in this sort of exercise. Seven years before the beginning of the excavation of Feddersen Wierde, Haarnagel set up a geobotanical division within his institute, following the example of the Biological-Archaeological Institute at Groningen and its founder, Prof. Dr. G. E. G. van Giffen.

Numerous excavations of wurten in the Netherlands and in the north-western German coastal area had already shown that there are both prehistoric and medieval wurten. Some were built up with clay alone, others with alternating layers of clay and refuse from animal stables. Some are agricultural settlements and others, such as one at Emden, were established as market settlements. Some of the wurten are today no longer inhabited, while others lie beneath modern cities and villages. Nevertheless, relatively little was known about the foundings of wurten, and about the lives of their occupants in the midst of a landscape that was continually modified through the centuries by the tides and floods that originally created it.

Various criteria operate in the selection of a wurt for investigation. It should be of prehistoric age. It should contain in addition to clay, ample stable refuse so that the plant remains may be recovered to provide ecological and economic information. The wurt should be uninhabited today, so that area excavation is possible. After many sample borings in the coastal area between the Ems and the Elbe, Haarnagel singled out the Feddersen Wierde, which is at a distance of about 100 km from the Institute in Wilhelmshaven.

Field Methods

As the excavation began, the geobotanical division organised a simple laboratory in a nearby farmhouse, consisting of two tables, a shelf for the samples, and running water for the water-sieving of soil samples. The working day for myself and my technician ran such that we spent several hours of each day at the excavation site in order to observe the exposed layers, to discuss all resulting questions with our archaeological colleague Dr. Peter Schmid, and to take soil samples at appropriate points. These samples were then examined in the field laboratory, and sub-samples were exhaustively analysed for macrofossils and diatoms. Other sub-samples taken primarily for pollen analysis, and all remains of wood were given to Wilhelmshaven for treatment during the winter months. The aim in the first year of excavation was essentially to become acquainted with the fossil plant remains.

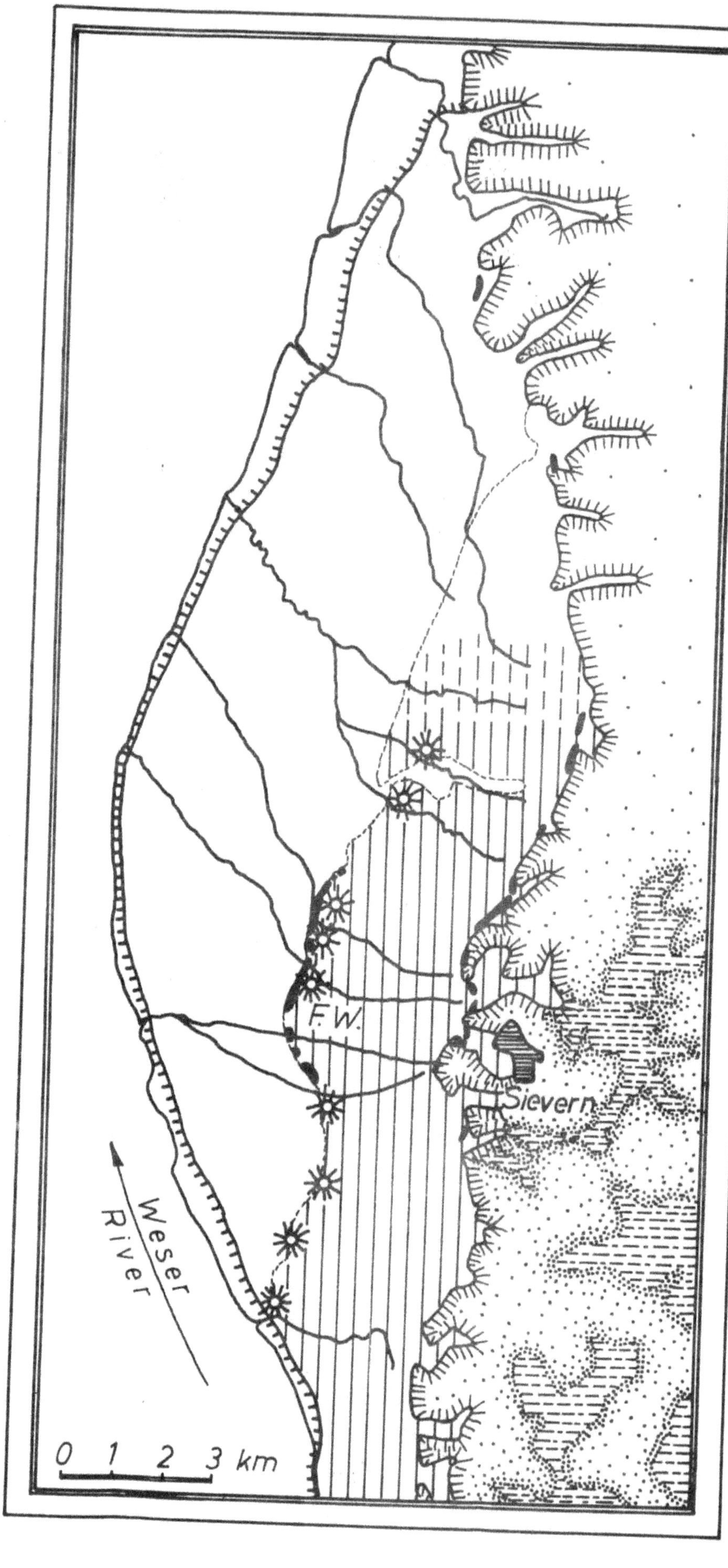
F.W.
Sievern
Weser
River
0 1 2 3 km

Legends

The prehistoric marshes at the time of about Birth of Christ.

Dwelling mounds with the Feddersen Wierde (F.W.)

Remains of two old beach walls; one wall along the west side of the prehistoric marshes and one wall along the west side of the Pleistocene sands ("Geest").

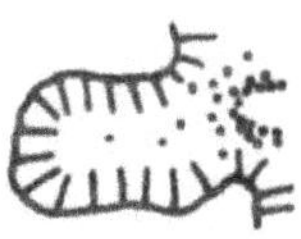

Pleistocene sand hills ("Geest") with peat bogs with the valleys.

Sea-dike near the modern coast-line.

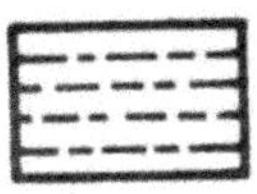

Peat bogs.

Fig. 13.2 The immediate environs of the Feddersen Wierde, showing the dwelling mounds and marshy areas in existence when the settlements began, in comparison with the modern coastline (after Haarnagel 1979).

Much time and care was devoted to the study of the present landscape; the strand meadows seawards of the dyke, the mud flats along the coast with its diatomaceous flora, and the zones of rushes and reeds around river mouths, particularly of the Elbe. In these areas the diatom and higher plant floras were studied in relation to the salt content of the water, the frequency of flooding, and the duration of each flood (Grohne 1959). Full use was made of past water level records in the study of flooding. In addition, the modern pollen rain in the marsh by the Feddersen Wierde was examined to assist the interpretation of fossil pollen in the layers of the wurt.

Since the object of the investigation was not only the wurt, but also the development of the surrounding landscape, select profiles of peat and clay deposits chosen from the numerous borings were analysed for pollen, diatoms, and general stratigraphy. The initial aim was to gain an idea of the appearance of the landscape at the time the settlement began. The changes up to the 5th century were to be determined, particularly those which might have given rise to the abandonment of this area.

Only by observing the modern vegetation in relation to the fossil remains could a valid interpretation of the identified plant remains be made. The co-ordinated observations of modern and fossil vegetation were undertaken over a period of six years, from May to October in each year. In the last two years a research trial was conducted using modern examples of the cultivated plants that had been detected in archaeological deposits from the Feddersen Wierde. This trial was carried out on a nearby unprotected plot of marshland lying seawards of the dyke. During the trial two midsummer storm floods flooded the small fields with salt water. As a result, it was possible to make observations and measurements of the relative tolerance of different economic plants to salt exposure.

The excavation was carried out such that each year a separate area of the wurt, of between 3500 and 4000 sq m was fully excavated down to natural. In this way, the relevant research questions could be discussed anew with each new excavation season. Every two weeks the current excavation surface was lowered by about 20-25 cm, so that in the course of a season 12-14 excavation surfaces were exposed. In addition, at key positions, profiles were left standing in order that the vertical construction sequence could be followed.

The development of the landscape from the Upper Sub-atlantic to the earliest settlement

In his book on the Feddersen Wierde Haarnagel (1979) describes the geological development of this area in the course of the Postglacial. The most crucial point he makes is that at the time of occupation of the area, the coastline ran in the immediate vicinity of the wurt, and not 4 km to the west as it does today (fig. 13:2). This coastline was formed by an embankment of gravel, approximately 40 m wide, whose origin Haarnagel assigned to the Dünkirchener Transgression, Stadium 1 (c. 500 BC). At various intervals along the 12 km ridge of this barrier beach lay seven more wurten, all contemporary with Feddersen Wierde. Cores through the neighbouring sediments demonstrated the existence of a wide arm of water running from the north about 1500 m inland, that is to the east, of the chain of wurten, and along the western margin of the geest. These sediments were dated by pollen analysis in order to determine whether the waterway had existed at the time of the settlement of the wurten. The diatoms and the sediments were analysed to determine whether the arm of water had held fresh or salt water.

In the settlement area itself and in its close surroundings several navigable channels were discovered by means of core sampling, and by the excavation itself.

About 1 m below the oldest settlement layer lies a clayey reed peat which was dated by pollen analysis to about 1000 bc. On the basis of diatoms it must have developed in a region of slightly brackish (oligohaline) water but under semiterrestrial conditions. The overlying sediments are at first unlayered, and of clay alone, but as one proceeds upward, they change into layered deposits of silt and fine sand. The evidence from macrofossils and the beginnings of root canals indicates that the area gradually accrued flood-lain sediments from marine inundations, and was finally colonised by salt marsh vegetation, (Strandwiesen-vegetation). Several decimeters of soil continued to build up during this last seral phase. Seed analysis showed that at the time when settlement of the land began, a Juncus gerardi vegetation existed in which the following halophytic species were detected; Spergularia salina Centaurium sp., Puccinellia distans, Glaux maritima, Triglochin maritimum and others.

A profile of the north-south flowing channel for geobotanical examination was selected from its margin, as the peat horizon had been eroded by the current in the centre of the channel. Pollen analysis suggested that the reed peat was contemporary with the reed peat found below the wurt. The formation of the higher peat layer in the upper part of the channel sediments could be dated to about the 6th or 7th century AD. The channel must have been open between these two periods, although by the latter period it was probably in its final phase before becoming land. The diatoms and sedimentological evidence showed that the channel had conducted slightly brackish waster, with inflows not only from the brackish marine coastal zone from the freshwater rivers of the geest. Analysis of the peat bogs on the geest has shown that prior to the settlement of the wurt, when the natural channel was most well developed, salt water diatoms and clay sediments had been washed far into these moors.

The earliest occupants of the wurt selected a 1.5-2 km wide strip of marshland for settlement, covered with a halophytic vegetation similar to the modern Juncetum gerardi community. It was protected by a barrier beach directly on the coast. They were not, incidentally, the first human occupants of the area. 12 cm below their upper surface, the marshland sediments mentioned above contain seeds of cultivation-dependent plants and skeletons of the diatom Hantzschia amphioxis, a lover of eutrophic conditions. The impressions of cattle hooves have since been discovered on earlier soil surfaces. Livestock must therefore have grazed here before the settlement of the wurt.

The nature of settlement on the Feddersen Wierde

In his book on the excavation of the Feddersen Wierde Haarnagel (1979) fully describes amongst other matters the construction of the wurt, the nature of the houses, and the settlement. Those aspects which are important for an understanding of the geobotanical investigations are summarised here.

The first dwellings were erected at ground level at about the end of the 1st century BC. The barrier beach with its slopes served as the construction site. This settlement stood for nearly 100 years and was renovated four times.

The building of artificial mounds began at the end of the 1st century AD. For each dwelling and its adjacent storehouse a long mound about 0.7 m high 20-25 cm long, and 8-10 m wide was constructed. Each mound was built up from a core of stable refuse, which had been painstakingly packed into layers. 'Stable refuse' is used here in a broad sense, to include the threshing debris that was intermixed with the actual manure and stable waste. Before the house was constructed, the core of manure and plant material was covered with a 30-50 cm layer of carefully packed clay sods. The structure of these sods indicates that they were taken from the immediate vicinity; their internal layers had remained completely intact. Deposits from storm floods on the slopes of the wurten show that during this period floods occasionally occurred over the surrounding land. At the Feddersen Wierde site, all such core-based wurten of this period together covered an area of about 13200 sq m.

During the 2nd century AD these wurt cores were systematically enlarged by additions of stable refuse and clay along their margins. The tops were also raised by about 50-80 cm.

The greatest extensions took place in the 3rd century. At that time the area of the wurten cores was substantially increased to obtain more room for the handling of harvests and household activities which were being performed on the wurten. The surfaces were raised by 20-50 cm at this stage. The widening of the separate wurten continued into the 4th century by which time they had merged into one continuous wurt. At this time a substantial village, arranged radially about a large central square, covered an area of about 3 ha. with a diameter of about 220 m in the NW-SE direction.

The excavation clearly showed that crop processing activities had taken place upon the wurt, or in the early phases, upon the separate core wurten, in the immediate vicinity of the storehouses. Here abundant remains of useful plants were to be found everywhere. Since the individual building stages could be dated by means of artefacts and the related stratigraphy, these stages formed a secure basis for the geobotanical investigations.

In the houses themselves the people and the livestock were sheltered under one roof. The largest of the houses included up to 16 double compartments for cattle which were so stalled in winter that their faeces ran into one liquid manure channel, lined with reed stems to assist drainage. This explains how in the layers of stable refuse whole portions could be found consisting entirely of cattle-faeces.

<u>Useful plants from the layers of stable refuse in the wurt</u>

Most noteworthy were the extensive layers of brown stems about 30-40 cm long and 2-3 mm thick which still possessed their roots and branches. In addition to these stems the valves of fruit capsules which had been removed in the process of threshing were regularly found. These plant remains could be identified as gold of pleasure, <u>Camelina sativa</u>. This species, whose seeds served as an important source of oil, was an important cultivar from the Iron Age to the beginning of the modern period.

Quite similar to the stems of these plants were those of flax, Linum usitatissimum, which were, however, thinner and shorter. The two species could be distinguished by microscopic examination of a cross-section of their stems. The stems of flax also retained their roots, indicating that, as today, the entire plants had been pulled up during harvest. The fruit capsules released by threshing were found in smaller or larger heaps, sometimes covering several square meters.

Also recovered from the excavated surfaces were layers of thick stems with longitudinal ribs and whose roots were also found intact. A thorough inspection of these layers revealed the impressions of individual pods of horsebean, Vicia faba, which were the only clue to the identity of these remains. Microscopic examination of the stem cross-section confirmed this identification.

Occurring in even greater quantities than the remains of gold of pleasure, flax, and horsebeans, were the threshing residues of cereals. They were spread over areas up to several square meters in extent, though fairly thinly. The long fragmented awns and rachis segments indicated that virtually all the residues were derived from barley, Hordeum vulgare, Only once were residues of emmer, Triticum dicoccum, also found, and these were in the carbonised state. Approximately 25% of all carbonised grains were of oats, but this cereal was entirely absent from the threshing residues. This may be because the thin-walled chaff of oats is not capable of preservation.

Broomcorn millet, Panicum miliaceum, must also be mentioned here. This occasionally occurred as very small piles of broken seed coats which were particularly noticeable by their yellow colour. It is suggested that they were the refuse from individual meals.

Finally, an important record at the Feddersen Wierde, of which there have been only a few prior examples, is of the plant woad, Isatis tinctoria, the ancient source of blue dye distributed throughout the Old World. Several times the inflorescences of this plant were found with fruits attached. This indicates that these plants must have been raised on the wurt.

Besides the uncharred remains of cultivated plants there also occurred charred grains and charred fruits and seeds of all the species mentioned above, with the exception of woad and millet. They lay in some of the houses scattered on the hearth among the ashes and burnt clay. Of these finds, one deserves special comment. In the oven of a burnt-down house, beneath the collapsed mud walls, lay an earth and ash free layer, 10 cm thick, of pure seeds, of cereals, flax, gold of pleasure and horsebean. The various species lay in carefully separated piles. The total weight of this find (sample No. 104) at 29 kg amounted to about double the sum total of all the other charred finds, even though these other finds did contain earth and ash impurities. Whether these stock piles were only being dried in the oven or were also being stored there could not be established. The total assemblage consisted of about 75% spelt wheat and 25% oats. Only one or two handfuls were of flax, gold of pleasure and horsebeans. A small number of grains of emmer, bread wheat, and naked barley were also found.

An interesting comparison may be made between the proportions of cultivated plants found in a carbonised state, and of those found in an uncarbonised

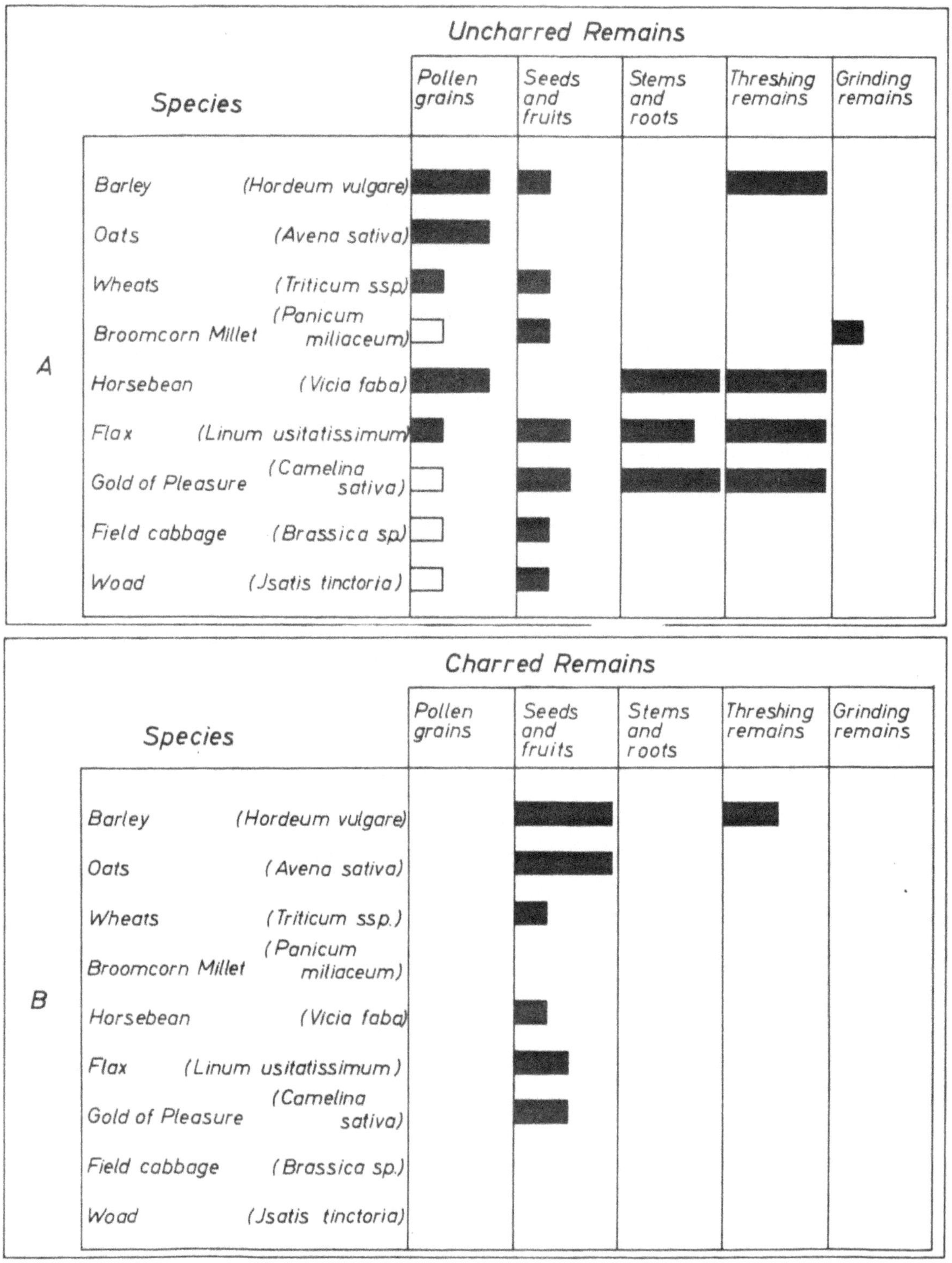
Uncharred Remains
Species
Pollen grains
Seeds and fruits
Stems and roots
Threshing remains
Grinding remains
Barley (Hordeum vulgare)
Oats (Avena sativa)
Wheats (Triticum ssp.)
Broomcorn Millet (Panicum miliaceum)
A
Horsebean (Vicia faba)
Flax (Linum usitatissimum)
Gold of Pleasure (Camelina sativa)
Field cabbage (Brassica sp.)
Woad (Jsatis tinctoria)
Charred Remains
Species
Pollen grains
Seeds and fruits
Stems and roots
Threshing remains
Grinding remains
Barley (Hordeum vulgare)
Oats (Avena sativa)
Wheats (Triticum ssp.)
Broomcorn Millet (Panicum miliaceum)
B
Horsebean (Vicia faba)
Flax (Linum usitatissimum)
Gold of Pleasure (Camelina sativa)
Field cabbage (Brassica sp.)
Woad (Jsatis tinctoria)

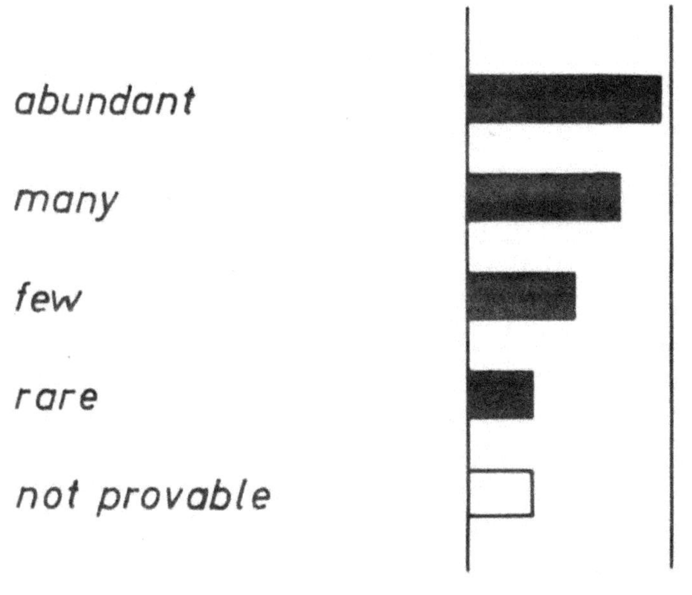

Fig. 13.3 Comparison of the occurrence of economic plant remains in uncharred and charred material from the site.

state. The site is useful in this respect in that both forms of preservation are well represented. Figure 13:3 shows how widely the two diverge. Three species present in the uncarbonised state were entirely absent from the carbonised record. It would appear that carbonised material will yield quantitative information about cereals alone. One should be aware of this bias on sites that produce only carbonised material. It is obviously preferable to recover material from both forms of preservation on a site.

The origin of the crops: The halophytic marshland or the inland geest?

A hypothesis suggesting that barley, oats, gold of pleasure, flax, and horsebeans were cultivated on brackish marshlands that would normally bear a *Juncetum gerardi* vegetation must arouse scepticism. The close-by flood-free geest, which was connected by navigable channels, and from which all wood for house-construction, tool manufacture, and fuel was undoubtedly derived, could theortically also have supplied cultivated crops, perhaps in exchange for livestock. In order to resolve this, the threshing remains were examined for seeds of weeds from cultivated fields, to establish whether these species were from sandy soils of the freshwater zone or were halophytes. The latter proved to be the case, accompanied however by certain non-halophytic species such as *Trifolium repens*, *Bromus mollis*, *Poa pratensis*, and *Poa trivialis*. Nonetheless, the species typical of the poor sandy fields on the geest were absent. The presence of roots still attached to the threshed plants of flax, gold of pleasure and horsebean also suggests they had not travelled far. In addition, sections of ploughed land were subsequently found. The inverted clods, which were produced not by a simple ard but by a turning plough, were analysed for seeds and humic content, the latter in the Peat Research Institute in Bremen.

It may be stated with certainty that agriculture in the form of summer cultivation had been practised in the immediate vicinity of the wurt. Its relative intensity, as reflected in the various settlement layers, was recorded and mapped during the 1956-9 seasons of excavation. The means by which such an agriculture could be conducted will be persued after various observations on the species of salt-marsh vegetation in relation to the recovered wild plant material have been made.

Wild plants from the layers of stable refuse in the wurt, and the inferred vegetation types

The layers of animal stable refuse in the wurt consisted of a closely pressed, brown to olive substance (plate 13:I). Layers of plant stems in whole or broken condition alternated with accumulations of pure cattle faeces. The rounded droppings of sheep or goats, and the faeces of horse and swine were also discernable, the latter recognisable by the presence of cleanly cut stalks, but these were rare. Usually the cattle faeces were found mixed with stem remains. However, there were also smaller, though sometimes extensive deposits consisting of unmixed layers of plant stalks. Of these, essentially four types could be distinguished:

1) Long, thick stalks of 5-8 mm diameter, with nodes, internodes and a bud at each node. When freshly cut, they appeared ochre in colour. These stems were identified as reed, *Phragmites*. Although in general they were

broken, they lay in the drainage troughs of livestock stalls still in bundles, just as they had been cut from the reed plants. The upper reed layer was permeated with the faeces of cattle, but the lower layers of up to 40 cm depth remained uncontaminated. Here were found, not uncommonly, entire fruit clusters of Lythrum salicaria, Sium erectum, Euphorbia palustris, Hypericum tetrapterum, and also fronds of the fern Dryopteris thelypteris.

2) Black stalks with white markings. These were about as thick as the reed stalks but were triangular in cross section and lacked nodes. Along their interiors cross-walls could be found at intervals of 5-20 cm. The presence of well preserved fruit clusters allowed these to be identified as the bulrush, Scirpus maritimus. These were not as plentiful as the reeds.

3) Very thin stalks of about 1 mm diameter, some with and some without nodes. Plentiful amongst these were the fruit clusters of Juncus gerardi, although other species such as Eleocharis palustria/uniglumis, Trifolium fragiferum, and Potentilla anserina, were also present.

4) Threshing remains of cultivated plants which have already been discussed.

As a consequence of the good condition of preservation in the tightly compressed layers of stable refuse, seeds, fruits, calyces, and sometimes even petals were preserved. Altogether, 86 soil samples were processed for seeds of wild plants.

Since these samples were obviously composed of a mixture of plants from various sources, including what came in with the cattle faeces, it was thought desirable to obtain some pure, uncontaminated samples. This was only possible because of the very large area exacavated. As mentioned above, the exposed surface under excavation was taken down by about 25-30 cm every two weeks. These surfaces could be thoroughly searched for preserved stalks. The exposed stable refuse was split apart into layers in order to obtain a good idea of its make-up. In this way, remains of hay were frequently found, in one case covering an area of several square metres. The stems and fruit clusters from these small 'pure' samples could be counted and compared with the relative proportions of seeds of the same species. Supplementary analyses of pollen were also carried out on these samples. As a result the following vegetation types could be deduced:

A) Layers of the thin short stems of Juncus gerardi with varying numbers of fruit clusters. These could be differentiated into three types of vegetation:

I) Layers composed of Juncus gerardi stalks alone, or with fruit clusters from Agrostis stolonifera, Triglochin maritimum, Atriplex hastata/patula Potentilla anserina, and a few others. This is a species-poor vegetation, with an average of 17 species per sample, in which Juncus gerardi dominates.

II) Layers of Juncus gerardi stalks, containing as characteristic components; Bromus mollis, Festuca rubra, Trifolium pratense, and Poa trivialis/pratensis. Leontodon autumnalis and Odontites rubra are also present in abundance. This is a richer vegetation with an average of 32 species per sample.

Plate 13.I Excavation of the Feddersen Wierde. Above: Excavation horizon with the remains of two houses in the background (dark areas are stable refuse, light areas are clay).
Below: Section through the deposits showing the vertical construction of the wurt (from Körber-Grohne 1967).

III) Layers of *Juncus gerardi* stalks containing a striking quantity of *Eleocharis palustris/uniglumis* fruit clusters. These were often more numerous than the fruit clusters of *Juncus gerardi* itself. *Oenanthe lachenalli* and *Juncus bufonius/ranarius* were typical components. In contrast with the above groups, this group also contained numerous non-halophytic species, such as *Juncus articulatus*, *Bidens tripartitus*, and others.

B) Layers composed of the thick long stems, usually of *Phragmites* sp., but sometimes of *Scirpus* sp.:

Ia) One sample composed of *Scirpus maritimus* stalks alone, together with the seeds of *S. tabernaemontani* and *Puccinella distans*.

Ib) A more common assemblage, with stalks of both *Scirpus maritimus* and *Phragmites* sp., together with seeds, particularly of *Carex disticha*. The average number of species per sample, at 55, was relatively high.

II) Layers of *Phragmites* stalks without *Scirpus*. In this group, entire fruit clusters of such tall herbaceous fresh-water plants as *Lythrum salicaria*, *Hypericum tetrapterum*, and others, were frequently found. *Eleocharis* and *Juncus gerardi* fruit clusters were also present, At 51, the average species number for this group was also large.

III) The final *Phragmites* assemblage was characterised by the presence, often in quantity, of fronds of the fern, *Dryopteris thelypteris*. This group also contained tall herbaceous species associated with non-halophytic conditions, such as *Iris pesudacorus*, *Mentha aquatica*, and *Solanum dulcamara*. The halophyte, *Juncus gerardi* was however also present in small quantities. The presence of leafy twigs of sallow, *Salix cinerea*, ratified the impression that this group was derived from the vegetation type farthest from the coast.

In order to determine the spatial arrangement of these various vegetation types in relation to the wurt, it was necessary to investigate the vegetation of the original soil surface beneath the wurt, as well as of the clay sods in the various layers of the wurt. This was done by means of seed and diatom studies.

The vegetation of the original clay surface and of the clay sods

The clay sods used in the construction of the wurt consisted of sedimentary material in which thick layers of clay and fine sand or silt were detectable. These layers were between 2 and 18 mm thick. Their irregular curved boundaries and numerous root canals support the idea that these were of the same alluvial origin as the material directly beneath the wurt. In most cases the root canals were small with diameters of only about 1 mm, and resembling those of *Juncus gerardi* and grasses in modern area lying outside the dykes. Only rarely were the black root crowns of *Scirpus maritimus* encountered. The results of seed analyses showed that a good match existed between vegetation type AI and a species-poor form of *Juncetum gerardii*, including *Glaux maritima*, *Triglochin maritimum* and others. Frequent occurrence of pioneers of open spaces, for example *Spergularia salina*, as well as plants associated with human settlement, such as *Plantago major* and *Polygonum aviculare* indicates that in places this *Juncus gerardi* marsh contained disturbed vegetation.

The original surface of the marsh lay directly beneath the construction layers of the wurt and above undisturbed soil. This surface was composed of a layer of humus and silty clay, 5-10 cm thick and completely unstratified, and containing refuse from the settlement. It was also penetrated by fine root canals. It clearly represents the soil surface contemporary with the origin of the wurt. Seed analyses from this layer revealed that a number of species appeared in this soil that were detectable neither in the undisturbed alluvial deposits nor in the clay sods. Besides a considerable number of ruderal plants, the presence of Leontodon autumnalis, Bromus mollis and Trifolium repens is particularly significant, as these are plants characteristic of group AII. It would seem then, that the wurt was constructed over a barrier beach vegetation.

Consequently, only two of the vegetation types were detected in the immediate vicinity of the wurt. Before attempting a reconstruction of the contemporary landscape, it is necessary to consider corresponding plant associations in the modern coastal vegetation, whose ecologies are well known.

Comparison of subfossil vegetation types with corresponding modern plant associations

The species-poor variant of the Juncus gerardi groups, group AI, matches well with the modern Juncetum gerardii. This association is encountered today on all coastal margins of the North Sea, as long as they are composed of marshy soils, have a certain salt content, and are situated at a certain height above mean high tide. This vegetation has been described by many authors, for example Gillner (1960) on the west coast of Sweden, Tüxen et al. (1957) on Neuwerk Island, and Beeftink (1962) at many places in the Netherlands. According to records of water levels on the coastal area between the Weser and Elbe rivers, this vegetation of the coastal marsh in the area outside the dykes lies sufficiently high above mean sea level that, with the exception of storm floods, it is only flooded by the highest tides. Around the Feddersen Wierde today, the storm floods in July often result in the farmers having their hay washed away.

The Juncus gerardi vegetation with Bromus mollis, Leontodon autumnalis, Trifolium repens, and Poa sp., group AII, occurs on the German coastal mud flats, (Wattenmeer-Küste) wherever a barrier has silted up to a sufficient height. Along the top of such a barrier beach, Bromus mollis grows, and along its landward slope, Trifolium pratense is occasionally found. Around the Feddersen Wierde, this vegetation lies only about 8-10 cm higher than the preceding vegetation type. However, this slight difference generally suffices to keep summer storm floods off the barrier beach. Hence, during these floods, the beach is left just projecting above the floodwaters. On rare occasions it does get flooded, but only at intervals of several years. In winter, when the storm floods tend to reach a greater height, it may become repeatedly submerged. However, floods at this time do not damage the non-halophytic plants growing on the barrier. In terms of elevation above the water table, the zones rich in Trifolium repens lie between vegetation groups AI and AIII. This vegetation type, group AII, has been described from various marshy coasts of the North Sea. Gillner (1975) considered Leontodon autumnalis to be promoted in this association by grazing, and that Odontites rubra disappeared under grazing. However, this could not be confirmed in the prehistoric vegetation around the Feddersen Wierde, since both plants occurred together in abundance. One might conclude that the influence of grazing was here secondary to the influence of other ecological factors.

The Eleocharis- rich variant of the Juncus gerardi vegetation, group AIII, may be compared with the Eleocharetum uniglumis of the lower Elbe (Kötter 1961). Similarities also exist with the Juncetum gerardii as a subassociation with Eleocharis uniglumis which Tüxen (1956) has described from Baltrum Island. This and other similar plant associations match the subfossil group AIII in its principal features; however, group AIII also has a strong Molinietelia component, comprising wet grassland plants of the non-halophytic zone, a character not reflected in the modern Eleocharis uniglumis association. The best match was with associations in the sand dune valleys of Terschelling Island, brought to my attention by W. van Zeist. This vegetation, a mixture of halophytes and non-halophytic species, is flooded by storms only a few times each year. As a result of the diking of the Dutch and German coasts, the unprotected forelands have become too small for this vegetation type to flourish. This vegetation therefore must have occurred in the recessed water-saturated land of the marshy coast, which was only occasionally reached by storm floods.

Group B vegetation has already been discussed in general, but a few further comments may here be made about group BIII in particular. Forms of vegetation that include the swamp fern, Dryopteris thelypteris on inland marshy soils have been described from various sites. From among these, the best match with group BIII is an association from the northern Netherlands, described by Meyer, de Wit et al. (1955). This association occurs in large non-halophytic swamps without any influx of salt water. A large number of species of the Molinietalla association that were present at these sites, for example, the mosses Calliergonella cuspidata and Brachythecium rutabulum, could also be identified in the fossil group BIII. Among the woody plants in the modern association were sallow, Salix cinerea, and alder, Alnus glutinosa. Leafy twigs of the former were identified in some of the group BIII samples. It would seem that BIII was the vegetation type which was at the greatest distance from the coast. The few occurences of Juncus gerardi in BIII samples show that even this vegetation received a few sea floods.

Interpretations of the wild plant assemblages with respect to the life of the occupants of Feddersen Wierde

The evidence suggests that the first occupants settled in a treeless marsh. Alder and sallow would have been restricted to fern thickets in the vicinity of the inland geest. The settlement itself would have been laid out on a barrier beach grassland of the Juncetum gerardii type, with clover and grasses. They key questions are: where did the fields lie?. and how was it possible to manage a flourising agriculture for centuries on a halophytic littoral grassland?

The only place within this zone that agriculture would have been possible is the barrier beaches, which were generally only submerged by storm floods in the winter. The agriculture of the occupants of the Feddersen Wierde must have depended upon summer cultivation, with the exception of woad, which must have been raised on the wurt. In order to establish whether, and in what manner, agriculture is possible in an undiked coastal marsh under such con-

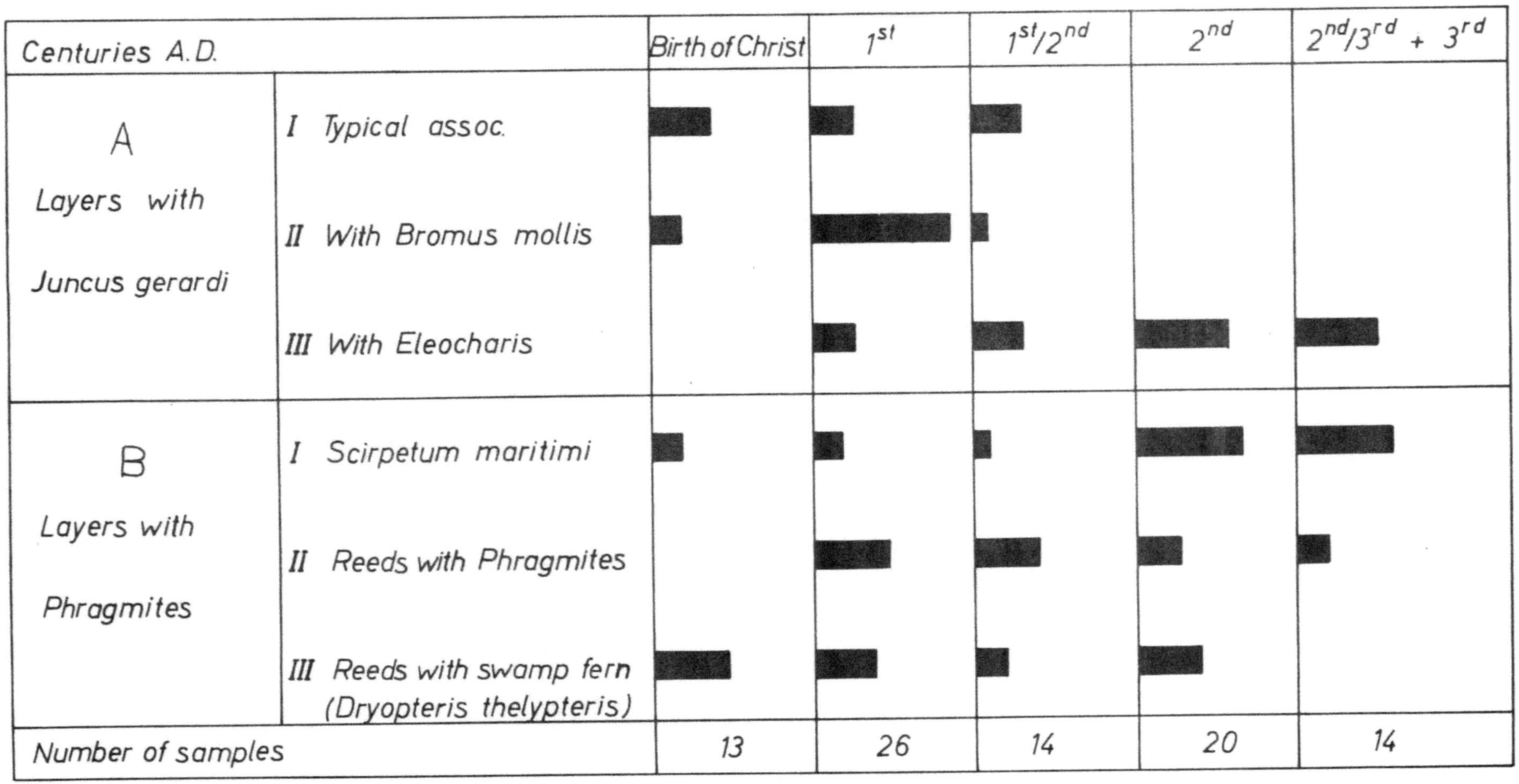

Fig. 13.4 The change in occurrence of the principal vegetation groups discussed on pp. 299-302 during the first three centuries AD.

ditions, an agricultural trial was carried out in the same region for two summers. Those cultivated plants which were recovered as fossils were planted on a plot of land lying outside the dike near Cappelersiel (Fig. 13: 1) in a raised zone with Juncus gerardi grassland containing Trifolium repens and Poa pratensis. In spite of the slightly salty water they grew well, until in each summer the plots were washed over by a storm flood. During this period of high standing salt water in front of the dike the fields remained under water for several hours. Damage was considerable, but interesting results were obtained with regard to the sensitivity and tolerance of the individual species. Trials were also undertaken using stable refuse as a fertiliser, Altogether, these experiments demonstrated that it is possible to raise crops on an undiked marsh of sufficient elevation. The productivity is naturally greater if the floods are confined to the winter. Clods of cultivated soils have in fact been exposed on many of the raised areas in the vicinity of the wurt. The fields definitely could not have been situated in the hinterlands of the coast, as these areas were far too wet.

A further question arises of whether there was any change in the vegetation through the centuries. To determine this, the quantitative proportions of these vegetation types were ascertained by mapping the deposits at every excavated surface exposed over a two year period. However, this was only possible for layers from the first three centuries AD, as plant remains were not preserved in the upper layers of the wurt. The results are presented in fig. 13:4. At the beginning of the 1st century AD the typical Juncus gerardi vegetation, AI, and the fern thickets, BIII, were the most abundant types, but the barrier beach vegetation, AII, and the brackish water reed thicket, BI, were also present. The ratios remained more or less the same until the end of the 1st century. In the 2nd century a noticeable alteration occurred; the Juncetum gerardii and the barrier beach vegetation disappeared. In their places, the Eleocharis-rich Juncus gerardi grassland, AIII, and the brackish water reed thicket became the most abundant forms. At the beginning of and during the 3rd century these and the Phragmites reed thickets became the only detectable forms of vegetation. The fern thicket ceased to be detectable. This indicates that the area became increasingly wetter from the 2nd century onwards. These shifts could be explained by the coastline having advanced and the wurt thereby coming to lie further and further into the marshy hinterland. These advances must have increased considerably in the 4th and 5th centuries, as the sediments forming at this period were thick layers of silty clay (Haarnagel, 1979). There was also evidence of considerable erosion of the older layers at this time. From this time on, the increased activity of the sea made habitation of the area impossible.

A series of changes may thus be seen in the conditions of human life in the coastal area, changes which are reflected in the plants and sediments, and which influence the way of life of several generations of inhabitants. The series began with a secure residence directly upon the level ground of an ideally suited coastal strip. It was followed by a period in which rising storm floods became a key factor, stimulating the construction of low mounds for each farmstead. Finally, changes in the coastal area lead to increased and severe periods of flooding, bringing to an end all habitation on the Feddersen Wierde.

Acknowledgements

I would like to thank Mr. Charles E. Carter, University of Hohenheim, and Mr. Martin Jones, University of Oxford, for translating the German text into English.

Bibliography

Beeftink, W. G. (1962). Conspectus of the Phanerogamic Salt Plant Communities in the Netherlands Biologisch Jaarboek 30. Jg. 325-362. Gent.

Gillner, V. (1952). Die Gürtelung der Strandwiesen und der Wasserstandswechsel an der Westküste Schwedens. Svensk Botanisk Tidskrift 46 393-428.

Gillner, V. (1960). Vegetations, und Standortsuntersuchungen in den Strandwiesen der Schwedischen Westküste Acta Phytogeographica Suecica 43 1-198.

Grohne, U. (1959). Die Bedeutung der Dietomeen zum Erkennen der subfossilen Vegetation höherer Pflanzen in Marschablagerungen.- Zeitschrift Deutscher Geologischer Gesellschaft 111 13-28.

Haarnagel, W. (1979). Die Grabung Feddersen Wierde, Methode, Hausbau Siedlungs- und Wirtschaftsform sowie Sozialstruktur In: Feddersen Wierde. Die Ergebnisse der Ausgrabung der vorgeschichtlichen Wurt bei Bremerhaven in den Jahren 1955-1963, Bd. 2 Wiesbaden.

Körber-Grohne, U. (1967). Geobotanische Untersuchungen auf der Feddersen Wierde In: Feddersen Wierde. Die Ergenisse der Ausgrabung der vorgeschichtlichen Wurt bei Bremerhaven in den Jahren 1955-1963 Bd. 1 Wiesbaden.

Kötter, F. (1961). Die Pflanzengesellschaften der Unterelbe Archiv Hydrobiologie/Suppllement 26 106-184.

Meijer, W., de Wit, R. J. u. Mitarb. (1955). Kortenhoef. Een veldbiologische Studie von een hollands verlandungsgebied.- Commissie voor de Vecht en het oostelijk en westlijk Plassengebied. Amsterdam.

Tüxen, R. (1956). Vegetationskarte der ostfriesischen Inseln: Baltrum- Bundesanstalt für Vegetationskartiarung, Stolzenau/Weser.

Tüxen, R. u. Mitarbeiter (1957). Die Pflanzengesellschaften des Aussendeichslandes von Neuwerk.- Mitteilungen Floristisch-soziologischer Arbeitsgemeinschaft/Niedersachsens 3 171-184.

14. SOIL AND BOTANICAL STUDIES OF THE "DARK EARTH"

R. I. Macphail

Introduction

The "Dark Earth", which is a dark coloured anthropogenic deposit, is commonly present on urban archaeological sites between Roman and medieval levels. It is usually poorly stratified and features are difficult to identify as they are commonly of a similar dark colour (Roskams 1980, pers. comm.). It may be 0.5 to 1.5 metres thick, and often contains reworked Roman pottery and merges with, or is contaminated by, medieval or post medieval debris and features. Opinions differ as to what this material represents, and therefore this enigmatic deposit is difficult, and some might say unproductive, to excavate. Nevertheless, careful examination of the "Dark Earth" and its relationships to the underlying and overlying levels and environmental studies in general, have begun to show how urban land use, in the period in question, varied both in time and space, and that information which is archaeologically pertinent is available if sought.

The importance of fully excavating the "Dark Earth" was realised relatively recently, mainly as a result of "Rescue's" survey of London in 1973 (Biddle et al.) which highlighted the dearth of urban archaeological information in the period 5th to 11th centuries AD. In London Saxon remains were elusive, firstly because of their destruction by medieval activity, and secondly because of the insubstantial nature of their structures. However, it was considered that these factors did not fully account for the lack of Saxon archaeology in the City of London—much may have been missed by poor excavation technique. In fact many of the pre-medieval and post-Roman deposits were not excavated but removed in bulk to reveal the Roman levels.

Such deposits of "Dark Earth" can be identified from sections drawn of excavations of Roman London at the beginning of this century (Norman and Reader 1912) as "Made Earth" or "Dark Made Earth" burying Roman surfaces and abutting Roman walls. In the nineteen sixties, the "Dark Earth" was certainly well known from excavations in London but was given little attention apart from the Saxon structures it contained (Merrifield 1965; Grimes 1968). Yet more recently, the London and Middlesex Archaeological Society's Special Paper (1976) on current problems of the archaeology of London appeared to ignore the whole question of the "Dark Earth".

However, the "Dark Earth" is an important deposit, commonly representing accumulation of material over seven centuries, and is not exclusively restricted to London or the United Kingdom. Archaeologically, it is significant because its occurrence between Roman and medieval structures represents a marked break in the continuity of normal urban land use. Moreover, the "Dark Earth" is not related soley to the Anglo-Saxon period, for it may be 3rd century Roman

(Roskams and Scholfield 1978) or Viking (Addyman, et al. 1976), or merge with the medieval levels. Consequently, a deposit known to archaeologists since the early excavations of Roman London has been generally neglected because of difficulties in interpreting its amorphous unstratified nature and long deposition period.

In the past, there was some doubt that the "Dark Earth" was an anthropogenic deposit—Kenyon (1959) working at Southwark believed it to be a flood sediment. Subsequently this view has been rejected after consideration of the nature of the material it contains (Sheldon 1978: 40) and Southwark "Dark Earth" is believed to have been brought in for use as a growing medium for a form of market gardening (Sheldon 1978). Contrary to expectations for an agriculturally used soil, no increased organic matter content was found in the upper part of the "Dark Earth" during detailed excavations at Milk Street, London (Roskams, unpublished report B); neither was there enough humus in "Dark Earth" from Southwark to suggest it had been cultivated (Taylor 1978).

If the material had been imported and dumped, such events might be illustrated by particle size analyses of sequences through the "Dark Earth" but Taylor (1978) reported little or no variation through columns of the "Dark Earth" at Chaucer Street, Southwark. Thus, earlier views on the nature of the "Dark Earth" seemed contradictory to the evidence. It was therefore reasoned that these questions could only be resolved by investigating "Dark Earth" from a number of sites in London and other localities in the United Kingdom.

SOIL AND POLLEN STUDY

During 1979 and early 1980 samples of "Dark Earth" were examined from four sites in London, and one each from York and Carlisle (Macphail 1980). Also one rural Roman/post-Roman site provided a contrast (Macphail 1980). In order to check uniformity of identification and character, samples were studied for colour, pH, extractable phosphate content, loss on ignition and alkali soluble humus. Grain size and coarse inclusion analyses were also carried out on samples from selected sites. Thin section were prepared of "Dark Earth" from GPO 75 (a site near St. Paul's Cathedral, London) as a comparison for thin sections from Southwark already produced by Mr. T. P. Taylor, and to enable past and present pedogenic processes to be identified more clearly. Complementary botanical studies were carried out, i.e. phytolith (opalised plant remains) analysis of samples from GPO 75, St. Bartholomew's Hospital (both London) and the Bedern (York) as pollen had been reported to be scarce. Nevertheless, an attempt was made to extract soil pollen from one level at GPO 75 (Scaife 1980a), but this was only partially successful so a second attempt was made on wetter samples from the Bedern, to provide a comparable source of data.

Methods

Soil colour, both wet and dry, was determined using a Munsell Chart. pH, loss on ignition and grain size analyses followed standard procedures; the test for alkali soluble humus is based on Cornwall's method (1958). Coarse inclusions (larger than 2 mm) were washed, identified and calculated as percentages, and extractable phosphorous content was estimated by the method

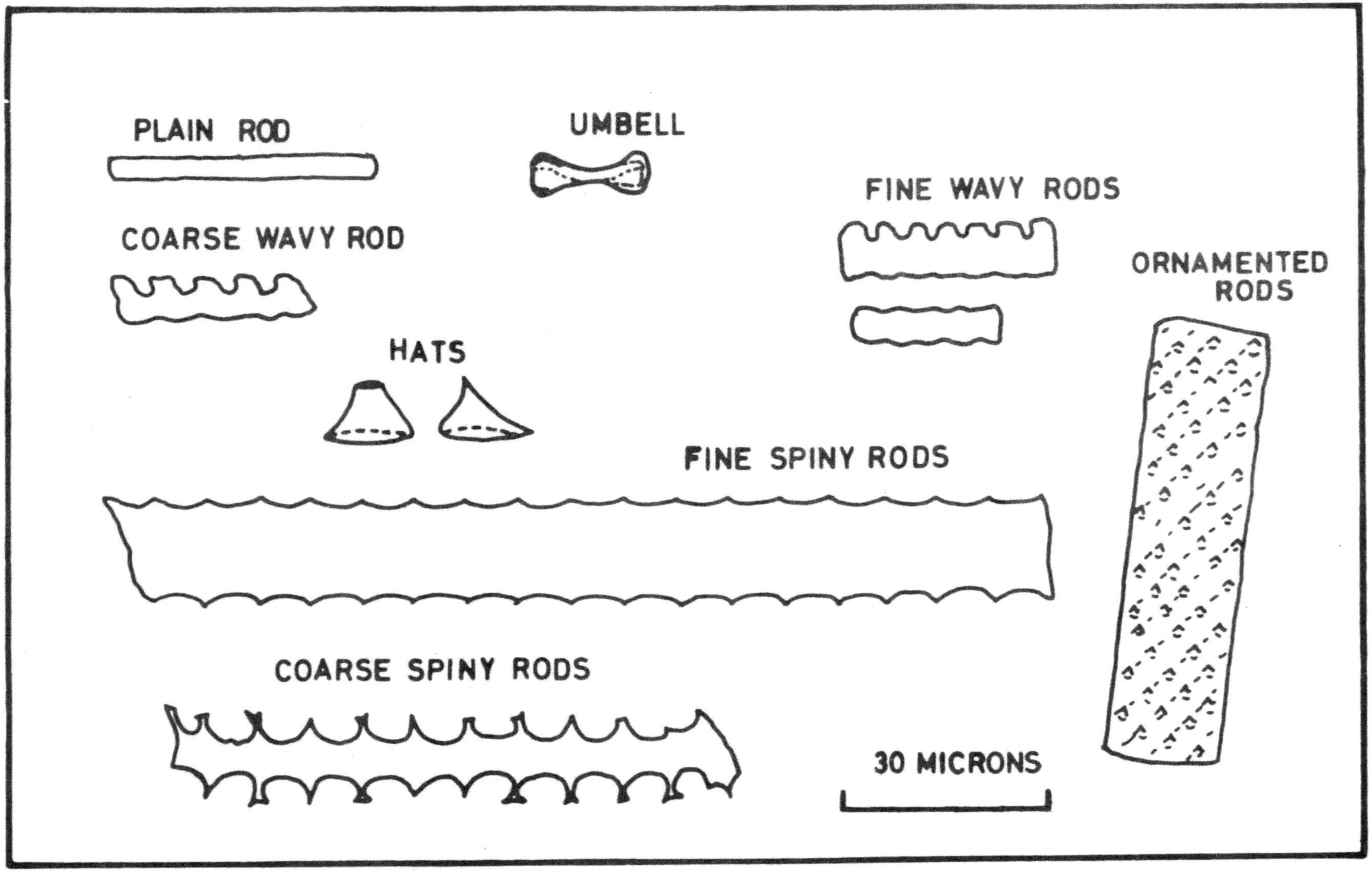

Fig. 14.1 Phytoliths.

of Schwartz (1967). Thin section manufacture was also based on standard methods. Soil pollen analysis followed standard concentration procedures (Faegri and Iversen 1974; Moore and Webb 1978).

Phytolith analysis was carried out on fine soil (less than 1 mm) after pre-treatment with HCl, and the combustion of non-opalised organic material. Three samples from each level were mounted on glass slides using Naphrax (Refractive Index 1.74). Phytoliths were counted under the microscope using three or more transects on each slide under phase contrast, and total counts per level varied from 460 to 680. Phytoliths themselves are constructed of biogenic silica or opal (Smithson 1958; Geiss 1973) which may strengthen cell walls, infill cell cavities, or occur in special cells and although there are a large number of problems relating to their use in palaeobotanical studies, as listed by Twiss et al. (1969), selected identification and grouping allows the differentiation of particular phytolith suites. These not only permit the grouping and comparison of the Gramineae present in each level, but also occasionally individual types may be recognised.

Phytoliths from the "Dark Earth" were counted in broad categories with descriptive names as follows: Plain Rod, Rectangular Rod, Crooked Rod, Fine Rod, Ornamental Rod, Fine Wavy-Rod, Coarse Wavy-Rod, Fine Spiny-Rod, Coarse Spiny-Rod, Spiny Shape, Hat and Dumb-bell, (Fig. 14:1) as these were types both common to the soil samples and to the 30 grass and cereal species examined at the Institute of Archaeology, London.

The Archaeological Sites

The rural site at Lechlade comprises small (generally less than 10 metres across) dark soil zones associated with Roman occupation areas (Miles 1979, pers. comm.). Samples were taken from the present day brown Ap horizon and the laterally merging dark Ap horizon (Table 1). Samples were also taken from "Dark Earth" above the Watling Street at Southwark (Dean 1979, pers. comm.); from "Dark Earth" above a Roman floor at St. Bartholomew's Hospital, London, hereafter referred to as Bar 79 (Bentley 1979, pers. comm.); from "Dark Earth" above Antonine levels at two sites near St. Paul's Cathedral, London, hereafter referred to as POM 79 and GPO 75 respectively (Norton 1979, pers. comm.; Roskams 1979, pers. comm.); and from "Dark Earth" overlying a 2nd century wooden and 3rd century stone building at Keays Lane (Macarthy 1979, pers. comm.). Lastly, "Dark Earth" lying between Roman walls and medieval deposits at the Bedern was received from Andrew Jones of the York Environmental Archaeological Unit.

RESULTS

Before discussing the laboratory results it is useful to consider a typical site description of the "Dark Earth" from GPO 75, as follows:-

> "black humic silt, frequent flakes of shell, moderate flecks of burnt daub and charcoal, moderate cobbles-pebbles, occasional flecks of brickearth". (Roskams, unpublished report A).

A similar description of the "Dark Earth" is available from the Milk Street publication:-

" a dark grey silt with inclusions of cultural material such as pottery, tile, and oyster shell, most of which is severely abraded" (Roskams and Schofield 1978).

In this present study the "Dark Earth" has been found to vary from wet to dry in its modern state and can sometimes be characterised by earthworm casts, yet the descriptions above are typical and show how basically "Dark Earths" have comparable attributes:-

1. "Dark Earth" has a strongly uniform colour throughout the sites (Table 2) especially when wet. However, much of the London "Dark Earth" had dried out to similar greyish browns. Brickearth and the Clayden Pike Ap horizon obviously differ.

2. pH is remarkably consistent, at neutral to alkaline, but this is not unexpected in deposits which are commonly rich in preserved bone and shell.

3. Extractable phosphorus levels vary greatly from Trace to Strong (see Table 2). Both the dark coloured and brown coloured areas of the soil at Clayden Pike have equal amounts, presumably from contemporary agricultural practices. However, as "Dark Earth" from London (i.e. samples 3-15) is often completely dried out, and as phosphorus is known to be irreversibly absorbed on drying (Peter Loveland of the Soil Survey of England and Wales, pers. comm.), the wetter deposits from Carlisle and York may give a better indication of original phosphorus levels. Variations may relate to the known inhumation at Bar 79 (samples 5 and 6), or be associated with leaching and earthworm activity from an overlying medieval cemetery as at GPO 75 (sample 12), or reflect "Dark Earth" containing fragmented bone or cess material.

4. Loss on ignition and alkali soluble humus are discussed together as the latter indicates amounts of organic carbon in the "Dark Earth", while slow ignition (at less than 400° C) includes carbonised material in its estimate of organic matter. Knowledge of organic carbon status is useful in ascertaining the agricultural potential of the deposit. Loss on ignition is an additional parameter for comparing homogeneity of levels in the "Dark Earth; it measures immobile material as well as the mobile organic carbon. The samples from Clayden Pike are interesting because they show amounts of alkali soluble humus in a modern cultivated soil, and reveal little difference in loss on ignition between the dark Ap horizon and the brown Ap horizon, suggesting homogenisation of the surface soil during cultivation.

In London generally, the alkali soluble humus content of the "Dark Earth" is lower than in contemporary soils at Clayden Pike, and tends to be similar at each individual site. Loss on ignition may vary to a much greater extent than, and apparently independently of, the alkali soluble humus content. Organic carbon is more easily disseminated throughout the "Dark Earth" due to pedologic, faunal and floral activity, while loss on ignition tends to reflect the original heterogeneity of the deposit which may result from additions of dumped material.

Samples from Carlisle and York exhibit a similar pattern, but contain far more alkali soluble humus than is found in London "Dark Earth" or reported by Taylor (1978) from Chaucer Street, Southwark. In the latter case these

TABLE 1

Site		Sample
Clayden Pike, Lechlade, Oxon.	1	Contemporary brown Ap
(Oxford Archaeological Unit)	2	Contemporary black Ap
Arcadia Buildings, Southwark	3	Above Watling St; "Dark Earth"
(Southwark Archaeological Unit)	4	to medieval.
St. Bartholomew's,London (Bar 79)	5	Late 2nd C. to 4th C.; resting on
(Department of Urban Arch-	6	on opus signinum floor.
aeology)	7	Late Roman or Saxon
	8	Probable Saxon
	9	Approx. 11th C.
GPO Middle (POM 79), London	10	Medieval
(Department of Urban Arch-	11	Possible "Dark Earth"
aeology)		
GPO 75, London	12	"Dark Earth" approx. 4th C.
(Department of Urban Arch-	13	"Dark Earth" " "
aeology)	14	"Dark Earth", " "
	15	Brickearth
Keays Lane, Carlisle	16	"Dark Earth" to medieval
(Planning Department,	17	"Dark Earth" to medieval
Carlisle City Council)		
The Bedern, York	18	"Dark Earth"
(York Archaeological Trust)	19	"Dark Earth"

TABLE 2: ANALYTICAL DATA

Sample		pH	Phosphate spot test	Alkali soluble humus	% loss on ignition	Colour Moist	Colour Dry
1	Clayden Pike, Oxon	7.5	Pos-W	108.0	10.3	10YR3/4	5YR3/4
2	Clayden Pike, Oxon	7.5	Pos-W	109.0	9.9	10YR3/1	10YR3/2
3	Arcadia Buildings,	7.0	W	56.0	3.96	10YR3/1	10YR6/2
4	Southwark	7.0	Tr	56.0	4.49	10YR3/1	10YR6/2
5	Bar 79, London	7.4	Pos	24.0	2.34	10YR3/1	10YR4/2
6	Bar 79, London	7.4	Pos	27.4	2.04	10YR3/1	10YR5/2
7	Bar 79, London	7.4	W	25.0	2.26	10YR3/2	10YR4/2
8	Bar 79, London	7.4	Pos	31.0	2.60	10YR3/1	10YR4/2
9	Bar 79, London	7.5	W	54.0	2.04	10YR3/1	10YR4/2
10	POM 79, London	7.6	Pos	92.0	6.81	10YR3/1	10YR4/2
11	POM 79, London	7.1	Pos	86.0	5.37	10YR3/1	10YR4/2
12	GPO 75, London	7.3	Pos	54.0	2.70	10YR3/1	10YR4/2
13	GPO 75, London	7.5	W-Pos	58.0	15.01	10YR3/1	10YR4/2
14	GPO 75, London	7.6	Tr	59.0	2.70	10YR3/1	10YR4/2
15	GPO 75, London	8.4	W-Pos	5.9	1.49	10YR5/4	10YR7/4
16	Keays Lane, Carlisle	6.9	Pos	118.0	5.16	10YR3/1	10YR4/2
17	Keays Lane, Carlisle	6.7	Pos	144.0	5.53	7.5YR2/0	10YR4/1
18	The Bedern, York	7.2	W	200.0	12.33	10YR3/0	10YR5/1
19	The Bedern, York	7.0	Str	212.0	10.30	10YR3/1	10YR5/1

N.B. i) Alkali soluble humus, mgms. per 100 gms. dry soil

ii) Phosphate: Tr, Trace, approx. 0.08 - 0.15% P_2O_3

W, Weak, approx. 0.15 - 0.4% P_2O_3

Pos, Positive, approx. 0.4 - 0.8% P_2O_3

Str, Strong, approx. 0.8% or more P_2O_3

After Schwartz (1967)

low amounts coincided with the small quantities of alkali soluble humus measured from the overlying post-medieval and immediately underlying Roman levels. The wetter nature of the "Dark Earth" from Carlisle and York apparently provides good evidence that lower alkali soluble humus contents of London result from severe oxidation of the organic matter content, because as noted previously, in London the "Dark Earth" is often completely dried out and contains evidence of previous intense biological activity. This evidence is derived from soil micromorphology and soil pollen studies and will be described later. Thus, the London "Dark Earth" may have originally contained similar amounts of organic matter as at Clayden Pike, i.e. levels equivalent to that of an agricultural soil.

5. Particle size analysis shows the "Dark Earth" at GPO 75 to be very poorly sorted, varying between a sand and a loamy sand, and is moderately stony (Table 3). Clay content is relatively low; variations in silt content may depend upon how much of the local "silty" brickearth has been admixed, as shown in thin sections. The "Dark Earth" shows only small variation in texture with depth as noted by Taylor (1978) at Chaucer Street, Southwark.

Much of the coarse material (i.e. larger than 2 mm in diameter) at GPO 75 consists of anthropogenic debris (Table 4), such as bone, "tile", mortar, pottery, charcoal and, in some cases, shell (mussel). At York, a similar suite of inclusions is present, so much of the "Dark Earth" is made up of finely fragmented anthropogenic debris, mixed with soil material and organic matter. At GPO 75 stones, and the mainly sandy matrix, must be regarded as exotic in an area dominated by a silty parent material; some coarse grains may derive from abraided pottery. The physical composition of the "Dark Earth" was examined in more detail in thin sections.

6. Three large slides of the central horizon of the "Dark Earth" at the GPO were examined under the microscope (see Micromorphology). Generally, the "Dark Earth" has a dark brown to opaque (Plane Polarised Light, PPL), agglomeroplasmic (after Brewer 1964) fabric with areas which are more porhyroskelic in nature. Cursory point counting showed the material to be quite dense with good channel structures. Skeletal material comprises much sand and silt, but also inclusions of ceramics and calcite/aragonite shell occur. In addition, recognisable charcoal fragments are present, and it seems likely that more finely comminuted material may be attached to the intimately mixed iron-clay complexes in such a way as to make them opaque. Indeed, evidence of earthworm activity which would account for this dense agglomeroplasmic fabric is commonplace. For example, the thin sections contain pedotubules (after Brewer 1964) comprised of silts without any dark colouration showing how earthworms have burrowed through the "Dark Earth" into the "brickearth" beneath, probably to aestivate. This evidence indicates areas where the "Dark Earth" has yet not been intimately mixed with the "fines" in the natural. Such mixing was apparent in the textural analysis of the site.

In addition, two slides manufactured by T. P. Taylor of "Dark Earth" from Chaucer Street, Southwark were examined in detail and showed the same kind of dark brown to opaque (PPL) agglomeroplasmic fabric. They also exhibit pedotubules relating to fauna importing "non-Dark Earth" material within the matrix of the "Dark Earth" proper. Consequently it appears that the

TABLE 3: % PARTICLE SIZE - GPO 75

Sample	Clay	Fine Silt	Medium Silt	Coarse Silt	Fine Sand	Medium Sand	Coarse Sand	(Gravel)
12	12	5	3	2	27	39	12	(19)
13	13	4	4	7	29	32	11	(20)
14	12	3	8	2	32	34	9	(17)

TABLE 4: % INCLUSIONS, LARGER THAN 2 mm

Sample		Bone	"Tile"	Mortar	Stones	Pottery	Charcoal	Mussel Shell
12	GPO 75	1.2	24.0	43.0	13.0	3.7	0.09	-
13	GPO 75	18.0	17.0	3.5	45.0	7.0	0.06	-
14	GPO 75	12.0	10.0	3.4	70.0	0.2	0.64	3.8
18	The Bedern	1.4	8.0	31.0	33.0	26.0	0.50	-
19	The Bedern	2.4	10.5	36.0	33.6	5.7	11.60	-

TABLE 5

London G.P.O. 75, Sample (13)	Pollen
Sinapis type	4
Chenopodium type	2
Papilionaceae indiff.	2
Centaurea cf. *scabiosa* type	1
Taraxacum type	10
Gramineae	5
Cerealia type	1
Dryopteris type	1
Pteridium aquilinum	1

Total pollen count = 35 grains

materials are similar and that faunal activity, especially of earthworms, has been important in both. The base-richness of the "Dark Earth" and obviously active faunal population may account for the apparent losses of organic matter.

Summary. Soil studies indicate that the "Dark Earth" is a neutral to alkaline deposit originally relatively rich in organic matter, made up of local and exotic mineral material, mixed with a variety of cultural debris (which in the case of charcoal has become intimately associated with the soil fabric) and therefore apparently not natural to the area. In London, at least, it would have provided a useful growing medium, containing quantities of phosphorus, bone, organic matter, and, in some cases, cess, in a freely draining environment. It is difficult using soil studies alone to differentiate between material which may be medieval in origin and "Dark Earth" which is late Roman to pre-medieval.

BOTANICAL STUDIES

As stated earlier, phytolith and soil pollen analyses were carried out, the latter by R. G. Scaife (1980a). Sample 13 from GPO 75 was examined; pollen was found to be sparse and a diminutive total of 35 pollen grains were identified. These are listed in Table 5 and show a predominance of grains with thick exines—Taraxacum type, (Liguliflorae) and Sinapis type. Where grains with thinner exines were identified such as Graminaea, a high degree of degradation was apparent. Crumpling, heavy oxidation and microbial attack resulted in poorly preserved pollen. Consequently differential pollen preservation may be an important feature of this spectrum creating a total biased in favour of more robust pollen types.

In view of the above, it can be said that types represented are of predominantly ruderal nature. Pollen of Cerealia type and Sinapis type (4 grains) are most likely to have been introduced onto the site from urban waste rather than from on site cereal production. Differential pollen preservation has almost certainly caused the under-representation of other weeds such as Urtica and Gramineae identified in later deposits at Broad Sanctuary, Westminster (Scaife 1980b) where a marked species diversity exists.

The poor pollen preservation further supports the hypothesis that the organic matter content of the "Dark Earth" in London may originally have been greater. Cerealia type and Sinapis type pollen could be secondary and associated with a cess input.

Calculation of percentage types of phytoliths from samples 6, 7 and 8 from Bar 79 and samples 12, 13 and 14 from GPO 75 showed these London "Dark Earth" to be dominated by Festucoid grass types (Twiss et al. 1969; Rovner 1971), which are represented by Fine Wavy-Rods, Coarse Wavy-Rods and by many of the Spiny and Hat types (Figures 14:1 and 14:2). There are only minor quantities of the diagnostic Dumb-bell phytolithis (Table 6) which relate to small additions from Panicoid grass types. The pollen data clearly suggested a waste-ground environment, and the phytolith suite present conforms to this.

TABLE 6: PERCENTAGE PHYTOLITH TYPES FROM ST. BARTHOLOMEW'S HOSPITAL, GPO 75 AND THE BEDERN

		Plain-Rods			Total Rods		Wavy-Rods		Spiny-Rods		Spiny			Total
Sample	Plain	Rect.	Crook.	Fine	Pl. Rods	Orn.	Fine	Coarse	Fine	Coarse	Shape	Hats	Dumb-bells	Count
St. Bartholomew's Hospital (London)														
6	10.7	3.5	2.4	1.6	18.2	19.8	22.5	31.8	4.8	1.4	-	2.3	0.4	541
7	26.8	3.5	3.5	-	33.8	13.1	23.5	10.7	5.6	2.9	-	3.7	0.7	571
8	30.6	2.1	1.5	-	33.2	14.0	28.5	6.1	6.1	4.6	-	5.4	0.2	477
GPO 75 (London)														
12	20.2	0.3	1.0	1.6	22.9	14.1	30.0	6.3	12.1	6.3	1.6	4.1	1.6	569
13	19.6	0.4	1.5	1.7	24.2	23.3	33.9	1.7	8.7	3.9	0.6	3.3	0.8	460
14	12.7	-	0.9	0.8	14.4	21.7	40.0	4.6	8.6	4.6	1.5	2.6	1.7	544
The Bedern (York)														
18	12.2	-	0.7	3.9	16.8	14.3	24.4	4.4	9.8	18.9	4.7	1.4	4.5	680
19	13.4	-	1.3	1.9	16.6	14.4	25.6	5.0	7.9	9.7	3.6	0.9	16.2	617

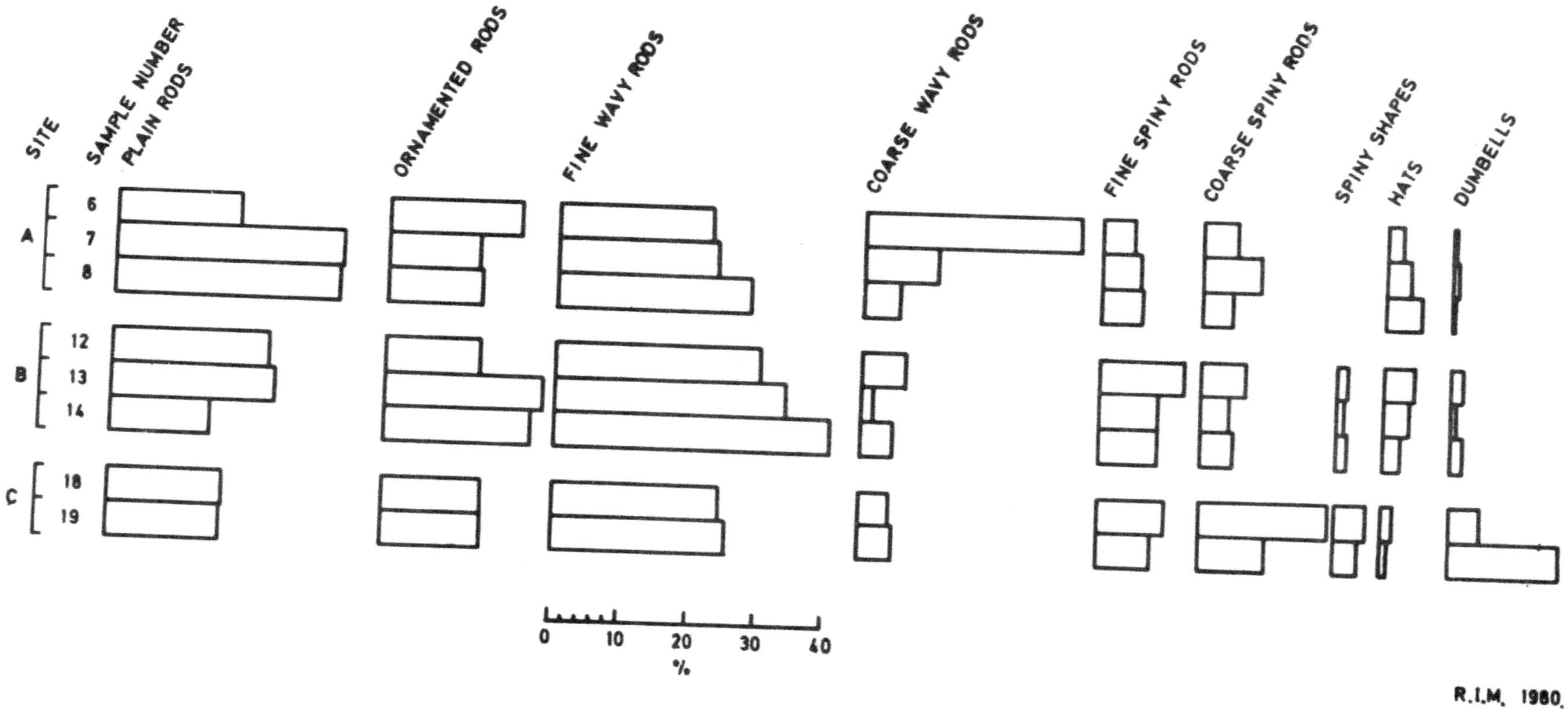

Fig. 14.2 Percentage Phytoliths from (a) GPO 75 (b) St. Bartholomews Hospital (c) The Bedern.

At GPO 75 the three levels sampled show minor variations in their respective phytolith suites and thus, despite disturbance by earthworms, seem to reflect slight change in flora (either local or imported). At Bar 79 the two lower levels (i.e. samples 6 and 7) are far more alike than the Saxon level above (i.e. sample 8; see Table 1). The latter also differs from the suites present at GPO 75 (see Fig. 14:2) being characterised by many Coarse Wavy-Rods typical of grass genera such as Cynosurus and Poa, which could indicate a sharp change in usage of or input into the "Dark Earth" at this time.

The two samples from York were examined for pollen by R. G. Scaife and the results were encouraging—absolute pollen frequencies were higher than at GPO 75 enabling a basic count of 100 grains to be made of sample 18 (see Table 1) and 200 grains were counted from sample 19 (the lower level of the Bedern "Dark Earth"). Pollen preservation was also notably better, although a divergence was found in the pollen spectrum of the upper level.

On one hand a number of types including Taraxacum and Anthemis type with thicker exines, showed a degree of degradation similar to that at GPO 75. The bulk of the remaining pollen including Gramineae, Cerealia type and Centaurea cyanus (Table 7), was in a moderate to good state of preservation.

The pollen taxa are dominated by ruderal and pastoral pollen types of which Gramineae are the most important. In addition to these, however, are Cerealia type, Centaurea cyanus and Sinapis type which are usually indicative of arable agriculture. These pollen types are likely to be of secondary-derived origin in such urban contexts. Analyses of medieval cess material show that high concentrations of cereal pollen and macro remains are present (Scaife 1980b). It seems likely, therefore, that arable pollen types are derived from human faecal waste deposited on or in the soils. Further evidence of this comes from the presence, albeit in low numbers, of parasite eggs (Trichuris spp.) and these have been reported elsewhere from urban deposits (Pike and Biddle 1966). Whilst these may be from animal sources, especially pigs, a human origin is likely.

The dual nature of the preservation noted, suggests that two phases of pollen influx are present:

a) The degraded pollen of robust exine types may represent autochthonous growth of these plant types. These may therefore pre-date the well preserved pollen.

b) Well preserved pollen may have been contemporaneous and in material dumped on the site including the ruderal and arable pollen types discussed above.

Phytolith analysis of the York samples again showed a basically Festucoid suite, but with far greater quantities of Panicoid grass types than found in London. Interestingly, sample 18 contains a high proportion (19%) of Coarse Spiny-Rods (see Fig. 14:2) many of which were very large indeed (approximately 140 microns by 20 microns). Study of the Institute of Archaeology (University of London) phytolith collection indicated that many grass and sedge types could not be represented by these particular phytoliths. Even taxa with Coarse Spiny-Rods such as Cyperacae spp. and Nardus spp., seemed unlikely candidates because they lack the multiple elements present on the

TABLE 7

York, Bedern, Sample (18)

	Pollen
Alnus	2
Calluna	2
Sinapis type	2
Chenopodium type	1
Plantago lanceolata	2
Taraxacum type	7
Anthemis type	5
Bidens type	1
Centaurea cyanus	5
C. nigra type	1
Gramineae	58
Cerealia type	7
Cyperaceae	5
Unidentified pollen	2
Dryopteris type	1
Parasite eggs (*Trichuris* spp.)	3

Total pollen count = 100 grains

TABLE 8

York, Bedern, Sample (19)

		%TP
Betula	2	1
Alnus	3	1.5
Corylus	9	4.5
Calluna	91	35.5
Ranunculus	1	0.5
Silene type	2	1.0
Chenopodium type	2	1.0
Trifolium type	1	0.5
Papilionaceae indiff.	1	0.5
Umbelliferae	1	0.5
Polygonum convolvulus	2	1.0
Rumex	2	1.0
Scabiosa	1	0.5
Bidens type	1	0.5
Anthemis type	1	0.5
Taraxacum type	17	8.5
Centaurea nigra type	1	0.5
Gramineae	43	21.5
Cerealia type	14	7.0
Cyperaceae	4	2.0
Unidentified pollen	1	0.5
Sphagnum	1	0.5
Pteridium aquilinum	3	1.5

A.P.F. = 14000 grains per gramme

species of these Coarse Spiny phytoliths. However, reference phytoliths of Triticum aestivum, both from the collection and from archaeological material, are clearly similar to the Coarse Spiny types identified from York (Hubbard, pers. comm.). As shown by the soil pollen, there may have been increments of secondary cereal pollen into this deposit, together with arable indicators, further supporting the contention that cereal material was included in the "Dark Earth". Armitage (1975) found well preserved phytoliths on the teeth of ungulates at Baynards Castle, and this argues that phytoliths at York could be of secondary origin. On the other hand, they may relate to waste from, for example, threshing.

In the lower level at York pollen preservation was good, possibly due to a higher moisture content. The dominant types are Calluna (46% TP), Taraxacum type (9% TP), Gramineae (22% TP) and Cerealia type (7% TP), (Table 8). These occur in association with a diverse ruderal assemblage similar to that of York Sample 18. Calluna is the notable exception to this and poses the question of its origin. Natural transportation cannot be the case because although Calluna produces large quantities of pollen, its dispersion is poor due to the high specific gravity of its pollen tetrads. Calluna must have been collected and utilised for one or more reasons of floor cover, animal bedding, fodder or brushwood. Indeed, evidence of Calluna and specific beetle remains has been similarly interpreted for an Anglo-Danish site elsewhere in York (Buckland 1974). The presence of high Gramineae totals may also be noted in this respect. Such material may have been dumped onto the soils of waste ground. The presence of cereal pollen as in preceding samples may similarly result from kitchen waste or human faecal material being dumped.

Phytoliths included 16% Dumb-bells (see Fig. 14:2), revealing a relatively high composition of Panicoid grass types. Two typical Panicoid grasses are Sieglingia sp. and Molinia sp., and a review of the reference slides, together with Smithson's (1958) findings from North Wales, suggested that both of these may have contributed towards the Panicoid grass types represented in sample 19. It is likely that the Calluna was collected from local moorland areas and Sieglingia and Molinia may have easily become incorporated in the same way, whether by design or accident it is difficult to say; alternatively these particular phytoliths may have been brought in by stock which had been grazing on the moor.

Summary

Urban areas are highly complex depositional environments, which causes problems when interpreting both phytolith and pollen data because the majority of pollen and phytoliths found result directly or autogenically from human activity. This is illustrated by the relatively low pollen frequencies of arboreal types transported naturally from rural areas. The assemblages therefore appear to reflect:-

a) The growth of ruderals on waste ground in urban areas, e.g. Plantago lanceolata, Compositae spp., Rumex spp.

b) Plants introduced/brought in from rural areas for various uses, e.g. Calluna sp. and Gramineae.

c) Refuse material discarded onto waste or open areas, with which are associated the arable pollen types and related weeds, derived from domestic waste, and possibly parasite eggs.

It must be remembered however, that some of the secondary material may have been imported with the "Dark Earth" which is not directly related with its use.

DISCUSSION

Soil and botanical studies showed the "Dark Earth" as apparently fulfilling varying but similar roles and comprising mainly rubbish material. As Grimes (1968) pointed out for the medieval period, in urban areas there was probably a general accretion of waste matter, not easily disposed of and likely to have been spread over open ground in gardens and elsewhere. Evidence of such additions of material is provided by the pollen, phytolith, loss on ignition and inclusions data. The uniformity of grain size described by Taylor from Chaucer Street, Southwark (1978), which was also noted at GPO 75, does not in fact preclude successive dumping as a means of accumulating "Dark Earth", especially if it all comes from a similar source. Particle size analysis, with the micromorphology, also revealed how extraneous material has been mixed in with the local parent materials. There is also clear indication of dumping, from West to East at GPO 75, in the form of tip lines, zones with much higher proportions of inclusions, or layers of building material (Roskams, unpublished report A). In contrast, the "Dark Earth" from Milk Street, London appeared to be homogeneous throughout. This is thought to have resulted from its more intensive use, obscuring evidence of dumping (Roskams, unpublished report B). This intensive cultivation perhaps accounts for the severe abrasion of pottery, and the position of a 9th century sherd 60 cms down into the "Dark Earth" (Roskams and Schofield 1978). Yet "Dark Earth" was apparently tipped onto an undisturbed mosaic at Milk Street (Roskams and Schofield 1978), in the same way as "Dark Earth" seems to have been dumped onto an opus signinum floor at St. Bartholomew's Hospital.

"Dark Earth" may have also included "street sweepings" (Sheldon 1978) which were deposited in market gardens in the early 17th century and would tend to produce layers of uniform texture, mixed with anthropogenic debris. Both Orton (1978) and Wilson, et al. (1979), looking at the distribution of cultural debris in "Dark Earth" from Southwark and Abingdon respectively, have found the progressive introduction of later pottery up through the deposit, together with mixing of relic pottery from earlier periods. This suggests successive accumulation and reworking of "Dark Earth".

The inclusion of anthropogenic debris would increase fertility and produce freely draining, neutral to alkaline "soils", in some cases phosphorus rich, effectively base rich brown earths. The soil pollen may suggest a cultivated soil, although a more organic upper soil, expected from agriculture, has been obscured by the general oxidation of the "Dark Earth" and by earthworm activity, as well as by continued reworking as described above.

At York, increments of material to the "Dark Earth" seem to be similar to other areas, although at the Bedern more evidence is preserved of botanical additions other than than the local ruderal flora. Here, elements of waste occur, similar to that from later crowded Medieval cities when it was common to use any spare ground as an open sewer (Ziegler 1969), as illustrated by the 15th to 17th century stream-fill from Broad Sanctuary, Westminster. At this site Scaife (1980) has shown very large urban ruderal pollen input into a narrow riverine deposit which was probably rich in cess; this assessment agrees with findings from pre-Norman York (Addyman et al. 1976).

CONCLUSIONS

Study of samples of "Dark Earth" has revealed the numerous components which produce this deposit, which may result purely from rubbish dumping or was purposely used for market gardening. "Dark Earth" is important to the history of the sites and cities where it occurs, because the simple model of Roman abandonment and Dark Ages decay may not be wholly correct.

London was thought to have been occupied as a city almost to the end of the 5th century; after an unspecified time, walls collapsed and ruins were deliberately demolished and subsequently overlain by material representing the Dark Ages (Merrifield 1976). However, evidence from Staines indicates a decline in the 4th century (Crouch 1976), and nearer the City, at Southwark, mid 2nd century buildings at the Borough High Street were destroyed and overlain by "Dark Earth" containing 3rd to 4th century pottery (Southwark and Lambeth Archaeological Excavation Committee 1978). General findings from Southwark show earlier buildings being buried by "Dark Earth" by the mid 4th century and the "spreading process" continuing as late as the end of the 4th century (Sheldon 1978). Previously built-up areas were given over to market gardening towards the end of the Roman period, presumably to meet local requirements, and the "Dark Earth" continued to accumulate into the medieval and post-medieval periods when many of the sites in question were known to be gardens (Sheldon 1978). At Staines, the "dark soil" was used for agriculture during the 9th-12th centuries as evidenced by relic ridge and furrow (Crouch 1976).

Evidence therefore exists of a dramatic change from an urban existence to cultivation of artificial soils burying these previously built-up areas in the "suburbs". At the end of the 4th century, London was still the seat of the Provincial Treasury (Biddle et al. 1973) and the City's bastions had only been recently completed (Grimes 1968). Even so, at St. Bartholomew's Hospital the opus signinum floor is buried by "Dark Earth" or between 2nd-4th century in age (Bentley 1979, pers. comm.), while the lowest "Dark Earth" at GPO; 75 dates from the 4th century (Roskams, unpublished report A). At GPO 75 the late Roman deposits had been cut away at different levels, sometimes removing the underlying Antonine deposits completely, while much residual building material had been loosely deposited before the "Dark Earth" accumulated (Roskams, unpublished report A). Further, at Milk Street an absence of destruction debris seems to show systematic dismantling of the Roman at the end of the 2nd century, whilst the actual deposition of the "Dark Earth" commenced in the early 3rd century (Roskams and Schofield 1978). Again, as at Southwark, there seems to have been a radical change in land use, with a lack of 4th century finds at Milk Street indicating a non-domestic occupation of the area (Roskams, unpublished report B).

The possible implications of this are that late Roman London had a sparser population, with farming within the walls, and that there had been a dramatic change from the early commercial expanding town with its entrepreneurial class (Roskams and Schofield 1978). Seemingly, built-up areas had been dismantled and now were possibly used for agriculture based on dumped "Dark Earth", while the legal, government and military functions of the city continued. Therefore, the "Dark Earth" in London, seems to represent the changing nature of the City during the late Roman period rather than its sudden abandonment and decay relating to Roman withdrawal and the arrival of the Dark Ages. This then is relevant to questions of urban continuity during the late Roman and Saxon periods as posed by Biddle (1976).

In the case of York, the Anglo-Saxons inherited buildings, while later the walled town was also utilised by the Danes, giving the city a continuity of use up to the Norman period, and this intensive occupation produced in many cases dark deposits which arecommonly organic in nature (Addyman et al. 1976). This better preservedtype of"Dark Earth" as described above and analysed in this paper, cannot be directly compared with the more oxidised deposits associated with the changing urban environment of late Roman London. In contrast, they seem to indicate continued accumulation of domestic waste in restricted areas over many centuries.

We may then have two types of "Dark Earth", one relating to a sparser urban population in London, as an initially purposely dumped artificial garden soil, and the other a deposit concomitant with continued urban refuse disposal in a more densely occupied environment as at York, although the York deposit is not related to the immediate post-Roman period (Addyman, pers. comm.). The next step is to examine a wet "Dark Earth" site from London so that it can be more accurately compared with material from York. In addition, it would be worthwhile to look at "Dark Earth" from other towns to see if there are other instances of supposed late Roman "Dark Earth".

Acknowledgements

I should like to thank H. C. M. Keeley and R. M. Rees who commented upon the text, and S. Roskams for his comments and for generously making unpublished reports available. I would also like to thank R. G. Scaife for his contributions to this paper and B. Lambert for typing the script.

Bibliography

Addyman, P. V., Hood, J. S. R., Kenward, H. K., MacGregory, A. and Williams, D. (1976). Palaeoclimate in urban environmental archaeology at York, England: problems and potential. World Archaeology 8 2 220-32.

Armitage, P. L. (1975). The extraction and identification of opal phytoliths from the teeth of ungulates. Journal of Archaeological Science 2 187-97.

Biddle, M., Hudson, D. and Heighway, C. (1973). The Future of London's Past: a survey of the archaeological implications of planning and development in the nation's capital. Rescue; a trust for British Archaeology. Rescue Pub. 4.

Biddle, M. (1976). Towns. In: The Archaeology of Anglo-Saxon England (ed Wilson, D. M.) pp. 99-150.

Brewer, R. (1964). Fabric and Mineral Analysis of Soils. London: John Wiley and Sons Inc.

Buckland, P. C. (1974). Archaeology and environment of York. Journal of Archaeological Science 303-16.

Cornwall, I. W. (1958). Soils for the Archaeologists. London: Phoenix House.

Crouch, K. (1976). The archaeology of Staines and the excavation at Elmsleigh House. Transactions of the London and Middlesex Archaeological Society 27 71-124.

Faegri, K. and Iversen, J. (1974). Textbook of Pollen Analysis. 3rd ed., Oxford: Blackwell.

Geiss, J. W. (1973). Biogenic silica in selected species of deciduous angiosperms. Soil Sci. 116 113-9.

Grimes, W. F. (1968). The Excavations of Roman and Medieval London. London: Routledge and Kegan Paul.

Hubbard, C. E. (1968). Grasses. 2nd ed., Harmsworth, Pelican Books Ltd.

Kenyon, K. M. (1959). Excavations in Southwark (1945-57). Research Papers Surrey Archaeological Society 5, Guildford.

London and Middlesex Archaeological Society, (1976). The Archaeology of the London Area: Current knowledge and problems. Special Paper 1. London and Middlesex Archaeol. Soc.

MacPhail, R. I. (1980). Report on soil colour variation of rural Roman and perhaps later occupation areas at Lechlade (Oxon.), and the relationship to the "Dark Earth" of comparable age at urban sites. Unpub. Ancient Mon. Lab. Report 3054.

Macphail, R. I. (1980). Soil report on "Dark Earth" at GPO Middle (POM 79), London. Unpublished Ancient Mon. Lab. Report 3055.

Macpahil, R. I. (1980). Soil report on "Dark Earth" at the Arcadia Buildings, Southwark, London. Unpublished Ancient Mon. Lab. Report 3057.

Macpahil, R. I. (1980). Soil report on "Dark Earth" at Keays Lane, Carlisle. Unpublished Ancient Mon. Lab. Report 3059.

Macphail, R. I. (1980). Soil report of the "Dark Earth" at GPO 75, London. Unpublished Ancient Mon. Lab. Report 3060.

Macphail, R. I. (1980). Soil report on "Dark Earth" from the Bedern excavation at York. Unpublished Ancient Mon. Lab. Report 3061.

Merrifield, R. (1965). The Roman City of London. London: Ernst. Benn Ltd.

Moore, P. D. and Webb, J. A. (1978). An Illustrated Guide to Pollen Analysis. London: Hodder and Stoughton.

Norman, P. and Reader, F. W. (1912). Further discoveries relating to Roman London, 1906 12. Archaeologia LXIII 257-344.

Orton, C. (1978). Sequence of Medieval and Post-Medieval pottery. In: Southwark Excavations 1972-74. Joint Publication No. 1. Southwark and Lambeth Archaeological Excavation Committee, London and Middlesex Archaeological Society, Surrey Archaeological Society, 140-6.

Pike, A. W. and Biddle, M. (1966). Parasite eggs in Medieval Winchester. Antiquity 40 293.

Rovner, I. (1971). Potential of opal phytoliths for use in palaeoecological reconstruction. Quaternary Research 1 343-59.

Roskams, S. and Schofield, J. (1978). The Milk Street Excavation, Part 2. London Archaeologist 3 9 227-34.

Roskams, S. The GPO excavation. Unpublished report (A).

Roskams, S. The Milk Street excavation. Unpublished report (B).

Scaife, R. G. (1980a). Pollen analysis of some dark earth samples. Unpublished Ancient Mon. Lab. Report 3001.

Scaife, R. G. (1980b). Pollen analytical investigation of Broad Sanctuary, Westminster. Unpublished Ancient Mon. Lab. Report 3070.

Sheldon, H. L. (1978). The 1972-74 Excavations: Their Contribution to Southwark's History. In: The Southwark Excavations 1972-74. Joint Publication No. 1. Southwark and Lambeth Archaeological Excavation Committee. London and Middlesex Archaeological Society, Surrey Archaeological Society.

Smithson, F. (1958). Grass opal in British Soils. Journal of Soil Science 9 1 148-54.

Southwark and Lambeth Archaeological Excavation Committee. 1978. Southwark Excavations 1972-74. Joint Publication No. 1. London and Middlesex Archaeological Society, Surrey Archaeological Society.

Taylor, T. P. (1978). Preliminary report on the black earth deposits at Chaucer Street, Southwark. Unpublished Ancient Mon. Lab. Report 2493.

Twiss, P. C., Suess, E. and Smith, R. M. (1969). Morphological classification of grass phytoliths. Soil Science Society of America. Proceedings 33 109-114.

Wilson, B., Thomas, R. and Wheeler, A. (1979). Sampling a profile of town soil-accumulation, 57 East St. Helen's Street, Abingdon. Oxoniensia xliv.

Ziegler, F. (1969). The Black Death. Ch. 9, London: Hygiene and the City. London: Collins.

Micromorphology, "Dark Earth" sample 13 at G.P.O. 75

An undisturbed sample of the "dark earth" was air-dried and impregnated with an Araldite mixture, and thin sections (3 mm x 1.5 mm) were examined under the petrological microscope.

The "dark earth" is heterogenous; black (opaque) to dark brown (P.P.L. —Plane Polarised Light), and black with a medium lustre under reflected light (R.L.). It comprises fine rounded accommodated primary peds forming in most areas more coarse, rounded blocky secondary peds. Total voids equal 20-25%, with mostly compound packing voids which are mainly metavughs or well formed channels both within and between larger peds.

There is approximately 18%skeletal material comprising sub-rounded medium sized (0.2 - 0.6 m) sand grains and gravel sized (2 - 6 mm) grey shell, and dark red brown (P PL) ceramic material. The "dark earth" contains nearly 20% silt. The shell fragments, which are most probably oyster, exhibit extreme birefringence colours concomitant with a calcite/aragonite composition.

Varying quantities of silt and fine sand sized (0.06 - 0.2 mm) charcoal fragments are present. Reflected light suggests both soil humus and finely divided charcoal are intimately mixed with the iron and clay complexes forming the matrix.

Both manganic and ferric glaebules can occur. Generally, fabric is silasepic; agglomeroplasmic-porphyroskelic.

Pedological features include pedotubules; one of which is 5 mm is diameter and has three zones. The outer edge is a greenish colour, most probably due to gleying and demarcates the non opaque soil of the pedotubule from the black, carbon-rich "dark earth". The outer zone is orange-brown (PPL) and silasepic. The inner zone is grey-brown and asepic (free of silt or sand), whilst the centre is void space. It seems likely that this is a faunal burrow (earthworm) which has brought "natural material" (brickearth-silasepic) into "dark earth". The grey-brown iron-clay material in the centre is likely to be fines which have been washed into the burrow through some soil disturbance.

Elsewhere in the "dark earth" there are plasmic separations both in the form of non-birefringent iron-clay channel linings and rare grain and channel argillans indicating some movement of material through disturbance other than faunal.

Thus we have a soil environment where wetting and drying are important processes. In addition, faunal activity is evident through obvious pedotubules and the nature of the soil aggregates or peds, which tend to occur in high base status brown earths. Therefore it is likely that the intimate mixing of charcoal and organic matter with the soil mineral content is liable to have occurred through faunal as well as wetting and drying activity.

It is likely that soil material that has been mobilised and redeposited has originated from the underlying natural brickearth, for these "fines" tend to have no "dark earth" content and thus were most probably not flocculated and thus more readily moved.

Overall, the micromorphology of GPO "dark earth" is similar to that of Southwark "dark earth" except for perhaps there being more large charcoal fragments at Southwark which notwithstanding is characterised by an agglomeroplasmic fabric and pedotubules.

CONCLUDING REMARKS

Geoffrey Dimbleby

So wide was the range of papers presented at the Conference that it is safe to say that everyone present will have met facts, theories and points of view that are new to them. Certainly this applied to me, and I would therefore find it very difficult to attempt a balanced summary of the proceedings even if I had been commissioned to do so. Fortunately my brief was to draw together whatever thoughts occurred to me on hearing, and subsequently reading, all the papers.

What do we mean by the Environment of Man? Peter Reynolds is right in saying that environmental archaeology has just grown, with little direction, but he goes on to take the restricted view that it is "an integral part of excavation and needs to be viewed as such and no more". This is, I suspect, a common view among archaeologists, who think in terms of sites rather than in terms of 'social landscapes' (David Miles). Differences in the way 'environment' is understood have parallels today. The Natural Environment Research Council likes to exclude man as an environmental factor, even though, as we know, he has had a recognisable influence on the landscape at least since mesolithic times. At the other extreme planners and architects sometimes create what they call environments which contain no living thing except man. And of course we have our monumental Department of the Environment under whose authority environments can be either conserved or destroyed. I do not think it is useful to attempt a definition of 'the environment of man' because of the subjective attitudes we hold; the important thing is to recognise that we do not necessarily mean the same thing when we use the term. Our organiser and editor, Martin Jones, has expressed this well by using distinct headings in the arrangment of the contents of this volume, and by doing so makes clear how 'environment' can be related to archaeology, either quite indirectly (Hubert Lamb) or very closely (e.g. the papers under Ecology of Settlement).

Though climate is beyond any influence of man (at least in the period we are concerned with) the impact of man on the environment may have some effect on our reading of the climate of earlier times. As long as we use evidence which is itself subject to an element of human pressure there is a possibility of bias. For instance, taken on its own, the evidence of contemporary water-tables, flooding and alluviation is a measure not only of rainfall but also of land use—or misuse—through its effect on run-off and water collection. Microclimate, too, is a factor of direct impact on man's way of life and can be drastically modified without any change in overall climate.

It would be possible to write a book on the environment of man in Britain over the ages without drawing upon any evidence from archaeological sites (*pace* Peter Reynolds). It might not give the complete story, but it would present a massive body of evidence based on pollen analysis, faunal analysis, palaeosols, geomorphology, etc. from sites which are archaeologically sterile. In this volume the only contribution from this field of research is that by Judith Turner, and her paper serves to remind us of how much of our background knowledge of contemporary landscapes comes from this type of research. This applies especially to the Highland zone, and the point was well made that our conference papers, with the exception of Judith Turner's and Allen Hall's,* all dealt almost exclusively with sites south of the Trent.

It was interesting to meet the attitude that archaeological sites can be treated as potential sources of information about the wider environment, even to the extent of excavating a site primarily for its environmental potential. Indeed that is how I myself first became involved—illicitly, I fear—in archaeology, and on occasion I can still turn a jaundiced eye on the archaeological gems of metalwork, pottery and worked flint whilst thrilling to the challenge of an iron pan, a stone-free horizon or the possibility of extracting pollen or molluscs.

However, by the time we start investigating sites, particularly in a period as late as the one we are dealing with, our concept of environment is usually beginning to change. We are now turning to analyses of crop and weed remains and the bones of domestic (and some wild) animals. Interpretation is not now in the area of broad environment but in terms of the economy of the site; the emphasis is now on agronomy rather than ecology. To the archaeologists the material is still 'non-archaeological' and has to be analysed by biologists, but interest centres on such things as the introduction of species, breeding, processing, etc., related to direct use by man. Certainly the development of crops and animals and the way they are raised will have an impact on the wider environment, but this is no longer the prime agronomical, and therefore archaeological, interest.

At this stage, too, we find evidence of the exploitation of resources from the surrounding countryside, such as fuel, thatching material and even minerals. Their collection can bring about significant changes in local ecosystems and lines of movement may be set up which can be linked with other activities.

To interpret the evidence of such a period from its contemporary sites calls for a different kind of background knowledge from that of the 'landscape archaeologist'. It requires a more detailed knowledge of farming, of the crops and animals in use and how they would be managed and distributed. Several contributors made derogatory remarks about the standard of interpretation of environmental evidence from archaeological sites. These were aimed at both archaeologists and some specialists. The question was asked whether we must have a *cognoscenti* or whether an archaeologist can cope, at least as a knowledgeable co-ordinator. Whilst I am often impressed with the ability of some archaeologists to handle environmental evidence, I doubt whether, with our present system of training, such people will ever form more than a minority. One factor which mitigates against this is that the great majority of students reading archaeology (and it is true of students generally

* No paper submitted to volume.

today) are townsmen by upbringing and have only the most superficial knowledge of country practices. Furthermore, modern agricultural practices are often little guide to those of earlier times.

The environmentalists, too, have limitations. They specalise—in pollen, seeds, bones, molluscs and so on. Any report that is produced can only be as comprehensive as the knowledge of the specialist who makes it. What a fortunate thing it is that the botanist who worked on the Feddersen Wierde was not just a pollen analyst or a seed specialist. Professor Körber-Grohne was able to bring to bear the whole range of a botanical training, so enabling her to recognise a wide range of plant material as well as to ask, and answer, questions that would not have occurred to anyone with a narrower background. When one sees this level of proficiency it has to be admitted that archaeologists who have acquired a specialism such as seed or mollusc analysis, without the benefit of a broader training in botany or zoology, are working under a severe handicap. This must mean that at times the material they are working on is not made to reveal its full potential of information. To some extent this limitation can be overcome by using a team of specialists, as at York or in the Oxfordshire Unit, which brings the additional advantage of interdisciplinary intercourse.

Turning from personnel to the sites they are studying, it was refreshing to hear the view expressed by an archaeologist that sites are not all unique but are part of a 'population', some similar, others more or less different. This raises the question of why a site should be selected for excavation. Just because a site is threatened is no longer an adequate reason for digging it. We have already seen that selection might be made on environmental grounds; there may also be valid archaeological grounds, such as the filling of a gap in a sequence of types, areas or time. A site stands in relationship to other sites (assuming that we can define 'site') and this may provide a good reason for excavation, providing comparative evidence of archaeological, agronomical or environmental relevance. Yet it is not uncommon to find archaeologists turning their faces resolutely against comparative studies, particularly environmental, restricting discussion in their report strictly to the site in question. In this attitude they are supported by the fact that official funding is given strictly on a site basis. Comparison and synthesis are of the essence of ecological work, and environmental studies, whether or not in an archaeological context, are ecological and demand the methodology of ecology. In this conference the value of such inter-site correlation has been well demonstrated in papers dealing with bone analysis (e.g. Mark Maltby) as well as in papers dealing with crops and weeds (Martin Jones; Frank Green).

Most of what I have said so far could be said about any period, so is there anything which is of particular significance to the Iron Age - Anglo-Saxon period? One archaeological situation which appears for the first time in Britain in this period is the town. As many of our present towns and cities are situated on the same sites as those of Roman, Viking and Anglo-Saxon age rescue archaeology has unearthed levels of these periods. Moreover, as these levels are often waterlogged the preservation of organic remains has often been remarkable, permitting a whole new line of study—the urban 'environment'. If it is true that our interpretation of rural environments has been simplistic it is not surprising that that of urban settings also has at times been less sophisticated. But through critical re-appraisal of our assumptions more valid reconstructions are being created. The impressive work at York, London and Winchester is being co-ordinated on an interdisciplinary basis and is shedding unexpected light on contemporary life, commerce and even foreign trade. Moreover, towns are now being seen not simply as discrete urban centres but as focal points in the surrounding countryside, another example of the complementary nature of inter-site studies.

Taking the wider view of the Iron Age - Anglo-Saxon landscape, several contributors thought we were not yet in a position to present a reliable picture of the countryside at this time, though we do know enough to be able to refute some opinions widely held until quite recently. As already said, our papers do not give an even coverage over the country as a whole. From the pollen evidence presented by Judith Turner we have some knowledge of northern landscapes in Roman times, which were much more forested than southern ones at that time (Mark Robinson). Yet both are part of the same 'social landscape', connected by the Roman road system, which emphasises the importance of comparative environmental studies covering the widest possible geographical range.

Earlier in this commentary reference was made to the fact that the landscapes of our period had resulted from man-induced change going back even to pre-agricultural times. This concept of change is fundamental and is perhaps the greatest justification of comparative field research. Though, as I say, changes can be detected in much earlier periods, there is evidence that in our period they were becoming more marked. Not only this, but it seems that the changes could in some cases be looked upon as adverse; that is, there was environmental deterioration. This is familiar in earlier times in regions of more fragile ecosystems (poor soils, high altitude, cool and wet climate) but here we are seeing it in the lush environment of southern England. Soil exhaustion, especially of nitrogen, was referred to by one or two contributors as a possible explanation of observed changes. Increased alluvial deposition in the Thames Valley in this period as compared with the Bronze Age was noted by Mark Robinson, an observation which could link with Martin Bell's remarkable demonstration of hillwash associated with intensified land use.

The papers collected in this volume represent the raw material of what was a most stimulating and enjoyable conference, which owed its success not only to the enthusiasm of all the participants but more particularly to those who had the foresight to envisage a conference on this topic and then planned it through all its stages. I hope the spirit of enquiry, challenge and healthy doubt which brought the conference to life will to some degree be transmitted through this volume. I am sure all readers will agree that our knowledge of this period has been extended by the research described, for which we thank all who took part and shared their learning so willingly.

www.ingramcontent.com/pod-product-compliance
Ingram Content Group UK Ltd.
Pitfield, Milton Keynes, MK11 3LW, UK
UKHW061845041026
15591UKWH00015B/911

9 780860 541288